T0258884

Physical Characterization
of Pharmaceutical Solids

DRUGS AND THE PHARMACEUTICAL SCIENCES

A Series of Textbooks and Monographs

edited by

James Swarbrick
AAI, Inc.
Wilmington, North Carolina

ADDITIONAL VOLUMES IN PREPARATION

Modern Pharmaceutics: Third Edition, Revised and Expanded, *edited by Gilbert S. Banker and Christopher T. Rhodes*

Physical Characterization of Pharmaceutical Solids

edited by
Harry G. Brittain

Ohmeda, Inc.
Murray Hill, New Jersey

informa
healthcare

New York   London

Informa Healthcare USA, Inc.
52 Vanderbilt Avenue
New York, NY 10017

© 2007 by Informa Healthcare USA, Inc.
Informa Healthcare is an Informa business

No claim to original U.S. Government works

10

International Standard Book Number-10: 0-8247-9372-2 (Hardcover)
International Standard Book Number-13: 978-0-8247-9372-2 (Hardcover)

Library of Congress Cataloging-in-Publication Data

Physical characterization of pharmaceutical solids / edited by Harry G. Brittain.
    p. ; cm. -- (Drugs and the pharmaceutical sciences ; 70)
    Includes bibliographical references and index.
    ISBN-13: 978-0-8247-9372-2 (alk. paper); ISBN-10: 0-8247-9372-2 (alk. paper)
    1. Solid dosage forms. 2. Pharmaceutical chemistry. I. Brittain, H. G. II. Series: Drugs
    and the pharmaceutical sciences ; v. 70. [DNLM: 1. Chemistry, Pharmaceutical--methods.
    2. Chemistry, Physical--methods. 3. Powders--chemistry. 4. Tables--chemistry.
    W1 DR893B v. 70 1995 / QV 744 P5779 1995]

RS201.S57P49 1995
615'.19--dc20   DNLM/DLC for Library of Congress 95-4852 CIP

Visit the Informa Web site at
www.informa.com

and the Informa Healthcare Web site at
www.informahealthcare.com

Preface

It is evident even to the casual observer that the vast majority of pharmaceutical products are administered as solid dosage forms, which are in turn produced by the formulation and processing of powdered solids. All too often characterization of raw materials and products has centered on aspects of chemical purity, with only passing attention being given to the physical properties of the solids. However, every pharmaceutical scientist knows of at least one instance in which a crisis arose due to some variation in the physical properties of input materials, and in which better characterization would have prevented the problem.

The continued refinement of manufacturing processes results in increasing degrees of automation. In this streamlined operation style, raw materials are mixed, granulated, dried, tableted, coated, and packaged in a continuous operation. The key requirement associated with these manufacturing procedures is that all ingredients be totally characterized, since an unattended operation functions only if its materials fall within a specified range of material properties. Any lot-to-lot variation in physical properties has to be detected before the material is used, or else the entire manufacturing run is compromised.

The pharmaceutical community is rapidly becoming aware of the need to obtain proper physical characterization of raw materials, drug substances, and formulated products. The present volume seeks to address these concerns by outlining a comprehensive program for this work. The modern and highly regulated industry cannot tolerate the inconsistent practices of the past, when the only physical properties documented were those that a particular lab knew how to determine. Proper physical characterization must be systematic in its approach, and it should follow a protocol that is rationally designed to obtain all needed information.

In the present work, such a systematic approach to the physical characterization of pharmaceutical solids is outlined. Techniques available for the study of physical properties are classified as being associated with the *molecular* level (properties associated with individual molecules), the *particulate* level (properties pertaining to individual solid particles), and the *bulk* level (properties associated with an ensemble of particulates). Acquisition of this range of physical information yields a total profile of the pharmaceutical solid in question, whether it is an active drug, an excipient, or a blend of these. The development of a total profile is a requirement for successful manufacture of any solid dosage form.

Each author has provided an introduction that serves to initiate the reader into the topic. Where appropriate, a brief exposition of associated theory is presented, but the essence of each chapter lies in the practical examples used to illustrate each topic. It is anticipated that even though the range of presented examples is not necessarily comprehensive, sufficient information is given to allow the reader to understand the strengths, advantages, and limitations of each technique. It is important that workers in the field have a good feel for what a given method cannot yield, as well as what it can provide.

Harry G. Brittain

Contents

Contributors

Gregory E. Amidon, Ph.D. Senior Scientist, Drug Delivery R&D—Pharmaceutics, The Upjohn Company, Kalamazoo, Michigan

Harry G. Brittain, Ph.D. Director, Pharmaceutical Development, Pharmaceutical Products Division, Ohmeda, Inc., Murray Hill, New Jersey

David E. Bugay, Ph.D. Senior Research Investigator II, Analytical R&D, Bristol-Myers Squibb Pharmaceutical Research Institute, New Brunswick, New Jersey

David J. W. Grant, M.A., D.Phil., D.Sc. Professor of Pharmaceutics, Department of Pharmaceutics, University of Minnesota, Minneapolis, Minnesota

Mark J. Kontny, Ph.D. Associate Director, Pharmaceutics Department, Boehringer Ingelheim Pharmaceuticals, Inc., Ridgefield, Connecticut

James A. McCauley, Ph.D. Senior Investigator, Analytical Research, Merck Sharp & Dohme Research Laboratories, Rahway, New Jersey

Ann W. Newman, Ph.D. Senior Research Investigator, Materials Science Group, Analytical R&D, Bristol-Myers Squibb Pharmaceutical Research Institute, New Brunswick, New Jersey

Cynthia S. Randall, Ph.D. Senior Investigator, Pharmaceutical Technologies, SmithKline Beecham Pharmaceuticals, King of Prussia, Pennsylvania

Raj Suryanarayanan, Ph.D. Associate Professor and Director of Graduate Studies, Department of Pharmaceutics, College of Pharmacy, University of Minnesota, Minneapolis, Minnesota

Adrian C. Williams, Ph.D. Lecturer in Pharmaceutical Technology, School of Pharmacy, University of Bradford, Bradford, West Yorkshire, United Kingdom

George Zografi, Ph.D. Professor of Pharmaceutics, School of Pharmacy, Univeristy of Wisconsin—Madison, Madison, Wisconsin

1

Overview of Physical Characterization Methodology

Harry G. Brittain
Ohmeda, Inc., Murray Hill, New Jersey

I. INTRODUCTION

The normal route of administration for the majority of pharmaceutically active therapeutic agents is through solid dosage forms [1], and these units are conventionally produced by the formulation and processing of powdered solids [2]. The priority of regulatory bodies has always been to focus primarily on concerns of safety and efficacy, which in turn has led to an almost total emphasis on aspects of chemical purity. This consideration has resulted in the situation that often only passing attention is given to the physical properties of the solids that comprise a dosage form.

Ignoring the physical aspects of a given formulation can be disastrous, since the stability of the drug entity can be strongly affected by its matrix [3]. A wide variety of reactions are known to take place in the solid state [4], the pathway of which can be dramatically different when compared with how the same reaction would proceed in the liquid or gaseous phase [5].

It is assumed during the course of drug development that any concerns related to the physical properties of the substances being formulated will be adequately researched at the appropriate moment. Unfortunately, this work is often not conducted until a crisis situation develops due to some variability in the physical properties of input materials. It is perhaps a truism that most of these problems could have been avoided had the materials received more balanced characterization. The economics of drug development, however, often interferes with the desire of the formulator to fully understand his system, and performance of the proper background work can become a casualty.

Nevertheless, the acquisition of a sufficiently detailed body of physical information can allow a formulator to go far beyond the mere ability to cope with crises when they develop at unexpected times. For a well-understood system, it is theoretically possible to design an automated or semi-automated manufacturing scheme for which the processing variables would be appropriately controlled so as to minimize the possibility of batch failure. Materials passing the hurdles of physical test specifications would be totally predictable in their performance, and they could therefore be blended, granulated, dried, compressed, and delivered into containers without operator intervention.

The need for physical characterization becomes even more crucial for the use of excipients in formulations. These materials have historically been characterized solely by the criteria of the National Formulary, which only rarely includes any mention of physical testing. Every formulation scientist knows that the present situation allows for the existence of wide variability in excipient properties, and such lot-to-lot variations can often be the cause of significant processing difficulties. To avoid problems during drug development, the physical characterization of bulk drugs, excipients, and blends of these should become part of the normal process. The degree of physical testing would necessarily vary with the particular formulation, but it would include any and all test methods deemed appropriate.

It may be envisioned that a protocol for the complete physical character- ization of a solid material could easily be developed. At the early stages in drug development, each lot of active drug, excipients, and formulated blends would be characterized as fully as possible. A feedback loop would be established after each formulation run, in which the physical characteristics of the input materials were correlated with the quality of formulated product. Out of these studies would come an understanding of what particular properties were essential to the production of an acceptable formulation.

As the maturity of the process increased, only the key parameters would require continued monitoring. Ultimately, the data collected on these properties would permit the generation of material specifications. If the work had been performed properly, then it would be possible to specify limits for the appropriate bulk drugs and raw materials that would ensure that the final product always was satisfactory. These guidelines would naturally apply only to the specific formulation, but their implementation would enable manufacturers to deliver their products with a greater degree of security than is now possible.

With these concerns in mind, it is appropriate to outline a comprehensive program for the physical characterization of pharmaceutical solids. A modern industry cannot tolerate the inconsistent practices of the past, where the only physical properties that might be documented were those that could be conve- niently measured. It is of extreme importance that investigators measure the parameters that need to be measured and not merely collect the type of data that is convenient to obtain. Proper physical characterization must be systematic in its approach, and it should follow a protocol that is rationally designed to obtain all needed information.

A systematic approach to the physical characterization of pharmaceutical solids has been outlined [6], and it will be filled out in significantly more depth in the chapters of the present work. Within this system, physical properties are classified as being associated with the *molecular* level (those associated with individual molecules), the *particulate* level (those pertaining to individual solid particles), or the *bulk* level (those associated with an assembly of particulate species).

II. PROPERTIES ASSOCIATED WITH THE MOLECULAR LEVEL

Molecular properties may be defined as those material characteristics which theoretically could be measured for a small ensemble of individual molecules. Due to the minimal sample requirements, molecular properties can be determined at the earliest stages of drug development. Though the molecular level techniques are all spectroscopic in nature, substantial information of great use to formulators can be obtained from appropriately designed experiments. For example, a screening of stressed materials can be carried out on the microgram level using infrared microscopy [7], and the results of such work would aid the preformulation characterization of a new chemical entity.

A.   Ultraviolet/Visible Diffuse Reflectance Spectroscopy

With the exception of single-crystal transmission work, most solids are too opaque to permit the conventional use of ultraviolet/visible (UV/VIS) electronic spectroscopy. As a result, such work must be performed through the use of diffuse reflection techniques [8–10]. Important work has been conducted in which UV/VIS spectroscopy has been used to study the reaction pathways of various solid state reactions. Other applications have been made in the fields of color measurement and color matching, areas which can be of considerable importance when applied to the coloring agents used in formulations.

It was recognized some time ago that diffuse reflectance spectroscopy would be a very useful tool for the study of interactions among various formulation components, and the technique has been successfully used in the characterization of many solid state reactions [11]. Investigations conducted under appropriately designed stress conditions have been useful in the study of drug–excipient interactions, drug degradation pathways, and alterations in bioavailability owing to chemisorption of the drug onto other components in the formulation.

Connors and Jozwiakowski have used diffuse reflectance spectroscopy to study the adsorption of spiropyrans onto pharmaceutically relevant solids [12]. The particular adsorbants studied were interesting in that the spectral characteristics of the binary system depended strongly on the amount of material bound. As an example of this behavior, selected reflectance spectra obtained for the adsorption of indolinonaphthospiropyran onto silica gel are shown in Fig. 1. At low concentrations, the pyran sorbant exhibited its main absorption band around 550 nm. As the degree of coverage was increased the 550 nm band was still observed, but a much more intense absorption band at 470 nm became prominent. This secondary effect is most likely due to the presence of pyran–pyran interactions, which become more important as the concentration of sorbant is increased.

The perception of color is subjectively developed in the mind of an individual, and consequently different people can perceive a given color in various ways. Such variability in interpretation causes great difficulty in the evaluation of color-related phenomena, leading to problems in making objective judgements. The development of quantitative methods for color determination was undertaken to eliminate the subjectivity associated with visual interpretative measurements.

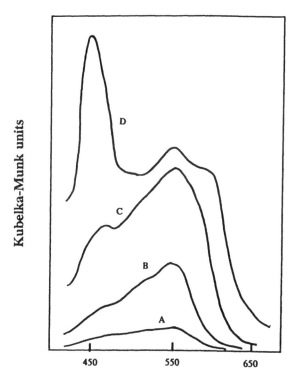

Wavelength (nm)

Fig. 1 Absorption spectra, obtained through the Kubelka–Munk transformation of diffuse reflectance spectra, of indolinonaphthospiropyran adsorbed onto silica gel. Spectra are shown for coverages of (A) 2.35, (B) 9.49, (C) 34.2, and (D) 46.7 $\mu g/m^2$. (Data adapted from Ref. 12.)

The most successful quantitative expression of color is that known as the CIE (Commission Internationale de l'Éclairage) system [13]. This methodology assumes that color may be expressed as the summation of selected spectral components (blue, green, and red hues) in a three-dimensional manner. The CIE system is based on the fact that human sight is trichromatic in its color perception, and that two stimuli will produce the same color if each of the three tristimulus values (X, Y, and Z) are equal for the two. Detailed summaries of the CIE and other quantitative systems for color measurement are available [14,15].

In a recent application, the appearance testing of tablets through measurement of color changes has been automated through the use of fiber optic probes and factor analysis of the data [16]. Good correlation between measured chromaticity parameters and visual subjective judgment was demonstrated, with samples of differing degrees of whiteness being used to develop the correlation. The methodology was complicated since surface defects on the analyzed materials could compromise the quality of the correlation.

B. Vibrational Spectroscopy

Infrared (IR) spectroscopy, especially when measured by means of the Fourier transform method (FTIR), is another powerful technique for the physical characterization of pharmaceutical solids [17]. In the IR method, the vibrational modes of a molecule are used to deduce structural information. When studied in the solid, these same vibrations normally are affected by the nature of the structural details of the analyte, thus yielding information useful to the formulation scientist. The FTIR spectra are often used to evaluate the type of polymorphism existing in a drug substance, and they can be very useful in studies of the water contained within a hydrate species. With modern instrumentation, it is straightforward to obtain FTIR spectra of micrometer-sized particles through the use of a microscope fitted with suitable optics.

The FTIR method makes simultaneous use of all the frequencies produced by the source, thus providing a large enhancement of the signal-to-noise ratio when compared with that of a dispersive instrument. Infrared spectra are best obtained on powdered solids through the use the diffuse reflectance method, and interpreted through the conventional group frequency compilations [18]. Three different spectral intervals are commonly identified: far-IR ($100-400$ cm^{-1}), mid-IR ($400-4000$ cm^{-1}), and near-IR ($4000-14,000$ cm^{-1}) regions, although most applications have been made in the mid-IR or near-IR regions.

When the vibrational modes of a compound are affected by fine details of molecular structure (i.e., polymorphism), the diffuse reflectance IR spectra of the polymorphs can be used for study of this behavior. For example, fosinopril sodium contains three carbonyl groups, the stretching frequencies of two of which are essentially equivalent (1600 versus 1598 cm^{-1}, and 1622 versus 1621 cm^{-1}) in the two known polymorphs [19]. This finding would suggest that the sidechains containing these groups are also equivalent in the two polymorphic structures. The stretching frequency of the third carbonyl group was found to be significantly different (1759 versus 1753 cm^{-1}) in the two forms, suggesting in turn that the polymorphism of fosinopril sodium was conformational in nature and associated with modified packing arrangements of the acetal sidechain [19].

Another technique of vibrational spectroscopy suited for the characterization of solids is that of Raman spectroscopy. In this methodology, the sample is irradiated with monochromatic laser radiation, and the inelastic scattering of the source energy is used to obtain a vibrational spectrum of the analyte [20]. Since most compounds of pharmaceutical interest are of low symmetry, the Raman spectrum will generally produce spectra equivalent to those obtained using the FTIR method. Differences in peak intensity can be observed, however, which may in turn provide details into the structural composition of the analyte and its host matrix.

In one application, Raman spectroscopy was used to identify and quantitate various drugs present in polymer matrices [21]. In Fig. 2, Raman spectra obtained within the fingerprint region for diclofenac, sodium alginate, and a 20% dispersion of diclofenac in sodium alginate are shown. It is evident in the spectra

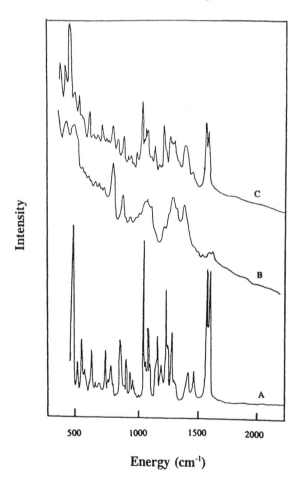

Fig. 2 Raman spectra obtained for (A) diclofenac sodium, (B) sodium alginate, and (C) a 20% dispersion of diclofenac sodium in sodium alginate. (Data adapted from Ref. 21.)

that the vibrational spectrum of the solute is easily distinguished in the mixture. The authors were able to perform a spectral subtraction of the data to reproduce the Raman spectra of the drug itself. In addition, they found that the bands at 1578 and 1603 cm^{-1} could be used to develop a quantitative method for the levels of drug in the polymer matrix.

The range of applications possible when using near-IR spectroscopy are numerous, but the range of problems addressable by this methodology are quite different than those just described [22]. Near-infrared spectra consist of overtone transitions of fundamental vibrational modes, and they therefore are not generally useful for identity purposes without the use of multicomponent analysis. The spectral features are of greatest utility in the detection and determination of functional groups that contain unique hydrogen atoms. For example, studies of water in solids can be easily performed through systematic characterization of the characteristic OH band, usually observed around 5170 cm^{-1}. The determination of hydrate species in an anhydrous matrix can easily be performed using near-IR analysis.

The near-IR technique has been used very successfully for moisture determination, whole tablet assay, and blending validation [23]. These methods are typically easy to develop and validate, and far easier to run than more traditional assay methods. Using the overtone and combination bands of water, it was possible to develop near-IR methods whose accuracy was equivalent to that obtained using Karl–Fischer titration. The distinction among tablets of differing potencies could be performed very easily and, unlike HPLC methods, did not require destruction of the analyte materials to obtain a result.

C. Magnetic Resonance Spectrometry

Yet another spectroscopic characterization tool that can be used to probe the solid state is that of nuclear magnetic resonance (NMR). The combination of magic-angle spinning and cross-polarization techniques now permit NMR spectra to be obtained in the solid state with only moderate difficulty relative to analogous solution-phase studies [24]. Although any NMR-active nucleus can be studied in the solid state, most of the work has focused on ^{13}C investigations. As was the case for FTIR spectroscopy, extensive compilations of ^{13}C resonances for various functional groups are available [25].

When the crystallography of compounds related by polymorphism is such that nuclei in the two structures are magnetically nonequivalent, it will follow that the resonances of these nuclei will not be equivalent. Since it is normally not difficult to assign organic functional groups to observed resonances, solid state NMR spectra can be used to deduce the nature of polymorphic variations, especially when the polymorphism is conformational in nature. Such information is extremely valuable at the early states of drug development when solved single crystal structures for each polymorph or solvate species may not yet be available.

As mentioned before, fosinopril sodium is known to be capable of existing in two polymorphic forms, and the diffuse reflectance IR spectra of the two forms indicated that the two structures differed in the conformation of one sidechain. The solid state ^{13}C NMR spectra obtained on both forms were found to confirm this hypothesis [19]. As may be seen in Fig. 3, the significant spectral differences were all associated with nuclei contained within the acetal sidechain.

III. PROPERTIES ASSOCIATED WITH THE PARTICULATE LEVEL

Particulate properties are defined as those material characteristics which theoretically could be determined by the analysis of a small number of particles. Since the sample requirements for these assay methods are extremely modest, these properties can be investigated as soon as a drug candidate is available in milligram quantities.

A. Optical and Electron Microscopies

The majority of materials having pharmaceutical interest consist of small microcrystals, which are often aggregated into much larger composite structures. Microscopy is the best method for study of such aggregate species, and both optical and electron microscopies are the methods of choice for such work [26]. Using the various techniques available, it is possible to study the optical crystallography of individual crystal particles, and to evaluate the nature of aggregate species. In addition, reliable information regarding particle sizes can be obtained. A variety of micromeritic parameters are related to the morphology of the constituent particles making up a given powder, and such information is easily obtained through a microscopic examination of the solid. The methodology for microscopic analysis is being standardized by the U.S. Pharmacopeia [27].

A variety of concerns require a determination of all possible crystalline forms of a given compound that might be encountered under different conditions. When only small amounts of material are available, crystallization from a variety of solvents can still be effected, and a full microscopic examination (paying critical attention to the optical crystallography of the sample) is used to observe any possible differences in crystal habit or structural class. Should a compound be capable of exhibiting polymorphism, hot-stage microscopy can be used to identify the most stable polymorph and provide determinations as to the physical interconvertibility of these [28]. After the performance of properly designed studies, no unexpected crystal forms should be encountered either during the scale-up of manufacturing or during stability studies.

Crystal polymorphs will usually exhibit different melting points, with the highest melting form being regarded as the most stable. The interconversion of

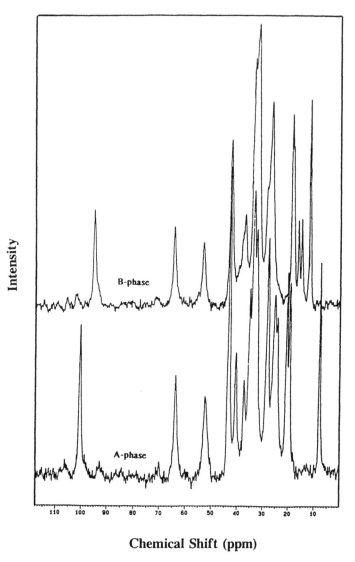

Chemical Shift (ppm)

Fig. 3   Solid state ^{13}C NMR spectra obtained on the A-phase and B-phase polymorphs of fosinopril sodium, illustrating the differences observed in the CH aliphatic and CC aliphatic regions of the spectrum. (Data adapted from Ref. 19.)

polymorphs is classified as being either enantiotropic or monotropic, according to whether the transformation of one modification into the other is reversible or not. Enantiotropic modifications interchange reversibly at the transformation point, and each form is characterized by having its own stability range of temperature. Monotropic substances are characterized by a hypothetical transition temperature, which is predicted to be higher than the melting points of both polymorphs. Monotropic polymorphs are characterized by the fact that one form is stable at all temperatures below its melting point, while the second form is metastable at all temperatures.

Photomicrographs of the two polymorphs of fosinopril sodium are shown in Fig. 4. The A-phase has been found to be the most stable, and it is normally

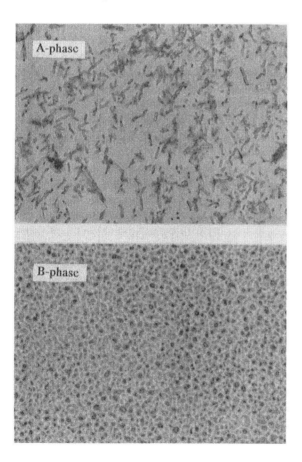

Fig. 4 Crystal morphologies as obtained by optical microscopy (600× magnification) of the A-phase and B-phase polymorphs of fosinopril sodium.

obtained as masses of fine needles. This material is ordinarily very crystalline, with individual needles being typically found to be 2–4 μm in width and 10–20 μm in length. The less stable B-phase is obtained as small tabular species, the individual crystals of which are substantially less crystalline compared with crystals of the A-phase.

B. Particle-Size Distribution

The particle-size distributions of drugs and excipients will exert profound effects on mixing phenomena, and on possible segregation in mixed materials [29]. It is generally accepted that in the absence of electrostatic effects, it is easiest to produce homogeneously mixed powders if the individual components to be mixed are of equivalent particle size. The distribution of particle sizes in a powdered material can affect the bioavailability of certain active drugs and certainly exerts a major effect on powder flowability. All pharmaceutical dosage forms must be produced in uniform units, and good content uniformity is only possible when the particle size of the active component is carefully controlled.

Different methods are available for the determination of the particle-size distribution of powdered solids [30]. These are optical microscopy (usually combined with image analysis), sieve analysis, laser light scattering of suspended particles, and electrical zone sensing.

The choice of sizing method should be made after taking into consideration the type of sample to be analyzed and the nature of information required. Since light scattering and electrical zone sensing are normally carried out on solids dispersed in an inert solvent medium, they are most suited for determinations of particle-size distribution in suspensions. Difficulties can arise when blindly using either technique to size dry powders, since the suspension process can grossly affect the size distributions obtained by these methods. The data will reflect what exists in the suspending medium, which may not necessarily indicate what had existed in the original powder sample. Microscopy and sieving are normally carried out on dry powder samples, and they are probably the most reliable indicators of the actual particle-size distribution existing for a powdered solid.

The only direct or absolute method of particle-size determination is that of microscopy, which becomes most efficient when combined with some form of image analysis [31]. Calibration of observed image is easily effected with the use of stage micrometers, and a given set of optics does not require recalibration. In the automated methodology, microscope parameters are adjusted so as to optimize the contrast between the background and particles to be sized. A video image of the powder is transmitted to a computer system, which then counts the number of pixels making up a particle. The size of each pixel is easily converted to micrometers, and the data is analyzed as desired by the analyst. Average particle sizes, full weight distributions, or shape information can be generated.

The advantage of the optical microscope method is that it provides direct and absolute information on the particles under characterization. Its chief disadvantage is that it can only provide data on the particles on the slide, and it can therefore be biased by the method used to prepare the slide.

In principle, sieve analysis represents the simplest method for the determination of particle sizes, and it is certainly one of the most widely used methods [32]. The methodology for analytical sieving is currently being standardized by the U.S. Pharmacopeia [33], but it may be summarized as follows. Particles are allowed to distribute among a series of screens (typically wire mesh), and the amount of material retained on each screen determined. The smaller particles that pass through a screen are termed the fines, while the larger particles remaining on the screen are the coarse particles. When using multiple screens, the intermediate size particles that pass through one or more screens (but that are retained on a subsequent screen) are called the medium fraction. A variety of assistance is normally provided during the sieving process, with vibration, ultrasound, or air suspension being employed to enable the particles to pass through the various screens. A proper size determination requires the use of five to six sieves, whose sizes are selected to obtain approximately equal amounts of powder on each screen and past the smallest sieve. The data are most commonly displayed as the percent of material retained on each sieve, the cumulative percent of sample retained, and the percent of sample passing each sieve. A general system for standardization of sieve data has been proposed [34].

Particle sizing may also be performed using electrical zone sensing (methodology based on the Coulter principle), in which the measurement of electrical pulses caused by passage of particles through a sensing zone is used to deduce size information [35]. One drawback to the Coulter method is that calibration using monodispersed particles of known diameter is required to assign the particle sizes of unknown species. The lowest size limits that may be measured are limited by thermal and electrical noise, and by the ability of the discrimination electronics to distinguish true signal pulses from the background.

For particles larger than 1 μm, Fraunhofer diffraction can be used to obtain particle-size distributions [36]. In this method the scattering of laser light (usually at low angles of incidence) is interpreted, and particle-size distributions are deduced from the scattering intensities. Calibration is not necessary, since the method does not require the use of standards for the conversion of signal response into a size distribution. The data are normally displayed as a volume distribution, with the percentages present in each size band being tabulated.

C. X-Ray Diffraction

The diffraction of x-rays by crystalline substances is of great analytical interest, since it is only by pure coincidence that two compounds would form crystals in which the three-dimensional spacing of planes was totally identical in all directions. In most instances, a powdered sample will present all possible crystal faces at a given interface, and the diffraction off this powdered surface will therefore provide information on all possible atomic spacings (the crystal lattice) [37]. The powder pattern consists of a series of peaks detected at various scattering angles. These angles, and their relative intensities, are correlated with computed d-spacings to provide a full crystallographic characterization of the powdered sample. After indexing all the scattered lines, it is possible to derive unit cell dimensions from the powder pattern of a substance.

The primary method for demonstration of the existence of drug polymorphs, or solvate species, is that of powder x-ray diffraction (XRD). Such measurements represent a specification of the internal structure within a crystal, and an evaluation of its lattice type. Since dissolution and subsequent drying can sometimes yield an undesired structure, it is also important to confirm crystal structures at each formulation stage during the beginning of the development process.

The technique is based on Bragg's law, which describes the diffraction of a monochromatic x-ray beam impinging on a plane of atoms [38]. Parallel incident rays strike the crystal planes and are then diffracted at angles that are related to the spacings between planes of molecules in the lattice.

To measure a powder pattern, a randomly oriented powdered sample is prepared so as to expose all the planes of a sample. The scattering angle is determined by slowly rotating the sample and measuring the angle of diffracted x-rays (with a scintillation counter) with respect to the angle of the incident beam. Alternatively, the angle between sample and source can be kept fixed while moving the detector to determine the angles of the scattered radiation. Knowing the wavelength of the incident beam, the spacing between the planes (identified as the d-spacings) is calculated using Bragg's law.

A very useful complement to ordinary powder x-ray diffraction is variable-temperature XRD. In this method, the sample is contained on a stage that can be heated to any desired temperature. The method is extremely useful in the study of thermally induced phenomena, and it is seen as a complement to thermal methods of analysis.

The XRD powder patterns of the anhydrate and trihydrate phases of ampicillin are shown in Fig. 5. The nonequivalence in the crystal structures is immediately evident from a comparison of the powder patterns. It is also known that amoxicillin trihydrate is isomorphous with ampicillin trihydrate [39], and a comparison of the corresponding powder patterns demonstrates the equivalence of the structures.

D. Thermal Methods of Analysis

Thermal analysis methods are defined as those techniques in which a property of the analyte is determined as a function of an externally applied temperature

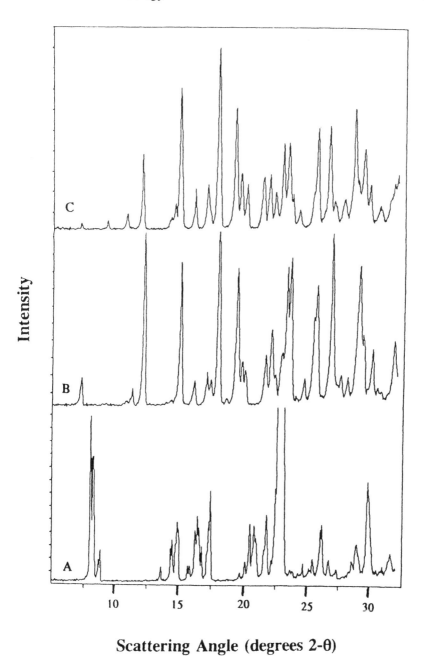

Fig. 5 X-ray powder patterns of (A) ampicillin anhydrate, (B) ampicillin trihydrate, and (C) amoxicillin trihydrate.

[40]. The sample temperature is increased in a linear fashion, while the property in question is evaluated on a continuous basis. These methods are used to characterize compound purity, polymorphism, solvation, degradation, and excipient compatibility [41]. Thermal analysis methods are normally used to monitor endothermic processes (melting, boiling, sublimation, vaporization, desolvation, solid–solid phase transitions, and chemical degradation) as well as exothermic processes (crystallization and oxidative decomposition). Thermal methods can be extremely useful in preformulation studies, since the carefully planned studies can be used to indicate the existence of possible drug–excipient interactions in a prototype formulation [7].

Differential thermal analysis (DTA) is the monitoring of the difference in temperature between a sample and a reference as a function of temperature [42]. Differences in temperature between the sample and reference are observed when changes occur that require a finite heat of reaction. If ΔH is positive (endothermic reaction), the temperature of the sample will lag behind that of the reference. If the ΔH is negative (exothermic reaction), the temperature of the sample will exceed that of the reference. Differential thermal analysis is not normally used for quantitative work; instead it is used to deduce temperatures associated with thermal events.

Differential scanning calorimetry (DSC) is similar to DTA and is the most widely used method of thermal analysis. In the DSC method, the sample and reference are kept at the same temperature and the heat flow required to maintain the equality in temperature is measured [43]. This is achieved by placing separate heating elements in the sample and reference cells, with the rate of heating by these elements being controlled and measured. The plots from DSC are obtained as the differential rate of heating (in units of watts/second, calories/second, or Joules/second) against temperature. The area under a DSC peak is directly proportional to the heat absorbed or evolved by the thermal event, and integration of these peak areas yields the heat of reaction (in units of calories/second · gram or Joules/second · gram).

When a compound is observed to melt without decomposition, DSC analysis can be used to determine the absolute purity [44]. This method can therefore be used to evaluate the absolute purity of a given compound without reference to a standard, with purities being obtained in terms of mole percent. Unfortunately, the method is limited to reasonably pure compounds that melt without decomposition, since the assumptions justifying the methodology fail when the compound purity is below approximately 97 mole%.

Thermogravimetry (TG) is a measure of the thermally induced weight loss of a material as a function of the applied temperature [45]. Thermogravimetric analysis is restricted to studies that involve either a mass gain or loss, and it is most commonly used to study desolvation processes and compound decomposition. The major use of TG analysis is in the quantitative determination of the total volatile content of a solid. When a solid can decompose by means of several discrete, sequential reactions, the magnitude of each step can be separately evaluated. Thermogravimetric analysis of compound decomposition can also be used to compare the stability of similar compounds. The higher the decomposition temperature of a given compound, the more positive would be the ΔG value, and therefore the greater would be the stability.

Differential scanning calorimetry can be extremely useful in the study of compound polymorphism. Suitably prepared films of 2,4-dinitrophenyl-2,4-dinitrobenzoate will exhibit phase transformations among all four polymorphs [46], as has been shown in Fig. 6. The complicated DSC thermogram contains

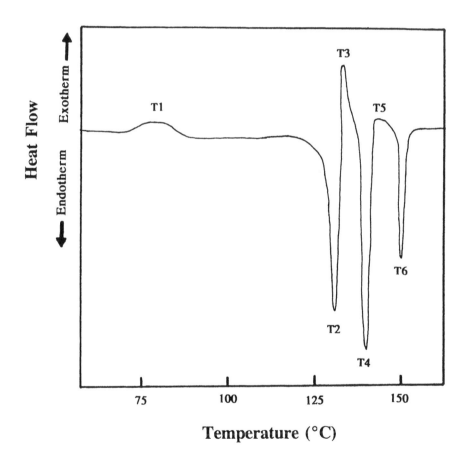

Fig. 6 Differential scanning calorimetry thermogram of 2,4-dinitrophenyl-2,4-dinitrobenzoate, illustrating the recrystallization of form IV into form III (T1), the melting of forms III and II (T2 and T4), the solidification of the melts produced by T2 and T4 (T3 and T5), and the melting of form I (T6). (Data adapted from Ref. 46.)

six phase transitions, but each of these have been assigned. The transition T1 corresponds to the recrystallization of form IV into form III, while transitions T2 and T4 represent the melting of forms III and II, respectively. Transitions T3 and T5 correspond to the solidification of the melts produced by T2 and T4, and finally T6 marks the melting of form I.

IV. PROPERTIES ASSOCIATED WITH THE BULK LEVEL

Bulk material properties may be conveniently defined as those characteristics of a solid which require a large ensemble of particles for measurements to be made. Once a solid formulation has reached the bulk manufacturing stage, these bulk physical properties are probably of the highest degree of importance. The ability to totally control a large-scale formulation process will be governed by the degree of knowledge amassed for the system in question, and the most pertinent body of knowledge concerns the properties that directly relate to those of the bulk powders involved. It is also true that reproducibility in all aspects of lot-to-lot behavior can be significantly improved through the implementation of properly designed specifications. The testing of raw materials is particularly important at this stage, with only those species passing appropriate specifications being used for the manufacturing process [47].

A. Micromeritics

When applied to powders, micromeritics is taken to include the fields that relate to the nature of the surfaces making up the solid. Of all the properties that could be measured, the surface area, its porosity, and the density of a material are generally considered to be the most pharmaceutically relevant parameters.

The surface area of a solid material is important in that it provides information on the available void spaces on the surfaces of a powdered solid [48]. In addition, the dissolution rate of a solid is partially determined by its surface area. The most reproducible measurements of the surface area of a solid are obtained by adsorbing a monolayer of inert gas onto the solid surface at reduced temperature and subsequently desorbing this gas at room temperature. The sorption isotherms obtained in this technique are interpreted using the equations developed by Brunauer, Emmett, and Teller, and therefore the technique is referred to as the B.E.T. method [49]. The surface area is obtained in units of square meters of surface per gram of material.

Any condensable, inert gas can be used for B.E.T. measurements, but the preferred gases are nitrogen and krypton. Nitrogen is used for most samples exhibiting surface areas of 2 m^2/g or greater, but materials with smaller surface areas should be measured using krypton. The gas to be adsorbed (the adsorbate) is mixed with an inert, noncondensable, carrier gas (usually helium). A range of 5 to 30% adsorbate in carrier gas is commonly used. The use of multiple adsorbate gas levels in a B.E.T. determination is recommended. A proposed compendial method has been proposed [50].

Relations between the internal surface area and tablet properties have been drawn through the characterization of a variety of lactose compacts [51]. For example, a given bulk sample of anhydrous α-lactose was sieved into selected size fractions, and compacts of these prepared through compression at 37.5 mPa. Nitrogen-gas adsorption was used to evaluate the surface area of the compacts, and the tablet crushing strengths of each were determined. As can be seen in Fig. 7, the crushing strength of the compacts was directly proportional (and almost linearly dependent) on the surface area of the compacted material. It was also found that compacts prepared from the coarser size fractions contained

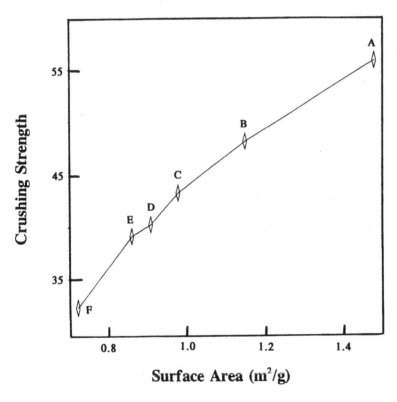

Fig. 7 Relation of tablet crushing strength and surface area of selected size fractions of anhydrous α-lactose, which had been compressed at 37.5 mPa. Data are shown for initial size fractions of (A) 24–32 μm, (B) 32–63 μm, (C) 63–100 μm, (D) 100–160 μm, (E) 160–200 μm, and (F) 250–315 μm. (Data adapted from Ref. 51.)

significantly lower surface areas than compacts prepared from the fine fractions [51].

Although a number of methods are available to characterize the interstitial voids of a solid, the most useful of these is mercury intrusion porosimetry [52]. This method is widely used to determine the pore-size distribution of a porous material, and the void size of tablets and compacts. The method is based on the capillary rise phenomenon, in which excess pressure is required to force a nonwetting liquid into a narrow volume.

Mercury, with its contact angle on glass of approximately 140°, is most commonly used as the intrusion fluid. The mercury is forced into the pores of the sample using an externally applied pressure, with the smallest pores requiring the highest pressures to effect the filling. The Washburn equation, as applied to circular pore openings, is used to relate the applied pressure and the pore size opening [52].

Measurements of particle porosity are a valuable supplement to studies of specific surface area, and such data are particularly useful in the evaluation of materials used in direct compression processes. For example, both micromeritic properties were measured for several different types of cellulosic-type excipients [53]. Surface areas by the B.E.T. method were used to evaluate all types of pore structures, while the method of mercury intrusion porosimetry used could not detect pores smaller than 10 nm. The data permitted a ready differentiation between the intraparticle pore structure of microcrystalline and agglomerated cellulose powders.

Another extremely important micromeritic parameter is that of powder density, defined as the ratio of mass to volume. Three types of density are normally differentiated, which differ in their determination of volume occupied by the powder. *Bulk* density is obtained by measuring the volume of a known mass of powder sample (that has been passed through a mesh screen) into a suitable volume-measuring apparatus [54]. When a graduated cylinder is used for the measurement, the volume is determined to the nearest milliliter. The bulk density is then obtained by dividing the mass of solid by the unsettled apparent volume. A measurement of *tapped* density is normally obtained at the same time, with the volume of the solid being measured after subjecting the system to a number of controlled shocks. The repeated mechanical stress causes the powder bed to pack into a smaller volume, and so it follows that the tapped density will always be higher than the bulk density.

The *true* density of a solid is the average mass per unit volume, exclusive of all voids that are not a fundamental part of the molecular packing arrangement [55]. This density parameter is normally measured by helium pycnometry, where the volume occupied by a known mass of powder is determined by measuring the volume of gas displaced by the powder. The true density of a solid is an intrinsic property characteristic of the analyte, and it is determined by the composition of the unit cell.

B. Powder Characterization

Evaluation of the mechanical characteristics of powdered solids is vitally important to the processing of these materials. Information can be obtained on bulk powders prior to their processing, during the compaction process through the instrumentation of tablet presses, or on tablet compacts after these have been compressed. In the first instance, work is generally centered on determinations of the degree of flowability associated with a given powder. Data obtained during the compaction of powdered solids can be an invaluable source of information to optimize the consolidation process. Measurements conducted on consolidated materials are also used during process optimization, and they can also be employed as part of the quality control testing. It should be recognized, however, that particle–particle interactions [56] are at the center of these investigations, and all the methods are designed to deduce such information.

It is generally agreed that one of the more important parameters of interest to formulators is the flowability of their powdered solids [57,58]. The process-ability of these materials is greatly affected by flowability concerns, since the materials invariably need to be moved from place to place. For example, when tablets are to be compressed at high speeds, the efficiency of the machine will only be suitable if the powder feed can be delivered at a sufficiently high rate.

When powders flow, they do so in either a steady controlled fashion or in an uncontrollable gushing manner. Since many pharmaceutical compounds are cohesive in nature, their flow characteristics tend to be undesirable [59]. One of the aims of granulation is to reduce the cohesive nature of the individual components, producing a uniformly blended material whose physical properties are more suitable for processing. The flowability of a given granulation is an exceedingly important characteristic, which should be maximized to permit use in high-speed tableting machines.

Carr has described a system that can be extremely useful in the evaluation of the flowability of powdered solids [60]. In his approach, Carr defined a number of parameters related to flow, which are scored after their measurement according to a weighting system. Powder flowability is evaluated using the *angle of repose* (defined as the angle formed when a cone of powder is poured onto a flat surface), the *angle of spatula* (defined as the angle formed when material is raised on a flat surface out of bulk pile), *compressibility* (obtained from measurement of the bulk and tapped material densities), and *cohesion* (relating to the attractive forces that exist on particle surfaces). The overall summation of these permits deductions regarding the degree of powder flowability.

When powders flow, they do so either in a steady controlled fashion (as in the case of dry sand), or in a uncontrolled gushing manner (as would damp sand, for which the entire bulk tries to move in a solid mass). This latter condition is termed *floodable* flow; it is most characteristic of the flow of cohesive, sticky powders. The floodability of a powder is determined by its overall flowability (the determination of which has just been described), the *angle of fall* (obtained as the new repose angle when the powder cone is mechanically shocked), the *dispersibility* (ability of a given powder to become fluidized), and the *angle of difference* (obtained as the numerical difference between the angle of fall and angle of repose). Carr has also detailed the procedure whereby indices are deduced for each floodability parameter, and how the summation of these indices yields a parameter indicative of the tendency of a powder to exhibit floodable flow [60].

To illustrate the utility of Carr's method in the evaluation of powder flow, full characterization of nine lots of nystatin was carried out [61]. Measurements of the actual mass flow rates were obtained, as well as each of the parameters specified by Carr. As evident in Fig. 8, the overall flowability index was indeed a reliable predictor of the relative degree of bulk powder flow. Detailed comparisons of the possible relation of each parameter with the mass flow rates were conducted with the aim of determining which parameter exerted the largest influence over the mass flow rate. It was found in this study that the powder flow rates were largely determined by the sample compressibilities and degrees of cohesion, with separate inverse linear relationships existing between these properties and the mass flow rates. Interestingly, no correlation between actual powder flow rates and the angle of repose was detected, in spite of the conventional wisdom that the repose angle may be used as an indicator of powder flow [62].

Studies involving instrumented compaction equipment can be extremely useful in the development of dosage forms, especially when the amount of drug substance is limited in quantity. Marshall has described a program in which dynamic studies of powder compaction can be used at all stages of the development process to acquire formulation information [63]. The initial experiments include a determination of the intrinsic compactability of the compound. In subsequent work, simple tablets are prepared, and tested for dissolution, potency, and content uniformity. Through studies of the compaction mechanism, it becomes possible to deduce means to improve the formulation under study.

When the work is carried out on a compaction simulator, it is possible to subject the formulation to high-speed tableting cycles in order to evaluate strain forces in the compacts. This work can be of utmost importance, since it appears that strain rates can exert strong influences on a variety of tablet properties [64]. The information gathered in this way can be used to aid in the transfer of the manufacturing process among different types of instrumentation, and to aid in the technology transfer process.

In one study, the formulation efficiency of several direct compression materials was evaluated using instrumented press methodology [65]. It was found that subtle changes in the structure of the component particles could lead to the

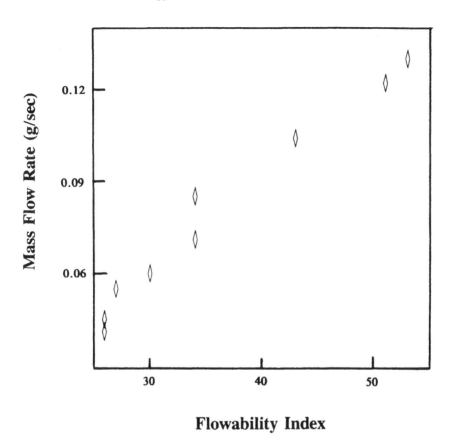

Fig. 8 Relation of the actual mass flow rate with Carr's flowability index, illustrated by data obtained on nine lots of nystatin. The mass flow rates were determined using a 11-mm aperture to govern the powder flow.

observation of significantly different behaviors upon compression. The tablet hardness and compressibility of differently sourced sucrose materials, obtained at comparable compressional forces, was found to vary significantly with the source of the compound.

The majority of active pharmaceutical agents are administered as tablets, and as a result the characterization of compact species is of great interest to formulators. During the compression step, a variety of particle–particle interactions take place, which ultimately lead to the formation of a stable entity. One may envision that the compaction process results in such consolidation of the input solids that the final density of the tablet approaches the true density of the component materials.

Most workers evaluate the quality of compacted materials through measurements of tablet hardness and friability. Hiestand and coworkers have developed several indices of tableting performance, which are indicative of the performance of materials during their compaction [66]. The *Bonding Index* is an estimation of the survival of tablet strength following the decompression that takes place after the tablet is ejected from the press. The *Brittle Fracture Index* is a measure of the brittleness of a material, and it provides a measure of the ability of a compact to relieve stress through plastic deformation. Finally, the *Strain Index* is an indicator of the relative strain that forms in a compact following its decompression.

Brittle fracture indices were obtained for a series of compacted direct compression excipients that had been exposed to various degrees of ambient relative humidity [67]. As illustrated in Fig. 9, compacts of the essentially

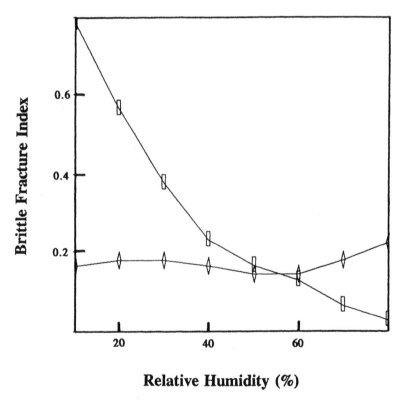

Relative Humidity (%)

Fig. 9 Effect of ambient relative humidity on the brittle fracture index of compacts made from Avicel PH 101 (◊) and Emdex (□). (Data adapted from Ref. 67.)

nonhygroscopic Avicel exhibited essentially the same degree of brittleness regardless of its exposure to relative humidity. On the other hand, compacts of the more hygroscopic Emdex (which is supplied as 90–93% dextrose and 2–5% maltose) exhibit a substantial decrease in tablet brittleness as the tablets are exposed to higher degrees of relative humidity. These phenomena were ascribed to the plasticizing effects of the adsorbed water.

C. Solubility

Evaluation of the solubility of a drug substance is of extreme importance to the drug development process, since it must become dissolved in a fluid medium for it to have its intended therapeutic effect. Essentially all aspects of this topic have been reviewed in the monograph of Grant and Higuchi [68]. The solubility of the drug entity is normally determined very early in the preformulation stage, since this property will play an important role in the formulation development and choice of dosage form. In fact, it has been stated that unless the aqueous solubility of the compound exceeds 10 mg/ml, problems may be encountered during adsorption [69]. For compounds having a solubility less than 1 mg/ml, the drug should be derivitized as a salt form to increase the intrinsic solubility.

One approach to the study of solubility is to evaluate the time dependence of the solubilization process, such as is conducted in the dissolution testing of dosage forms [70]. In this work, the amount of drug substance that becomes dissolved per unit time under standard conditions is followed. Within the accepted model for pharmaceutical dissolution, the rate-limiting step is the transport of solute away from the interfacial layer at the dissolving solid into the bulk solution. To measure the intrinsic dissolution rate of a drug, the compound is normally compressed into a special die to a condition of zero porosity. The system is immersed into the solvent reservoir, and the concentration monitored as a function of time. Use of this procedure yields a dissolution rate parameter that is intrinsic to the compound under study and that is considered an important parameter in the preformulation process. A critical evaluation of the intrinsic dissolution methodology and interpretation is available [71].

The other main approach to solubility is to measure the concentration of the drug substance after an equilibrium has been reached with the solvent in question. This work is also conducted very early during the development process, normally at the stage of preformulation characterization [7]. A full discussion of the various aspects of solution theory is beyond the scope of the present chapter, but it is available [68]. Only a few salient points will be addressed in the following paragraphs.

A solution can be defined as a homogeneous molecular mixture of two or more substances, with the component in excess being termed the *solvent* and the other being the *solute*. Since a solution is homogeneous, it must exist in one of the three fundamental states (gas, liquid, or solid). The properties of a pure substance depend on two of the three variables (temperature, pressure, and volume), with the third being fixed by the equation of state. For a solution, a fourth variable (composition) must also be specified.

The dissolution of a solute into a solvent perturbs the colligative properties of the solvent, affecting the freezing point, boiling point, vapor pressure, and osmotic pressure. The dissolution of solutes into a volatile solvent system will affect the vapor pressure of that solvent, and an *ideal solution* is one for which the degree of vapor pressure change is proportional to the concentration of solute. It was established by Raoult in 1888 that the effect on vapor pressure would be proportional to the mole fraction of solute, and independent of temperature. This behavior is illustrated in Fig. 10A, where individual vapor pressure curves are

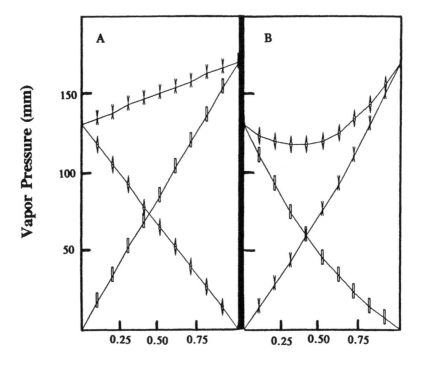

Mole Fraction, Solute

Fig. 10   Dependence of vapor pressure of a solution containing a volatile solute, illustrated for (A) an ideal solution and (B) a nonideal solution and shown as a function of mole fraction of the solute. Individual vapor pressure curves are shown for the solvent (◇) the solute (□), and for the sum of these (×).

shown. The overall vapor pressure of the system equals the sum of the individual components, and the ideality of the system is evident in the linearity of the composite curve. Simple relations can also be advanced in which the other colligative properties are specified by solution composition.

Whenever the solute and solvent exhibit significant degrees of mutual attraction, deviations from the simple relationships will be observed. The properties of these nonideal solutions must be determined by the balance of attractive and disruptive forces. When a definite attraction can exist between the solute and solvent, the vapor pressure of each component is normally decreased. The overall vapor pressure of the system will then exhibit significant deviations from linearity in its concentration dependence, as is illustrated in Fig. 10B.

The curves in Fig. 10 were drawn for the particular instance of a volatile solute dissolved in a volatile solvent, such as would exist for the acetone–chloroform system (whose diagram is very nearly like that of Fig. 10B). For many nonvolatile solutes, it not possible to trace smooth partial pressure curves across the entire range of mole fractions. This is especially true for aqueous salt solutions, where at a certain concentration of solute the solution becomes saturated. Any further addition of crystalline solute to the system does not change the mole fraction in the liquid phase, and the partial pressure of water thereafter remains constant, in accord with the phase rule. This phenomenon permits the use of saturated salt solutions as media to establish fixed relative humidity values in closed systems [72].

An ideal solution is defined as one for which the chemical potential of every component (μ_i) is related to its mole fraction by

$$\mu_i = \mu_i^* + RT \ln X_i \tag{1}$$

where μ_i^* is a function of temperature and pressure only. Beginning with an analysis of the properties of ideal solutions, Schroeder developed an equation relating solubility with several fundamental parameters [73]. The deductions derived from Schroeder's equation are that (1) the solubility of a solid increases with increasing temperature, (2) the lower the melting point of a solid, the greater its solubility will be, and (3) the smaller the latent heat of melting of a solid, the greater will be its solubility. Though these predictions were made on the supposition of ideal solutions, they have been found to be in general agreement with experience.

To deal with the properties of real solutions, most workers determine the magnitude of a pure number that can be multiplied by the mole fraction to bring the equation of state back to an ideal form:

$$\mu_i = \mu_i^* + RT \ln \gamma_i X_i \tag{2}$$

This quantity, γ_i, is known as the activity coefficient of the species in question, and it may be a function not only of temperature and pressure, but also of the mole fractions of all substances in the solution. The usual convention is to assume that the activity coefficient of the solute will approach unity at infinite dilution (the condition where $X_i = 0$), while the activity coefficient of the solvent becomes unity for the pure solvent. According to this convention, μ_i^* stands for the chemical potential of pure solute in a hypothetical liquid state corresponding to extrapolation from infinite dilution. A great effort has been placed on theories that permit the calculation of activity coefficients [74].

Much effort has been expended on models that can be used to predict the solubility behavior of solutes, with good success being attained using a semi-empirical, group contribution approach [75]. In this system, the contributions made by individual functional groups are summed to yield a composite for the molecule, which implies a summation of free energy contributions from constituents. This method has proven to be useful in the prediction of solubility in water and in water–cosolvent mixtures. In addition to the simplest methodology, a variety of more sophisticated approaches to the prediction of compound solubility have been advanced [68].

The properties of organic liquids relevant to their use as solvating agents have also been reviewed [76]. The ability of liquids to solvate a solute species depends mainly on their polarity and polarizability properties, ability to hydrogen bond, and cohesive electron density. These molecular properties are best measured by the Kamlet–Taft solvatochromic parameters, and the square of Hildebrand's solubility parameter.

No discussion of solubility would be complete without making reference to phase solubility analysis. This methodology entails the determination of the purity of a substance by the accurate measurement of its solubility as a function of the sample to solvent ratio [77]. The procedure consists of mixing increasing amounts of solute with a fixed volume of solvent, equilibrating the mixtures, and measuring the amount of solute dissolved in each instance. When the mass dissolved is plotted against the total weight of sample per unit volume of solvent, the presence of impurities in the analyte will cause to the curve to deviate from linearity. The method is not extensively used anymore owing to its tedious nature and lack of sensitivity (impurities less than 0.1% cannot normally be detected). It is, however, one of the few absolute methods available for determinations of compound purity, since a reference standard is not required. In addition, the number and types of impurities need not be known in advance, and they will in fact be disclosed from the analysis.

D. Sorption of Water

The sorption of water by a solid is of extreme interest to the pharmaceutical scientist since the stability of a drug entity in either its bulk or formulated state often critically depends on the nature of water contained within the solid [78].

Water may be considered as being either bound or unbound, with the unbound portion generally being responsible for reactions requiring moisture as a reactant. In addition to affecting the stability of the formulation, water can also cause strong perturbations in the overall physical properties of the solid.

Water is introduced into closed pharmaceutical systems either accompanying the input materials or in the headspace as relative humidity [79]. Whatever water is contained within the dosage form and its container will ultimately equilibrate among the components according to its affinity for the solid ingredients and the number of association sites. The Sorption–Desorption Moisture Transfer model has been used to evaluate the thermodynamically favored state that will result after the equilibration process is complete [79].

Zografi has considered in detail the sorption of water onto both crystalline and amorphous solid surfaces [80]. It has been concluded that the tendency for adsorption onto a solid surface is strongly dependent on the vapor pressure, the system temperature, and on the magnitude of the interfacial binding energy. It is generally agreed that the adsorption process takes place with the water forming hydrogen bonds with the hydrophilic sites on the surface of the solid. Water is unusual in its sorption properties in that it is capable of forming multiple layers during the sorption process, with these being able to form even before the entire surface is covered by a monolayer. Sorption at the defect sites existing on a nonideal crystal surface can take place preferentially, and the water bound at these sites may even become catalytically activated [81].

The sorption of water by excipients derived from cellulose and starch has been considered by numerous workers, with at least three thermodynamic states having been identified [82]. Water may be directly and tightly bound at a 1:1 stoichiometry per anhydroglucose unit, unrestricted water having properties almost equivalent to bulk water, or water having properties intermediate between these two extremes. The water sorption characteristics of potato starch and microcrystalline cellulose have been determined, and comparison of these is found in Fig. 11. While starch freely adsorbs water at essentially all relative humidity values, microcrystalline cellulose only does so at elevated humidity values. These trends have been interpreted in terms of the degree of available cellulosic hydroxy groups on the surfaces, and as a function of the amount of amorphous material present [83].

Since an understanding of the nature of water sorption by materials is of extreme interest to the formulator, an evaluation of the degree of hygroscopicity associated with a given material is crucial to the development process. A systematic approach for these types of studies has been outlined, in which the kinetics of water adsorption can be deduced [84]. This work should be performed at the preformulation stage, where the effect of water on the various components (and mixtures of these) needs to be addressed prior to any final decision as to the formulation composition [85]. It would be far better to discover any

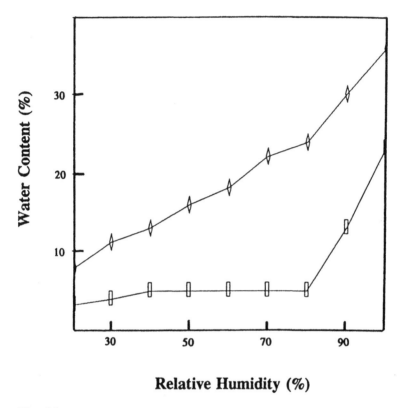

Fig. 11   Sorption isotherms for water onto potato starch ($\Diamond$) and microcrystalline cellulose ($\square$). (Data adapted from Ref. 83.)

moisture-related problems in the formulation at the early stages of development rather than later, when these can have profound effects on the quality of data being obtained for a regulatory filing.

The presence of hygroscopic excipients in a formulation containing a hygroscopic drug will lead to enhanced water sorption rates [86]. The percent weight gain associated with the exposure of two tablet formulations to 80% relative humidity is shown in Fig. 12. The hygroscopicity of formulation A was largely due to the water sorption characteristics of the drug entity, while the hygroscopicity of formulation B contained the additional effect of the starch excipient.

V.   SUMMARY

Each chapter in the present volume contains information that permits an entry-level understanding into a given method and provides the references that

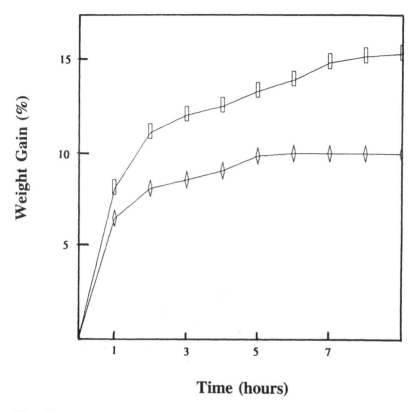

Time (hours)

Fig. 12 Percent weight gain associated with the exposure of tablet formulations to 80% relative humidity at 40°C. Formulation A (◇) was essentially a 1:11 blend of the drug entity and microcrystalline cellulose, while formulation B (□) was essentially 1:5.5:5.5 drug–microcrystalline cellulose–starch.

will lead into a deeper understanding. Each chapter also contains significant examples of how each technique is used, and what sort of information can be extracted from its use. The applications illustrate both the strengths and weaknesses of each technique, and they indicate where each method would fit in an overall characterization scheme.

A systematic Materials Science approach is now highly appropriate for the pharmaceutical field. It may be envisioned that a protocol for the complete physical characterization of a solid material could be easily developed following the approach just outlined. Ideally, each lot of active drug, excipients, or formulated blends of these would be characterized as fully as possible at the early stages of drug development. A feedback loop would then be established after each formulation run, in which the physical characteristics of the input materials were correlated with the quality of produced product. Out of these studies, we hope, would come an understanding as to the particular physical properties that would be predictive of success in the production of a given formulation.

As the maturity of the process increases, only these key parameters would require continued monitoring. Ultimately, the data collected on these properties would permit the generation of appropriate material specifications. If the work had been performed properly, then it would be possible to specify limits for raw-material properties that would ensure the final product will be satisfactory. These guidelines would naturally apply only to the specific formulation, but through continued use of the systematic Materials Science approach the more general trends would become apparent.

ACKNOWLEDGMENTS

Special thanks are due to my many coworkers at Bristol-Myers Squibb who helped to develop the systematic, multidisciplinary approach to the physical characterization of pharmaceutical solids described in this chapter and in the remainder of this book. They are Ann Newman, David Bugay, Kenneth Morris, Joseph DeVincentis, Clifford Sachs, Gary Barbera, Geoffrey Lewen, Susan Bogdanowich, Paul Findlay, Frank Rinaldi, Patricia Cortina, Glen Young, Ayman Ahmed, and Imre Vitez. Everyone has played a role in our Materials Science program, each making their special contribution, which in turn allowed the end result to become so impressive.

REFERENCES

1. S. R. Byrn, *Solid State Chemistry of Drugs*, Academic Press, New York, 1982.
2. L. Lachman, H. A. Lieberman, and J. L. Kanig, *The Theory and Practice of Industrial Pharmacy*, 3rd ed., Lea and Febiger, Philadelphia, 1986.
3. J. T. Carstensen, *J. Pharm. Sci.*, *63*, 1 (1974).
4. S. R. Byrn, *J. Pharm. Sci.*, *65*, 1 (1976).
5. D. C. Monkhouse and L. Van Campen, *Drug Dev. Ind. Pharm.*, *10*, 1175 (1984).
6. H. G. Brittain, S. J. Bogdanowich, D. E. Bugay, J. DeVincentis, G. Lewen, and A. W. Newman, *Pharm. Res.*, *8*, 963 (1991).
7. J. I. Wells, *Pharmaecutical Preformulation: The Physicochemical Properties of Drug Substances*, Halsted Press, New York, 1988.
8. G. Kortüm, *Reflectance Spectroscopy*, Springer-Verlag, New York, 1969.
9. W. W. Wendlandt and H. G. Hecht, *Reflectance Spectroscopy*, Interscience Pub., New York, 1966.
10. R. W. Frei and J. D. MacNeil, *Diffuse Reflectance Spectroscopy in Environmental Problem Solving*, CRC Press, Cleveland, Ohio, 1973.
11. D. G. Pope and J. L. Lach, *Pharm. Acta Helv.*, *50*, 165 (1975).

12. M. J. Jozwiakowski and K. A. Connors, *J. Pharm. Sci.*, *77*, 241 (1988).
13. *CIE Publication 15.2*, *Colorimetry*, 2nd ed., Central Bureau of CIE, Vienna, Austria, 1986.
14. D. B. Judd and G. Wyszecki, *Color in Business, Science, and Industry*, 2nd ed., John Wiley & Sons, New York, 1963, pp. 264–361.
15. F. W. Billmeyer, Jr., and M. Saltzman, *Principles of Color Technology*, Wiley-Interscience, New York, 1966.
16. M. Wirth, *J. Pharm. Sci.*, *80*, 1177 (1991).
17. K. Krishnan and J. R. Ferraro, in *Fourier Transform Infrared Spectroscopy*, Vol. 4, Academic Press, New York, 1982.
18. R. T. Conley, *Infrared Spectroscopy*, Allyn and Bacon, Boston, 1966.
19. H. G. Brittain, K. R. Morris, D. E. Bugay, A. B. Thakur, and A. T. M. Serajuddin, *J. Pharm. Biomed. Anal.*, *11*, 1063 (1993).
20. N. B. Colthup, L. H. Daly, and S. E. Wiberley, *Introduction to Infrared and Raman Spectroscopy*, 2nd edition, Academic Press, London, 1975.
21. M. C. Davies, J. S. Binns, C. D. Melia, P. J. Hendra, D. Bourgeois, S. P. Church, and P. J. Stephenson, *Int. J. Pharm.*, *66*, 223 (1990).
22. E. W. Ciurczak, *Appl. Spectrosc. Rev.*, *23*, 147 (1987).
23. B. F. MacDonald and K. A. Prebble, *J. Pharm. Biomed. Anal.*, *11*, 1077 (1993).
24. C. A. Fyfe, *Solid State NMR for Chemists*, CFC Press, Guelph, 1983.
25. H. Saito, *Mag. Reson. Chem.*, *24*, 835 (1986).
26. T. G. Rochow and E. G. Rochow, *An Introduction to Microscopy by Means of Light, Electrons, X-rays, or Ultrasound*, Plenum Press, New York, 1978.
27. G. E. Amidon, *Pharm. Forum*, *18*, 4089 (1992).
28. M. Kuhnert-Brandstätter, *Thermomicroscopy in the Analysis of Pharmaceuticals*, Pergamon Press, Oxford, 1971.
29. J. W. Carson, *Drug Dev. Indust. Pharm.*, *14*, 2749 (1988).
30. H. G. Barth, *Modern Methods of Particle Size Analysis*, John Wiley & Sons, New York, 1984.
31. B. H. Kaye, *Direct Characterization of Fine Particles*, John Wiley & Sons, New York, 1981.
32. M. E. Fayad and L. Otten, *Handbook of Powder Science and Technology*, Van Nostrand Reinhold, New York, 1984.
33. K. Marshall, *Pharm. Forum*, *19*, 4637 (1993).
34. H. G. Brittain, *Pharm. Forum*, *19*, 4640 (1993).
35. T. Allen, *Particle Size Measurement*, 3rd ed., Chapman and Hall, London, 1981.
36. B. Chu, *Laser Light Scattering*, Academic Press, New York, 1974.
37. M. M. Woolfson, *X-Ray Crystallography*, Cambridge University Press, Cambridge, 1970.
38. G. H. Stout and L. H. Jensen, *X-Ray Structure Determination: A Practical Guide*, Macmillan Co., New York, 1968.
39. M. O. Boles, R. J. Girven, and P. A. C. Gane, *Acta Cryst.*, *B34*, 461 (1978).
40. W. W. Wendlandt, *Thermal Analysis*, 3rd ed., John Wiley & Sons, New York, 1986.
41. D. Giron, *J Pharm. Biomed. Anal.*, *4*, 755 (1986).
42. M. I. Pope and M. D. Judd, *Introduction to Differential Thermal Analysis*, Heyden, London, 1977.

43. D. Dollimore, "Thermoanalytical Instrumentation," in *Analytical Instrumentation Handbook*, G. W. Ewing, ed., Marcel Dekker, New York, 1990, pp. 905–960.

44. R. L. Blaine and C. K. Schoff, *Purity Determinations by Thermal Methods*, ASTM Press, Philadelphia, 1984.

45. C. J. Keattch and D. Dollimore, *Introduction to Thermogravimetry*, 2nd ed., Heyden, London, 1975.

46. M. Kuhnert-Brandstätter and M. Riedmann, *Microchim. Acta (Wien)*, *II*, 107 (1987).

47. H. G. Brittain, *Pharm. Tech.*, *17*(7), 66 (1993).

48. S. Lowell and J. E. Shields, *Powder Surface Area and Porosity*, 2nd ed., Chapman and Hall, London, 1984.

49. S. Brunauer, P. H. Emmett, and E. Teller, *J. Am. Chem. Soc.*, *60*, 309 (1938).

50. D. J. W. Grant, *Pharm. Forum*, *19*, 6171 (1993).

51. H. Leuenberger, J. D. Bonny, C. F. Lerk, and H. Vromans, *Int. J. Pharm.*, *52*, 91 (1989).

52. J. M. Hayes and P. Rossi-Doria, *Principles and Applications of Pore Structural Characterization*, J. W. Arrowsmith, Bristol, 1985.

53. T. Pesonen and P. Paronen, *Drug Dev. Indust. Pharm.*, *16*, 31 (1990).

54. L. Augsburger and G. E. Amidon, *Pharm. Forum*, *20*, 6931 (1994).

55. M. S. Bergren, *Pharm. Forum*, *20*, 7221 (1994).

56. E. N. Hiestand, *J. Pharm. Sci.*, *55*, 1325 (1966).

57. B. S. Neumann, *Adv. Pharm. Sci.*, *2*, 181 (1967).

58. M. Deleuil, *S.T.P. Pharma*, *3*, 668 (1987); D. Duchene, *ibid.*, *3*, 794 (1987).

59. N. Pilpel, *Adv. Pharm. Sci.*, *3*, 173 (1971).

60. R. L. Carr, *Chemical Engineering*, 163 (January, 1965); *ibid.*, 69 (February, 1965).

61. H. G. Brittain, G. Lewen, and A. W. Newman, manuscript in preparation.

62. J. T. Carstensen, *Pharmaceutical Principles of Solid Dosage Forms*, Technomic Pubs., Lancaster, 1993, p. 209.

63. K. Marshall, *Drug Dev. Ind. Pharm.*, *15*, 2153 (1989).

64. M. Celik and D. N. Travers, *Drug Dev. Ind. Pharm.*, *11*, 299 (1985).

65. C. O. Ondari, C. E. Kean, and C. T. Rhodes, *Drug Dev. Ind. Pharm.*, *15*, 1517 (1988).

66. E. N. Hiestand and D. P. Smith, *Powder Tech.*, *38*, 145 (1984); *Int. J. Pharm.*, *67*, 217, 231 (1991).

67. S. Malamataris, P. Goidas, and A. Dimitriou, *Int. J. Pharm.*, *68*, 51 (1991).

68. D. W. Grant and T. Higuchi, *Solubility Behavior of Organic Compounds*, Wiley-Interscience, New York, 1990.

69. S. A. Kaplan, *Drug Metab. Rev.*, *1*, 15 (1972).

70. H. M. Abdou, *Dissolution, Bioavailability, and Bioequivalence*, Mack Pub., Easton, 1989.

71. M. Nicklasson, A. Brodin, and L. Sundelöf, *Acta Pharm. Suec.*, *19*, 109 (1982); *Int. J. Pharm.*, *15*, 87 (1983).

72. L. Greenspan, *J. Res. Nat. Bur. Std.*, *81A*, 89 (1977).

73. J. H. Hildebrand and R. L. Scott, *The Solubility of Non-Electrolytes*, Reinhold, New York, 1950, Chapter 17.

74. K. Denbigh, *The Principles of Chemical Equilibrium*, University Press, Cambridge, 1968.

75. S. H. Yalkowsky, R. Pinal, and S. Banerjee, *J. Pharm. Sci.*, *77*, 74 (1988).
76. Y. Marcus, *Chem. Soc. Rev.*, *22*, 409 (1993).
77. T. Higuchi and K. A. Connors, *Adv. Anal. Chem. Instr.*, *4*, 117 (1965).
78. J. T. Carstensen, *Drug Dev. Ind. Pharm.*, *14*, 1927 (1988).
79. M. J. Kontny, *Drug Dev. Ind. Pharm.*, *14*, 1991 (1988).
80. G. Zografi, *Drug Dev. Ind. Pharm.*, *14*, 1905 (1988).
81. M. J. Kontny, G. P. Grandolfi, and G. Zografi, *Pharm. Res.*, *4*, 104 (1987).
82. G. Zografi and M. J. Kontny, *Pharm. Res.*, *3*, 187 (1986).
83. T. C. Blair, G. Buckton, A. E. Beezer, and S. F. Bloomfield, *Int. J. Pharm.*, *63*, 251 (1990).
84. L. Van Campen, G. Zografi, and J. T. Carstensen, *Int. J. Pharm.*, *5*, 1 (1980).
85. G. Schepky, *Drug Dev. Ind. Pharm.*, *15*, 1715 (1989).
86. D. R. Heidemann and P. J. Jaroz, *Pharm. Res.*, *8*, 292 (1991).

2

Ultraviolet/Visible Diffuse Reflectance Spectroscopy

Harry G. Brittain
Ohmeda, Inc., Murray Hill, New Jersey

I. INTRODUCTION

Due to the opaque nature of most solids, investigations that use some type of electronic spectroscopy as a means of characterization cannot be performed in the transmission mode. Studies performed in the solid state therefore must make use of reflection techniques. Although the majority of workers would be most familiar with the solution-phase aspects, ultraviolet/visible (UV/VIS) techniques are readily adaptable for use in the characterization of pharmaceutical solids. Since most pharmaceutical agents are administered as solid dosage forms, the most appropriate forms of spectroscopy would employ reflectance techniques.

The most important UV/VIS applications have been in the fields of color measurement and color matching, areas of great importance to the dye, paint, paper, textile, and printing industries. The pharmaceutical industry has similar interests in that the use of coloring agents in formulations requires specification. Reflectance spectroscopy has been used, however, by a number of workers to study the kinetics and mechanisms associated with a variety of reactions that were found to take place in the solid state.

The scope of the present chapter will be exclusively concerned with investigations of diffuse reflectance work performed in the UV/VIS region of the spectrum, and with colors that can be perceived by the human eye. Much work has been conducted in the near-infrared region of the spectrum, but that aspect will be covered elsewhere in this book.

II. NATURE OF DIFFUSE REFLECTANCE

To study the ultraviolet or visible absorption spectroscopy of a solid material, the radiation reflected from the surface of the sample is detected and recorded as a function of the incident wavelength. The fraction of light reflected from a sample surface is given by

$$R = \frac{I}{I_0} \tag{1}$$

where R is the measured reflectance at some wavelength, I is the intensity of light reflected from the sample, and I_0 is the intensity of light reflected from the surface of a suitable standard. The choice of standard material is subject to the discretion of the investigator, but it is usually either MgO or $BaSO_4$. The most appropriate standard is one whose matrix permits the most useful data to be obtained on the system of interest.

The light reflected by a powdered solid will consist of a specular reflection component and of a diffuse reflection component. The specular component represents reflection of the incident light by the surfaces of the component particles, and it is characterized by a complete absence of light transmission through the interiors of the particles. By contrast, diffuse reflectance is associated with the radiation that penetrates into the particles to some extent and that then emerges from the bulk solid. This light will exhibit spectral characteristics that are modified from those of the incident beam by the electronic transitions that took place within the solid phase and at the boundaries of the component particles.

Owing to the complementary nature of specular and diffuse reflectance, it is essential to design experimental conditions for which only the diffuse reflectance is measured. High levels of specular reflectance are undesirable in this work, and both the collection optics and sample preparation are therefore optimized to minimize the effects of specular reflectance.

III. THEORY ASSOCIATED WITH DIFFUSE REFLECTANCE MEASUREMENTS

The theory associated with diffuse reflectance has been developed in great detail, and its full exposition is beyond the scope of this chapter. Interested readers are referred to the texts by Kortüm [1], and by Wendlandt and Hecht [2], where the general theory is presented in sufficient detail. The most generally accepted theory concerning diffuse reflectance was developed by Kubelka and Munk [3,4].

In the Kubelka–Munk model, four conditions are assumed. The first of these is that the extent of the horizontal layer is so large relative to its thickness that the light diffused horizontally out of the edges of the sample can be neglected relative to the light moving perpendicular to the layer front. The second assumption is that the material in the layer under study is homogeneous in its composition for the entire distance through which the light passes. The third assumption is that the light incident on the top of the layer is so perfectly diffused that all points on the surface receive equal irradiation. The final assumption is that the top of the sample has the same index of refraction as the medium (typically air) in contact with the sample.

The Kubelka–Munk theory treats the diffuse reflectance of infinitely thick opaque layers [4], a situation achieved in practice for UV/VIS spectroscopy through the use of powder path lengths of at least several millimeters. In this instance, the Kubelka–Munk equation has the form

$$\frac{k}{s} = \frac{(1-R'_\infty)^2}{2R'_\infty} \tag{2}$$

where R'_∞ is the absolute reflectance of an effectively infinitely thick layer, k is its molar absorption coefficient, and s is the scattering coefficient. The derivation of this equation is found elsewhere [5].

As mentioned earlier, it is common practice to measure reflectance values relative to that of a standard, thus defining

$$R_\infty = \frac{R'_\infty \text{ (sample)}}{R'_\infty \text{ (sample)}} \tag{3}$$

The substitution of Eq. (3) into Eq. (2) yields

$$\frac{k}{s} = \frac{(1 - R_\infty)^2}{2R_\infty} \tag{4}$$

The term on the left side of Eq. (4) is often termed the remission function (or the Kubelka–Munk function), and it is frequently denoted by $f(R_\infty)$. Equation (4) indicates that a linear relationship should exist between $f(R_\infty)$ and the sample absorption.

Take the logarithm of the Kubelka–Munk function yields

$$\log f(R_\infty) = \log k - \log s \tag{5}$$

We conclude that when $\log f(R_\infty)$ is plotted against the detection wavelength, the resulting diffuse reflectance spectrum will be identical to the transmission spectrum of the compound. The only difference will be a displacement in the ordinate by the magnitude of -log s.

It should be pointed out that this particular derivation is valid only for weakly absorbing systems without significant contribution from specular reflectance. In addition, the particle size of the powdered sample must be relatively small (around 1 μm). The theory for more complicated systems has been worked out, and the equations pertaining to a wide range of possibilities are available [4].

Another approach to the theory of reflectance spectroscopy has been made using the statistical single-particle theory. In this approach, a powdered sample is considered to be a collection of uniformly sized, rough-surfaced spherical particles. Using the Beer–Lambert equation, the Fresnel equation, and the Lambert cosine law, it is possible to derive an equation relating reflectance to the absorption coefficient (k), the relative index of refraction (n), and the average particle diameter (d) [6]. In its most general form, the equations for reflectance cannot be processed without the use of involved calculations. For the limiting case of weakly absorbing powder samples, however, the reflectance (R) is given by [7]

$$R = \exp{-2n \frac{kd}{3}}^{1/2} \tag{6}$$

IV. INSTRUMENTATION FOR THE MEASUREMENT OF DIFFUSE REFLECTANCE

Probably the first instrument designed for reflectance studies was that described by Nutting [8], and improved by Taylor [9]. Since that time, a large number of instruments have been proposed, and numerous modifications suggested to continually improve signal-to-noise ratios [1,2].

The only basic requirements needed for the construction of a diffuse reflectance spectrometer are a light source and monochromator (to obtain monochromatic radiation), an integrating sphere, and a detection and recording system. The heart of the system is the integrating sphere, whose inside surfaces are coated with highly reflective materials (either MgO or $BaSO_4$) to minimize absorption of the diffuse reflectance. The efficiency of the sphere is defined as the energy lost at the detector port divided by the total energy lost within the sphere. Detailed discussions are available regarding the design of and theory associated with integrating spheres [1,2,5]. For an ideal sphere, the intensity due to reflected light at any point would be independent of the spatial distribution within the sphere.

Detection of the light intensity at any point is therefore an accurate determination of the diffuse reflectance. The incorporation of a standard material surface (again, either MgO or $BaSO_4$) within the sphere thus yields the relative reflectivity, as required by Eq. (4). A variety of sample-handling devices can be incorporated into the integrating sphere so as to permit the performance of necessary experiments. For instance, the addition of a hot stage to the sphere allows the reflectance of materials to be obtained at elevated temperatures [10]. This advance has proved to be especially important to detect and follow the progress of solid state chemical reactions. With the combination of a gas evolution detection technique, the technique of high-temperature diffuse reflectance spectroscopy becomes particularly useful [11].

A schematic diagram illustrating the important points of a reflectance spectrometer is shown in Fig. 1. The incident energy from the source need not be diffuse in itself, and good results are obtained using direct irradiation of the sample. The light is rendered monochromatic by either a monochromator or filter arrangement, split, and allowed to fall on both the standard and sample faces. The light diffusely reflected by the reference and sample are collected by the internal reflections of the sphere, and the intensity ultimately is measured by the detector. The data are processed (usually according to the Kubelka–Munk equation) and then displayed on a suitable device.

To meet the requirements of the Kubelka–Munk theory, the average particle size of the sample should ideally be between 0.1 and 1 μm. At particle sizes below 0.1 μm, the scattering coefficient exhibits an unacceptably strong dependence on the frequency of the incident radiation. For particles larger than 1 μm, the specular reflectance becomes excessively large and depresses the diffuse reflectance spectrum. As an example of the effect of particle size on the intensity of diffuse reflectance, the reflection spectra of pure $KMnO_4$ samples of differing particle sizes are shown in Fig. 2. The loss of intensity at larger particle size diameters is evident in the data [13]. The sample thickness should be between 1 and 5 mm, ensuring that the path length is sufficiently long to meet the assumptions of the Kubelka–Munk theory.

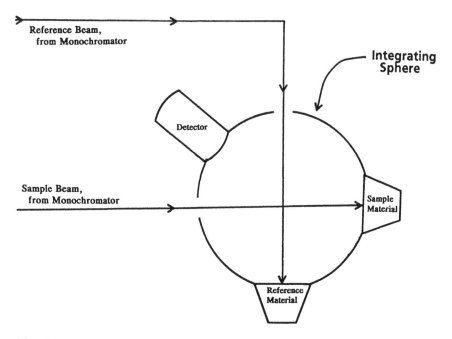

Fig. 1 Schematic diagram of the integrating sphere portion of a diffuse reflectance spectrometer, illustrating the key elements of the optical train. Although the detector has been placed in the plane of the sample and reference materials, in common practice it would be mounted orthogonal to the plane created by the intersection of the optical beams.

V. APPLICATIONS OF DIFFUSE REFLECTANCE SPECTROSCOPY TO THE STUDY OF PHARMACEUTICAL SOLIDS

Although diffuse reflectance spectroscopy in the UV/VIS region of the spectrum has not played a large role in the study of pharmaceutical solids, the technique has been used with success to follow the course of reactions for which color changes were directly related to stability-indicating parameters. The most important of these have concerned the color stability of agents used in solid dose formulations, where the reflectance spectra were used to evaluate the relative rates of decomposition among the accepted coloring agents. The other main area of investigation concerns the study of either drug–excipient or excipient–excipient reactions, and the technique has been particularly useful where color either develops or disappears as the reaction proceeds.

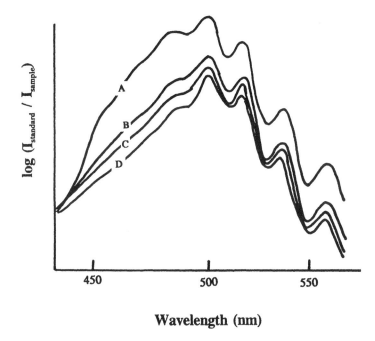

Wavelength (nm)

Fig. 2 Reflectance spectra of pure $KMnO_4$ samples of differing particle sizes, illustrating the effect of particle size on the intensity of diffuse reflectance. Spectra are reported for particle-size ranges of (A) 1–2 μm, (B) 42–50 μm, (C) 60–75 μm, and (D) 100–150 μm. (Data adapted from Ref. 12.)

A. Studies of Color Stability in Tablet Formulations

An initial use of reflectance spectroscopy in the characterization of pharmaceutical solids concerned studies of the stability of coloring agents in tablet formulations. With the description of a device that enabled the surface of intact tablets to be studied [14], the photostability of various dyes and lakes in tablets was followed [15,16]. Exposure of formulations to both normal and exaggerated light conditions was investigated, and the kinetics of the photodegradation evaluated. In most cases, the photoreactions appeared to follow first-order kinetics.

As an example of this work, the reflectance spectra of tablets formulated with FD&C Blue No. 1 and exposed to various light intensities are shown in Fig. 3. Under normal illumination (45 foot-candles), the decomposition kinetics are modest. When exposed to higher levels of illumination (550 foot-candles), complicated kinetics characterized by at least three different reaction pathways were observed [15].

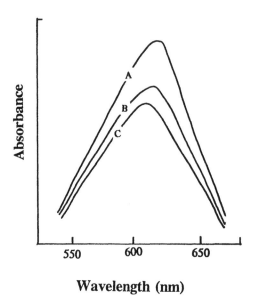

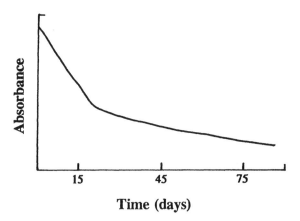

Fig. 3 Reflectance spectra of tablets formulated with FD&C Blue No. 1 and exposed to ordinary illumination for various time periods. Full spectra are shown in the upper box for tablets (A) as initially prepared, and after (B) 42 days and (C) 84 days of illumination. The dependence of dye absorbance on elapsed time is shown in the lower box. (Data adapted from Ref. 14.)

In general, the kinetics associated with the fading of the various dyes studied was found to vary over wide ranges. Lakes prepared from several dyes were found to exhibit less photostability than did the corresponding dyes [16]. In most instances, the initial rates of decomposition were significantly faster than the rates observed at later time points.

The ability of various protection schemes to retard the photocomposition was investigated by means of diffuse reflectance spectroscopy. Glasses of different colors were evaluated as to their ability to retard the fading of tablets colored with FD&C Blue No. 1 and D&C Yellow No. 10 [17]. The main conclusion of this work was that amber glass afforded the best protection against the effects of photoillumination, although any glass containing an effective UV-absorber could be equally effective. The incorporation of 2,4-di-hydroxybenxophenone into tablet formulations as an inherent UV-absorber was studied, and its effect on dye fading rates evaluated [18]. It was found that this compound could effectively protect against the photodecomposition reaction as long as its absorption spectrum matched reasonably well with the dye under question.

In a detailed study of the color stability of tablets containing FD&C Red No. 3 dye, the ability of a variety of sunscreen agents to retard the color loss was investigated [19]. The agents were incorporated in films coated upon the model tablet formulation, and the tablets subsequently exposed to light of 1000 foot-candles. The film-coating method was not found to yield acceptable stability when using either glyceryl *p*-aminobenzoate or salicylates as sunscreens. The greatest protection against fading was observed when 2-ethyoxyethyl *p*-methoxycinnamate was used as the protection agent.

The stability of a wide variety of certified dyes in tablets has been evaluated upon exposure to various illumination conditions [20–22]. In these works, more detailed studies of dye concentration, exposure time, and irradiation light intensity were performed, and the data analyzed more rigorously. The effect of varying tablet excipients was also investigated with respect to the decomposition kinetics. This work ended with the development of a systematic approach for the testing of color stability, and the investigators developed a ranking of colorant stabilities in tablets [22].

Diffuse reflectance spectroscopy was used to screen the possible interactions between a large number of adjuvants and several dyes [23]. It was concluded that supposedly inert excipients (such as starch or lactose) were capable of undergoing significant reactions with the dyes investigated (Red No. 3, Blue No. 1, and Yellow No. 5). For adjuvants containing metal ions (zinc oxide, or calcium, magnesium, and aluminum hydroxides), the degree of interaction could be considerable. It was concluded from these studies that dye–excipient interactions could also be responsible for the lack of color stability in certain tablet formulations.

B. Studies of Compound Interactions in Formulated Materials

It was recognized very early that diffuse reflectance spectroscopy could be used to study the interactions of various compounds in a formulation, and the technique has been particularly useful in the characterization of solid state reactions [24]. Lach concluded that diffuse reflectance spectroscopy could also be used to verify the potency of a drug in its formulation. In addition, studies conducted under stress conditions would be useful in the study of drug–excipient interactions, drug degradation pathways, and alterations in bioavailability owing to chemisorption of the drug onto other components in the formulation [24].

In a series of works, Lach and coworkers studied the solid state interactions of a variety of compounds with various adjuvants [25–27]. Working predominately with compounds containing conjugated aromatic systems (oxytetracycline, anthracene, phenothiazine, salicyclic acid, prednisone, and hydrochlorothiazide), it was deduced that the complexes formed with the adjuvants were of the donor–acceptor type. Large bathochromic and hyperchromic spectral shifts in absorption maxima upon complexation were inferred to be the result of strong charge transfer interactions. In the specific instance where the drug possessed a chelation center and the adjuvant contained a metallic center, site-specific complexation could be identified as the source of the change in the reflectance spectrum [27,28].

The stability of ascorbic acid formulations has been successfully studied using diffuse reflectance spectroscopy, since the degradation is accompanied by a color reaction and this color change is directly related to the potency of the active component [29,30]. Tablets stored at ambient conditions will gradually age from white to a yellowish-brown color, with the degree of stability being greatly affected by the excipients in the formulation. It was determined that materials containing metal ions (i.e., magnesium and calcium stearate, or talc) accelerated the color reaction, while metal-free excipients (such as stearic acid or hydrogenated vegetable oil) imparted maximal color stability [30]. The effect of temperature and relative humidity on the stability of the formulations was also followed using diffuse reflectance spectroscopy, and it was found that the effect of elevated humidity was more important than the effect of elevated temperatures.

The reaction of a primary amine with lactose is accompanied by a browning of the solids, and the path of such reactions is easily following by means of diffuse reflectance spectroscopy. For instance, the reaction of isonicotinic acid hydrazide (Isoniazid) with lactose could be followed through changes in the reflectance spectrum [31]. As may be seen in Fig. 4, a steady decrease in reflectance was noted as the sample was heated for increasing amounts of time. The spectral data were used to deduce the rate constants for the browning reaction at various heating temperatures, and these rates could be correlated with those

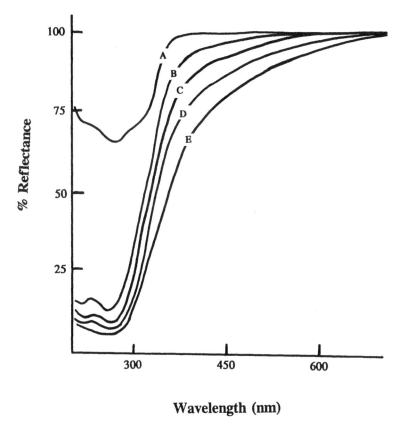

Fig. 4 Reflectance spectra of the reaction products formed when isonicotinic acid hydrazide (Isoniazid) reacts with lactose. Reflectance spectra are shown for (A) the initially prepared material, and for samples illuminated for (B) 4 hours, (C) 10 hours, (D) 21 hours, and (E) 44 hours. (Data adapted from Ref. 31.)

obtained through chemical analysis (formation of isonicotinoyl hydrazone of lactose) of the mixtures. In addition, the active was found to both chemisorb as well as physisorb onto magnesium oxide, with accompanying changes in the reflectance spectrum [31].

A similar study has been conducted in which the interaction of *d*-amphetamine sulfate with spray-dried lactose was investigated [32]. Upon storage at elevated temperatures, discoloration of the powder blends was noted, and the new absorption bands characterized. One maximum was noted at 340 nm, and this was attributed to the chemisorption of the amine onto the lactose particles. The other band appeared at 295 nm and was attributed to the new compound (*d*-amphetamine-hydroxymethylfurfural) formed as a result of the reaction between the two. Through the use of Arrhenius plots, the browning rate anticipated for a temperature of 25°C was predicted.

The interaction of tetracycline and its derivatives with excipients has been studied by diffuse reflectance spectroscopy [33,34]. Adjuvants containing metal ions were found to yield the largest degree of change in the reflectance spectra, with the effects of calcium being most prominent. Complexation of the metal ion with specific functional groups on the drug molecules was postulated as the origin of these effects. The complexation would be anticipated to influence the bioavailability of the compound, and therefore the technique is demonstrated to provide useful information regarding the degree of drug–excipient compatibility.

Compounds known to undergo changes in their absorption spectra upon sorption onto a solid surface are termed adsorptiochromic, and such effects would be ideally studied by means of diffuse reflectance spectroscopy. In one such study, the absorption of various spiropyrans onto many different solids was investigated [35]. For the compounds studied, the reflectance spectra were dominated by bands at 550 nm and in the range of 400–500 nm (most often at 472 nm). As an example, the reflectance spectra obtained for 6-nitrobenxospiropyran are shown in Fig. 5. When the difference spectrum was taken between the spectrum of the pure compound and that obtained after sorption onto silicic acid, the bands characteristic of the adsorbed species were clearly evident.

The interaction between drug compounds and excipients, as these influence drug dissolution, can be successfully studied by means of reflectance spectroscopy. In one study concerning probucol and indomethacin, it was deduced that hydrogen bonding and van der Waals forces determined the physisorption between the active and the excipients in several model formulations [36]. Chemisorption forces were found to play only minor roles in these interactions. These studies indicated that surface catalytic effects could be important during the selection of formulation excipients.

Although most often connected with investigations of solid dosage forms, diffuse reflectance spectroscopy can also be used to characterize alternative formulations. Through the use of a special sample cell, the technique has been used to study the stability of emulsions [37]. In this work, it was found that information could be obtained that pointed toward subtle changes in the emulsion microenvironment.

VI. QUANTITATIVE MEASUREMENT OF COLOR

Color is basically a perception that is developed in the mind of a given individual, and consequently different people can perceive a particular color in various fashions. Such variability in interpretation causes great difficulty in the evaluation of color-related phenomena, and it leads to subjective rather than objective judgements. For obvious reasons, the development of a quantitative method for

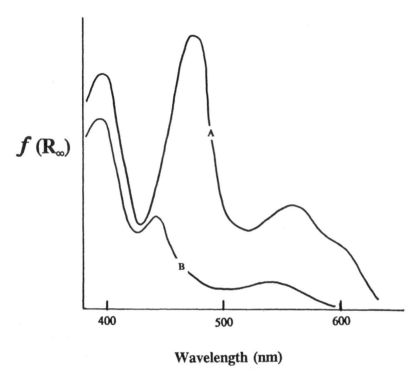

f (R$_\infty$)

Wavelength (nm)

Fig. 5   Diffuse reflectance spectra of 6-nitrobenxospiropyran (A) in pure form and (B) after its sorption onto silicic acid. (Data adapted from Ref. 35.)

color determination is highly desirable in that a good system would lead to elimination of the subjectivity associated with a visual interpretative measurement.

   Although several schemes have been developed for the quantitative expression of color [38], only the CIE (Commission Internationale de l'Éclairage) system has gained general acceptance [39]. This system assumes that color may be expressed as the summation of selected spectral components in a three-dimensional manner. The three primary colors normally added for such purposes are blue, green, and red. The CIE system is based on the fact that human sight is trichromatic in its color perception, and that two stimuli will produce the same color if each of the three tristimulus values (X, Y, and Z) arc cqual for thc two:

$$X = k \int S(\lambda) R(\lambda) \bar{x}(\lambda) \, d\lambda \tag{7}$$

$$Y = k \int S(\lambda) R(\lambda) \bar{y}(\lambda) \, d\lambda \tag{8}$$

$$Z = k \int S(\lambda) R(\lambda) \bar{z}(\lambda) \, d\lambda \tag{9}$$

where

$$k = \frac{100}{\int S(\lambda)\bar{y}(\lambda)\ d\lambda} \qquad (10)$$

In Eqs. (7)–(10), $S(\lambda)$ is the spectral power distribution of the illuminant, and $R(\lambda)$ is the spectral reflectance factor of the object. $\bar{x}(\lambda)$, $\bar{y}(\lambda)$, and $\bar{z}(\lambda)$ are the color-matching functions of the observer. In the usual practice, k is defined so that the tristimulus value, Y, for a perfect reflecting diffusor (the reference for $R(\lambda)$) equals 100. Using the functions proposed by the CIE in 1931, $\bar{y}(\lambda)$ was made identical to the spectral photopic luminous efficiency function, and consequently its tristimulus value, Y, is a measure of the brightness of objects. The X and Z values describe aspects of color that permit identification with various spectral regions.

The $\bar{x}(\lambda)$, $\bar{y}(\lambda)$, and $\bar{z}(\lambda)$ terms were derived by the CIE from data obtained in visual experiments where observers matched colors obtained by the mixing of the blue, green, and red primary colors. The average result for human observers were defined as the CIE 1931 2° standard observer, and the wavelength dependencies of these color-matching functions are illustrated in Fig. 6.

The proper implementation of the CIE system requires use of a standard illumination source for calculation of the tristimulus values. Three standard sources were recommended in the 1931 CIE system, and these may be presented in terms of color temperatures (the temperature at which the color of a black-body radiator matches that of the illuminant). The simplest source is an incandescent lamp, operating at a color temperature of 2856 K. The other two sources are combinations of lamps and solution filters designed to provide the equivalent of sunlight at noon, or the daylight associated with an overcast sky. The latter two sources are equivalent to color temperatures of 5000 K and 6800 K, respectively.

Since color is a perceived quantity, a strict mathematical relation relating tristimulus values to a concept of color is not possible. An equation has been proposed, however, that relates the perception of a color to these values. The trichromatic equation for tristimulus values is normally put into the general form

$$\mathrm{color}(C) \equiv x(X) + y(Y) + z(Z) \qquad (11)$$

where x, y, and z represent the chromaticity coordinates of C, obtained via the following relations:

$$x = \frac{X}{X + Y + Z} \qquad (12)$$

$$y = \frac{Y}{X + Y + Z} \qquad (13)$$

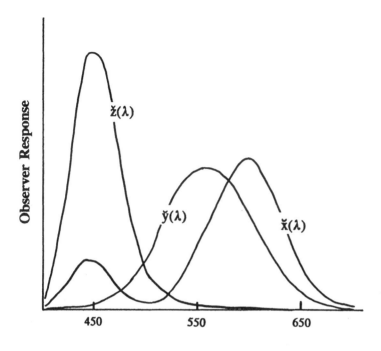

Wavelength (nm)

Fig. 6   Color-matching functions $\check{x}(\lambda)$, and $\check{z}(\lambda)$ of the CIE 1931 2° standard observer. (Data adapted from Ref. 40.)

$$z = \frac{Z}{X + Y + Z} \tag{14}$$

Only two of the three chromaticity coordinates need to be actually specified, since

$$1 = x + y + z \tag{15}$$

A chromaticity diagram can therefore be drawn presenting colors in terms of their x and y coordinates, which are often termed hue and saturation, respectively. A diagram of this type is presented in Fig. 7.

In terms of the chromaticity diagram, it can be said that two colors having the same chromaticities (x,y) and the same luminous reflectance (Y) are matched. Technically, no information is provided about the appearance of two matched colors, although subjective judgements about specific colors identified by a pair of (x,y) corrdinates are often made.

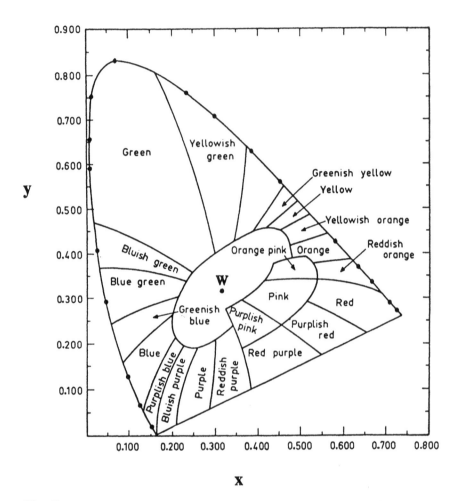

Fig. 7 Chromaticity diagram presented in terms of the x and y coordinates. (Data adapted from Ref. 39.)

The central point (marked W in Fig. 7) of a chromaticity diagram is termed the achromatic point, and represents the combination of x and y values yielding white light. A straight line passing through W would connect complementary colors. Combination of the colors specified at the endpoints of this line in the relative amounts indicated by the distances from their respective points to W will result in the generation of white light.

The actual determination of color is made with a photoelectric tristimulus colorimeter fitted with a CIE illuminating source. Measurements are made relative to a standard (usually magnesium oxide or barium sulfate) taken through each of the provided filters. The experimental readings are converted to tristimulus values by means of instrumental factors provided by the instrument manufacturer, and taking the filter specifications into account. The tristimulus values are converted to chromaticity coordinates using Eqs. (12)–(14). The accuracy of this approach is determined by how well the combination of detector and tristimulus filters approximates the characteristics of the CIE standard observer.

With the introduction of computers and microprocessor-controlled instrumentation, it has become possible to use spectrophotometry to obtain far more accurate determinations of color. The tristimulus values are obtained after integration of the data according to Eqs. (7)–(9). This degree of sophistication permits the use of more advanced methods of color quantitation, such as the 1976 CIE $L*u*v*$ system [41] or other systems not discussed in the present chapter.

Raff has provided a discussion summarizing how the CIE system could be applied to the characterization of materials having pharmaceutical interest [42]. At that time, the FDA was just becoming concerned about the stability of coloring agents and dyes, and the quantitative CIE system held a definite appeal for those seeking to eliminate subjectivity from such determinations. In this work, Raff provided the reflectance spectra of several pigments and attempted to study the changes in color that took place upon exposure to strong illumination.

In a subsequent work, Raff used the CIE system to quantify the colors that could be obtained when using FD&C aluminum lakes as colorants in tablet formulations [43]. He reported on the concentration dependence of the tristimulus values obtained when calcium sulfate dihydrate was compressed with various amounts of FD&C Blue No. 2 aluminum lake, and one example of the reported data is found in Table 1.

The Y tristimulus value may be taken as the relative lightness of the tablet

Table 1 Tristimulus Values Obtained on a Series of Tablets Colored with FD&C Blue No. 2 Aluminum Lake

% dye (w/w)	X	Y	Z	x	y
0.065	0.744	0.762	1.012	0.296	0.303
0.290	0.568	0.582	0.860	0.282	0.289
0.900	0.439	0.447	0.718	0.274	0.279
2.600	0.279	0.280	0.518	0.259	0.260
6.200	0.175	0.172	0.362	0.247	0.243
10.000	0.142	0.138	0.304	0.243	0.236

Source: Ref. 43.

surface, with a value of 1.0 being the maximum. It is evident that increasing colorant concentrations decrease the apparent brightness of the tablets. The chromaticity coordinates, x and y, indicate that the apparent color of the tablets shifts as the colorant levels are raised, and that no limiting color can be reached solely through the use of FD&C Blue No. 2 aluminum lake. Since the perceived color could not be saturated, careful control of the colorant level in this particular formulation would be required to ensure that all batches yielded equivalent appearances.

Goodhart and coworkers used the CIE parameters to evaluate the chromaticity coordinates and brightness of nearly 50 pharmaceutical colorants compressed with lactose [44]. The colorants were categorized as belonging to blue, black, brown, green, orange, red, violet, and yellow classes, and several iron oxide colorants were also studied. The aim of this work was to produce a data set that could be used to match any given color through the combination of pharmaceutically acceptable agents.

In an extension of their work, Goodhart and coworkers developed a system [45] whereby the final desired color of a compressed table formulation was first chosen from a standard color chart (such as the Munsell compilation [40]). This color was then analyzed as to its CIE parameters, and these parameters were in turn used to develop a colorant combination that would produce a match of the desired color. The ultimate end of this work was to produce a database of sufficient depth that the empirical nature of color matching could be eliminated.

Bogdansky has used a combination of the CIE and alternative color systems to deduce color parameters associated with tablet colorants [46]. In this work, thin color dispersions on chromatography paper were made, and the tristimulus values were determined through appropriate colorimetry. The materials studied were ordered through their perceived color, and these judgments correlated with the quantitative color parameters. Most importantly, an evaluation procedure was established for the acceptability criteria to be used in judging the range of chromaticity values that signified equivalence in the color of solid dosage forms produced as different lots. The compilation of tristimulus and chromaticity information was shown to provide a permanent description of a colored sample, and to define any difference between an analyte and its standard.

The effect of particle size, and hence dispersion, on the coloring properties of aluminum lake dyes has been studied through quantitative measurement of color in compressed formulations [47]. It was found that reduction in the particle size for the input lake material resulted in an increase in color strength, and that particles of submicron size contributed greatly to the observed effects. Analysis of the formulations using the parameters of the 1931 CIE system could only lead to a qualitative estimation of the effects, but use of the 1976 CIE $L^*u^*v^*$ system provided a superior evaluation of the trends. With the latter system, the effects of dispersion on hue, chroma, lightness, and total color differences were quantitatively related to human visual perception.

One intriguing application of color technology has been in the development of screened indicators [48]. Several reagents were prepared by use of complementary tristimulus data of different acid–base indicators and screening dyes. As illustrated in Fig. 8, the color change that takes place during the pH titration of an acid–base indicator is well described by the chromaticity parameters. It was shown that the color changes taking place with one pure acid–base indicator and different dyes (when represented in the complementary chromaticity diagram) lie on the same chromatic straight line. The equations defining this line from the color parameters were correlated with experimental data and used to provide the necessary compositional information required to obtain the optimal color change for the system. It was concluded that the optimum color change always occurs between two complementary colors with the same degree of gray proportion in the colors.

The appearance of tablets and powders during accelerated stability testing can be quantified using tristimulus colorimetry [29]. In this work, various formulations were stored under stress conditions, and the tristimulus parameters

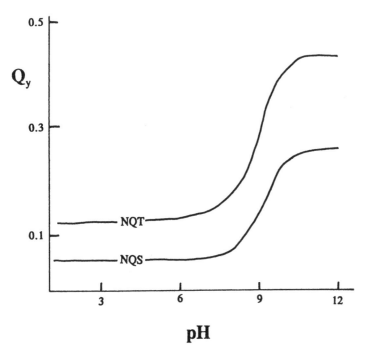

Fig. 8 pH dependence of the Q_y chromaticity parameters obtained during the potentiometric titration of 1,2-naphthoquinone-2-semicarbazone (NQS) and 1,2-naphthoquinone-2-thiosemicarbazone (NQT). (Data adapted from Ref. 48.)

of the tablets were measured at various time points during the storage time. Since the particular systems studied were characterized by changes in tablet color as the decomposition reaction proceeded, the X, Y, and Z parameters were judged useful in deducing the kinetics of the reactions. Both first-order and zero-order reactions were noted for the systems studied, and a formal Arrhenius treatment of the data could be used to predict shelf lives for the formulations.

Other workers have used the tristimulus parameters to study the kinetics of decomposition reactions. The fading of tablet colorants was shown to follow first-order reaction kinetics, with the source of the illumination energy apparently not affecting the kinetics [49]. The effect of excipients on the discoloration of ascorbic acid in tablet formulations has also been followed through determination of color changes [50]. In this latter work, it was established that lactose and Emdex influenced color changes less than did sorbitol.

The appearance testing of tablets through measurement of color changes has been automated through the use of fiber optic probes and factor analysis of the data [51]. Good correlation between measured chromaticity parameters and visual subjective judgement was demonstrated, with samples of differing degrees of whiteness being used to develop the correlation. It was pointed out, however, that surface defects on the analyzed materials could compromise the quality of the correlation, and that more sophisticated methods for data evaluation would be useful.

VII. SUMMARY

Although UV/VIS diffuse reflectance spectroscopy has not been used extensively in the study of pharmaceutical solids, its applications have been sufficiently numerous that the power of the technique is evident. The full reflectance spectra, or the derived colorimetry parameters, can be very useful in the study of solids that are characterized by color detectable by the human eye. It is evident that questions pertaining to the colorants used for identification purposes in tablet formulations can be fully answered through the use of appropriately designed diffuse reflectance spectral experiments. With the advent of newer, computer-controlled instrumentation, the utility of UV/VIS diffuse reflectance as a characterization tool for solids of pharmaceutical interest should continue to be amply demonstrated.

REFERENCES

1. G. Kortüm, *Reflectance Spectroscopy*, Springer-Verlag, New York, 1969.
2. W. W. Wendlandt and H. G. Hecht, *Reflectance Spectroscopy*, Interscience Pub., New York, 1966.
3. P. Kubelka and F. Munk, *Z. Tech. Physik*, *12*, 593 (1931).

4. P. Kubelka, *J. Opt. Sci. Am.*, *38*, 448 (1948).
5. R. W. Frei and J. D. MacNeil, *Diffuse Reflectance Spectroscopy in Environmental Problem Solving*, CRC Press, Cleveland, Ohio, 1973. The derivation is found as Appendix I, p. 209.
6. N. T. Melamed, *J. Appl. Phys.*, *34*, 560 (1963).
7. E. L. Simmons, *Opt. Acta*, *18*, 59 (1971).
8. P. G. Nutting, *Trans. Illum. Eng. Soc.*, *7*, 412 (1912).
9. A. H. Taylor, *J. Opt. Sci. Am.*, *4*, 9 (1919).
10. W. W. Wendlandt, P. H. Franke, and J. P. Smith, *Anal. Chem.*, *35*, 105 (1963); W. W. Wendlandt, *Pure Appl. Chem.*, *25*, 825 (1971).
11. W. W. Wendlandt and W. S. Bradley, *Thermochim. Acta*, *1*, 143 (1970).
12. G. Kortüm and J. Vogel, *Z. Physik. Chem.*, *18*, 230 (1958).
13. A. H. Norbury, *Laboratory Practice*, *18*, 754 (1990).
14. T. Urbanyi, C. J. Swartz, and L. Lachman, *J. Am. Pharm. Assoc.*, *Sci. Ed.*, *49*, 163 (1960).
15. L. Lachman, C. J. Swartz, T. Urbanyi, and J. Cooper, *J. Am. Pharm. Assoc.*, *Sci. Ed.*, *49*, 165 (1960).
16. L. Lachman, S. Weinstein, C. J. Swartz, T. Urbanyi, and J. Cooper, *J. Pharm. Sci.*, *50*, 141 (1961).
17. C. J. Swartz, L. Lachman, T. Urbanyi, and J. Cooper, *J. Pharm. Sci.*, *50*, 145 (1961).
18. L. Lachman, T. Urbanyi, S. Weinstein, J. Cooper, and C. J. Swartz, *J. Pharm. Sci.*, *51*, 321 (1962).
19. B. R. Hajratwala, *J. Pharm. Sci.*, *63*, 129 (1974).
20. M. E. Everhard and F. W. Goodhart, *J. Pharm. Sci.*, *52*, 281 (1963).
21. F. W. Goodhart, M. E. Everhard, and D. A. Dickcius, *J. Pharm. Sci.*, *53*, 338 (1964).
22. F. W. Goodhart, H. A. Leiberman, D. S. Mody, and F. C. Ninger, *J. Pharm. Sci.*, *56*, 63 (1967).
23. M. Bornstein, J. P. Walsh, B. J. Munden, and J. L. Lach, *J. Pharm. Sci.*, *56*, 1410 (1967).
24. D. G. Pope and J. L. Lach, *Pharm. Acta Helv.*, *50*, 165 (1975).
25. J. L. Lach and M. Bornstein, *J. Pharm. Sci.*, *54*, 1730 (1965).
26. M. Bornstein and J. L. Lach, *J. Pharm. Sci.*, *55*, 1033 (1966).
27. J. L. Lach and M. Bornstein, *J. Pharm. Sci.*, *55*, 1040 (1966).
28. M. Bornstein, J. L. Lach, and B. J. Munden, *J. Pharm. Sci.*, *57*, 1653 (1968).
29. J. T. Carstensen, J. Johnson, W. Valentine, and J. Vance, *J. Pharm. Sci.*, *53*, 1050 (1964).
30. R. B. Wortz, *J. Pharm. Sci.*, *56*, 1169 (1967).
31. W.-H. Wu, T.-F. Chin, and J. L. Lach, *J. Pharm. Sci.*, *59*, 1286 (1970).
32. S. M. Blaug and W.-T. Huang, *J. Pharm. Sci.*, *61*, 1770 (1972).
33. J. L. Lach and L. D. Bighley, *J. Pharm. Sci.*, *59*, 1261 (1970).
34. J. D. McCallister, T.-F. Chin, and J. L. Lach, *J. Pharm. Sci.*, *59*, 1286 (1970).
35. M. J. Jozwiakowski and K. M. Connors, *J. Pharm. Sci.*, *77*, 241 (1988).
36. D. C. Monkhouse and J. L. Lach, *J. Pharm. Sci.*, *61*, 1435 (1972).
37. M. J. Akers and J. L. Lach, *J. Pharm. Sci.*, *65*, 216 (1976).

38. D. B. Judd and G. Wyszecki, *Color in Business, Science, and Industry*, 2nd ed., John Wiley & Sons, New York, 1963, pp. 264–361.
39. D. B. Judd, *Nat. Bur. Stand. U.S. Circular 478*, 1950.
40. F. W. Billmeyer, Jr., and M. Saltzman, *Principles of Color Technology*, Wiley-Interscience, New York, 1966.
41. *CIE Publication 15.2, Colorimetry*, 2nd ed., Central Bureau of CIE, Vienna, Austria, 1986.
42. A. M. Raff, *J. Pharm. Sci.*, *52*, 291 (1963).
43. A. M. Raff, *J. Pharm. Sci.*, *53*, 380 (1964).
44. M. E. Everhard, D. A. Dickcius, and F. W. Goodhart, *J. Pharm. Sci.*, *53*, 173 (1964).
45. F. W. Goodhart, M. A. Kelly, and H. A. Lieferman, *J. Pharm. Sci.*, *54*, 1799 (1965).
46. F. M. Bogdansky, *J. Pharm. Sci.*, *64*, 323 (1975).
47. L. S. Wou and B. A. Mulley, *J. Pharm. Sci.*, *77*, 866 (1988).
48. E. Bosch, E. Casassas, A. Izquierdo, and M. Rosés, *Anal. Chem.*, *56*, 1422 (1984).
49. P. Turi, D. Brusco, H. V. Maulding, R. A. Tausendfreund, and A. F. Michaelis, *J. Pharm. Sci.*, *61*, 1811 (1972).
50. S. Vemuri, C. Taracatac, and R. Skluzacek, *Drug Dev. Indust. Pharm.*, *11*, 207 (1985).
51. M. Wirth, *J. Pharm. Sci.*, *80*, 1177 (1991).

3

Vibrational Spectroscopy

David E. Bugay
Bristol-Myers Squibb Pharmaceutical Research Institute, New Brunswick,
New Jersey

Adrian C. Williams
School of Pharmacy, University of Bradford, Bradford, West Yorkshire,
United Kingdom

I. INTRODUCTION

Most chemists tend to think of infrared (IR) spectroscopy as the only form of vibrational analysis for a molecular entity. In this framework, IR is typically used as an identification assay for various intermediates and final bulk drug products, and also as a quantitative technique for solution-phase studies. Full vibrational analysis of a molecule must also include Raman spectroscopy. Although IR and Raman spectroscopy are complementary techniques, widespread use of the Raman technique in pharmaceutical investigations has been limited. Before the advent of Fourier transform techniques and lasers, experimental difficulties limited the use of Raman spectroscopy. Over the last 20 years a renaissance of the Raman technique has been seen, however, due mainly to instrumentation development.

Today, the complete characterization of pharmaceutical solids (including bulk drugs, excipients, physical mixtures, formulated product, and placebo) is a requirement for the consistent, reliable, and safe development of drug products. One stage of the characterization of pharmaceutical solids must be at the molecular level. Ideally suited for this task are the various forms of molecular spectroscopy techniques such as infrared, Raman, and nuclear magnetic resonance (NMR). Infrared and Raman analysis provide a complete vibrational motion analysis of the molecule, whereas NMR provides insight into the local environment of each NMR active atom. This chapter focuses on the theory, experimental aspects, and selected pharmaceutical applications of vibrational spectroscopy. The next chapter provides insight into the use of solid state NMR techniques. Together, these forms of spectroscopy can provide a complete characterization of pharmaceutical solids at the molecular level.

Most solid state investigations of bulk drug material involve the identification and quantitation of polymorphic and pseudopolymorphic systems. Since different polymorphic forms of a drug substance exhibit different three-dimensional structures, the vibrational motion for each polymorphic form is potentially different, whence the ability to investigate polymorphism by vibrational spectroscopy techniques. Other areas of solid state analysis by vibrational spectroscopy include drug delivery systems, drug–excipient interactions, crystallization studies, mixing studies, particulate/contaminant characterization, and solid state transformations. Since the advent of Fourier transform techniques and advances in laser technology and computer systems, a myriad of solid state sampling techniques are available for the vibrational spectroscopist. Analysis can now be performed on single crystals, bulk material, and on process streams through the use of fiber optic probes. Additional advantages of vibrational analysis include the following: (1) It is typically nondestructive in nature with the ability to recover the material for further characterization; (2) It is a quantitative technique under proper sampling conditions; and (3) It is complementary to other physical characterization techniques such as powder x-ray diffraction (XRD), NMR, and thermal analysis.

II. VIBRATIONAL SPECTROSCOPY THEORY

A brief description of IR and Raman theory will be presented so that a common understanding of the techniques is available to the reader. A complete description of the underlying theory to IR and Raman spectroscopy is outside the scope of this chapter, but can be obtained from the literature [1–5].

A. Infrared

All molecules of pharmaceutical interest absorb some form of electromagnetic radiation. Within the electromagnetic spectrum (Fig. 1), infrared energy is a small portion; it is typically divided into three regions, the near-, mid-, and far-IR regions, with their respective energy/frequency limits. In IR spectroscopy, the energy unit wavenumber (cm^{-1}) is typically used. Wavenumber is the reciprocal of the IR wavelength expressed in centimeters.

When a broadband source of IR energy irradiates a sample, the absorption of IR energy by the sample results from transitions between molecular vibrational and rotational energy levels. A vibrational transition may be approximated by treating two atoms bonded together within a molecule as a harmonic oscillator.

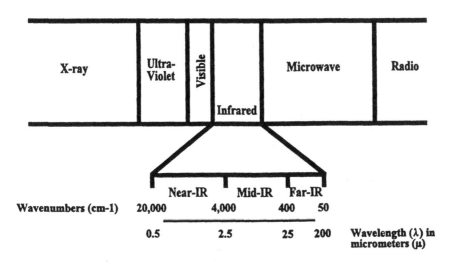

Fig. 1 A portion of the electromagnetic spectrum comparing infrared energy with other forms of radiation.

Based upon Hooke's law, the vibrational frequency between these two atoms may be approximated as

$$\nu = \frac{1}{2\pi} \sqrt{\frac{k}{\mu}} \tag{1}$$

where μ is the reduced mass of the two atoms—$\mu = m_1 m_2/(m_1 + m_2)$—and k is the force constant of the bond (dynes/cm). Quantum mechanical analysis of the harmonic oscillator reveals a series of equally spaced vibrational energy levels that are expressed as

$$E_n = (n + \frac{1}{2})h\nu_0 \tag{2}$$

where E_n is the energy of the nth level, h is Planck's constant, and ν_0 is the fundamental vibrational frequency. These energy levels may be graphically described in a Jablonski energy level diagram (Fig. 2). It must be noted that the

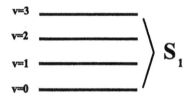

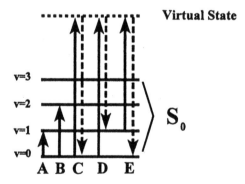

Fig. 2 Jablonski energy level diagram illustrating possible transitions, where solid lines represent absorption processes and dotted lines represent scattering processes. *Key*: A, IR absorption; B, near-IR absorption of an overtone; C, Rayleigh scattering; D, Stokes Raman transition; and E, anti-Stokes Raman transition. S_0 is the singlet ground state, S_1 the lowest singlet excited state, and ν represents vibrational energy levels within each electronic state.

fundamental vibrational frequencies in a polyatomic molecule do not necessarily correspond to the vibrations of single pairs of atoms, rather to vibrations from a group of atoms. The absorption of IR energy by a molecule corresponds to approximately 2–10 kcal/mole, which in turn equals the stretching and bending vibrational frequencies of most bonds in covalently bonded molecules. Thus, we get the correlation between IR spectroscopy and the ability to probe the vibrational motion of a molecule.

The number of fundamental vibrational modes of a molecule is equal to the number of degrees of vibrational freedom. For a nonlinear molecule of N atoms, $3N - 6$ degrees of vibrational freedom exist. Hence, $3N - 6$ fundamental vibrational modes. Six degrees of freedom are subtracted from a nonlinear molecule since (1) three coordinates are required to locate the molecule in space, and (2) an additional three coordinates are required to describe the orientation of the molecule based upon the three coordinates defining the position of the molecule in space. For a linear molecule, $3N - 5$ fundamental vibrational modes are possible since only two degrees of rotational freedom exist. Thus, in a total vibrational analysis of a molecule by complementary IR and Raman techniques, $3N - 6$ or $3N - 5$ vibrational frequencies should be observed. It must be kept in mind that the fundamental modes of vibration of a molecule are described as transitions from one vibration state (energy level) to another ($n = 1$ in Eq. (2), Fig. 2). Sometimes, additional vibrational frequencies are detected in an IR and/or Raman spectrum. These additional absorption bands are due to forbidden transitions that occur and are described in the section on near-IR theory. Additionally, not all vibrational bands may be observed since some fundamental vibrations may be too weak to observe or give rise to overtone and/or combination bands (discussed later in the chapter).

For a fundamental vibrational mode to be IR-active, a change in the molecular dipole must take place during the molecular vibration. This is described as the IR selection rule. Atoms that possess different electronegativity and are chemically bonded change the net dipole of a molecule during normal molecular vibrations. Typically, antisymmetric vibrational modes and vibrations due to polar groups are more likely to exhibit prominent IR absorption bands.

A transmission IR spectrum may be calculated by

$$\text{Transmission} = \frac{I}{I_0} \tag{3}$$

where I equals the intensity of transmitted IR radiation, and I_0 equals the intensity of irradiating IR energy. In this case, the transmission IR spectrum is represented by wavenumber (cm^{-1}) on the abscissa and transmission on the ordinate (Fig. 3). Infrared spectra may also be represented in absorbance units:

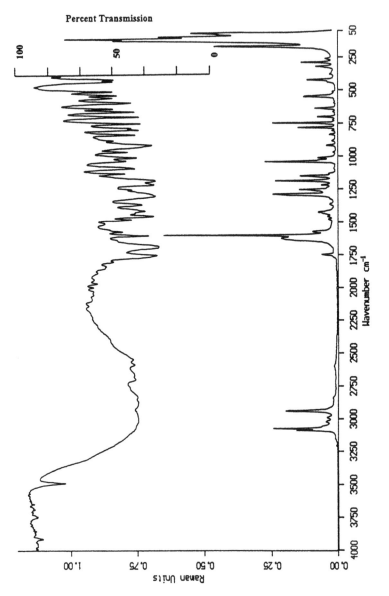

Fig. 3 Fourier transform (FT) infrared (upper) and FT-Raman (lower) spectra of aspirin. The left ordinate scale is representative of the Raman intensity, whereas the right ordinate scale represents IR transmission units.

$$\text{Absorbance} = \log \frac{I_0}{I} = abc \tag{4}$$

where a is absorptivity, b is the sample cell thickness, and c is concentration. The representation of IR spectra in absorbance units permits quantitative analysis.

As previously mentioned, IR spectroscopy is typically used for the identification of a molecular entity. This approach arises from the fact that the vibrational frequency of two atoms may be approximated from Eq. (1). If one assumes that the force constant (k) for a double bond is 10×10^5 dynes/cm, Eq. (1) allows one to approximate the vibrational frequency for C=C:

$$\nu = 4.12 \sqrt{\frac{k}{\mu}} = 4.12 \sqrt{\frac{10 \times 10^5}{(12 \times 12)(12 + 12)}} = 1682 \text{ cm}^{-1} \tag{5}$$

Fairly good agreement exists between the calculated value of 1682 cm^{-1} and the experimental value of 1650 cm^{-1}. Based upon the Hooke's law approximation, numerous correlation tables have been generated that allow one to estimate the characteristic absorption frequency of a specific functionality [3]. It becomes readily apparent how IR spectroscopy can be used to identify a molecular entity, and subsequently to physically characterize a sample or to perform quantitative analysis.

B. Near-Infrared

The near-IR region of the electromagnetic spectrum is generally from 13,300 to 4000 cm^{-1}. Typically, wavelength (in nanometers) is represented on the abscissa of a near-IR spectrum. Absorption bands in the near-IR region of the spectrum arise from overtones or combinations of overtones originating in the fundamental mid-IR region. Overtone absorption bands are the result of forbidden transitions arising from the ground vibrational energy level and where $n > 1$ in Eq. (2). Since overtone bands arise from forbidden transitions, they are typically 10–1000 times less intense than their corresponding fundamental absorption bands. The majority of overtone bands in the near-IR arise from RH stretching modes (OH, NH, CH, SH). Due to the large mass difference between the two atoms, large amplitude vibrations arise with high anharmonicity and large dipole moments. The large anharmonicity causes the frequency of the overtone (or combination) bands to be slightly less than the sum of the frequency of the participating bands. This is mathematically described by

$$\nu_v = v\nu_0 (1 - vx) \tag{6}$$

where ν_0 is the fundamental vibrational frequency, ν_v is the frequency of the (v – 1)st overtone, v is the vibrational quantum number (= 1 for fundamental, = 2 for overtone), and x is the anharmonicity factor that measures the deviation of the potential function from the parabolic function. Combination bands arise from the simultaneous changes in the energy of two or more vibrational modes and are observed at a frequency given by

$$\nu = n_1\nu_1 + n_2\nu_2 + n_3\nu_3 + \ldots \tag{7}$$

where ν_i = frequency of transition contributing to the combination band, n_i = integer). Again, since combination bands are forbidden transitions, the intensity of these bands within a near-IR spectrum are 10–1000 times less intense than the original fundamental vibrational bands. Typically, in pharmaceutical analysis, near-IR spectroscopy is used for the detection of water (strong OH combination band).

C. Raman

When a compound is irradiated with monochromatic radiation, most of the radiation is transmitted unchanged, but a small portion is scattered. If the scattered radiation is passed into a spectrometer, we detect a strong Rayleigh line at the unmodified frequency of radiation used to excite the sample. In addition, the scattered radiation also contains frequencies arrayed above and below the frequency of the Rayleigh line. The *differences* between the Rayleigh line and these weaker Raman line frequencies correspond to the vibrational frequencies present in the molecules of the sample. For example, we may obtain a Raman line at ± 1640 cm^{-1} on either side of the Rayleigh line, and the sample thus possesses a vibrational mode of this frequency. The frequencies of molecular vibrations are typically 10^{12}–10^{14} Hz. A more convenient unit, which is proportional to frequency, is wavenumber (cm^{-1}), since fundamental vibrational modes lie between 4000 and 50 cm^{-1}.

The Raman lines are generally weak in intensity, approximately 0.001% of the source, and hence their detection and measurement are difficult. Raman bands at wavenumbers less than the Rayleigh line are called Stokes lines, while anti-Stokes lines occur at greater wavenumbers than the source radiation. Generally the anti-Stokes lines are less intense than the Stokes lines because these transitions arise from higher vibrational energy levels containing fewer molecules, as described by the Boltzman distribution (Fig. 2). Hence, the Stokes portion of the spectrum is generally used. The abscissa of the spectrum is usually labeled as wavenumber shift or Raman shift (cm^{-1}) and the negative sign (for Stokes shift) is dispensed with (Fig. 3).

D. Comparison of Infrared and Raman Spectroscopies

Both infrared and Raman are vibrational spectroscopic techniques, and the Raman scattering spectrum and infrared absorption spectrum for a given species often resemble one another quite closely. There are, however, sufficient differences between the types of chemical groups that are infrared- and Raman-active to make the techniques complementary rather than competitive. This is illustrated in Fig. 3, where the infrared and Raman spectra of aspirin are shown.

Briefly, a vibrational mode is infrared-active when there is a change in the molecular dipole moment during the vibration, whereas a vibrational mode is Raman-active when there is a change in polarizability during the vibration. Consequently, asymmetric modes and vibrations due to polar groups (e.g., C=O, NH) are more likely to be strongly infrared-active, whereas symmetric modes and homopolar bonds (e.g., C=C or SS) tend to be Raman-active. The complementarity of the two techniques for molecular characterization for a range of antibacterial agents and related compounds has recently been demonstrated [6].

III. EXPERIMENTAL

A. Infrared Sampling Techniques for Solids

Sometimes, one of the greatest challenges for the IR spectroscopist is sample preparation. Since the development of the Fourier transform infrared spectrophotometer, with its inherent signal-to-noise and throughput advantages, an abundance of sampling techniques for solid state analysis have developed. In this experimental section, the general configuration of the IR spectrophotometer has been overlooked so that solid state IR sampling techniques can be briefly reviewed. A full description of the components of an IR spectrophotometer may be reviewed in the classic Griffiths and de Haseth book [7]. Although extensive, this section does not review all IR sampling techniques, just those widely used for solid state pharmaceutical problem solving and methods development.

1. Alkali Halide Pellet

The classic IR sampling technique is the alkali halide pellet preparation [8]. This technique involves mixing the solid state sample of interest with an alkali halide (typically KBr or KCl). The mixture is pulverized into a finely ground mixture, placed into a die (typically stainless steel), and subjected to approximately 10,000 psi of pressure for a period of time to produce a glass pellet. The pellet (with the sample finely dispersed throughout the glass) may then be placed into the IR spectrophotometer for spectral data acquisition. From the traditional view, this sampling technique is used for the preparation of samples for IR identity testing, which is required for every regulatory submission. The advantage of this technique is that only a small amount of sample is required (usually 1 mg) and a high-quality spectrum can be obtained in a matter of minutes. Disadvantages exist, such as solid state transformation of the sample due to the pressure requirements to form the glass pellet and possible halide exchange between KBr (or KCl) and the sample of interest [9,10]. These are critical disadvantages whenever IR spectroscopy is utilized for pharmaceutical polymorph investigations. Quantitation of mixtures utilizing the alkali halide pellet sampling technique has been attempted with fair success [11]. Due to the aforementioned disadvantages, it is suggested that this IR sampling technique be used only for simple compound identification assays.

2. Mineral Oil Mull

Another classical sampling technique for solids is the mineral oil mull preparation [8]. In this technique, a small amount of sample ($\sim$1 mg) is placed into an agate mortar. To this, a small amount of mineral oil is added, and the sample and oil are mixed to an even consistency. The mixture is then placed onto an IR optical window and sampled by the IR spectrophotometer. One advantage to this technique is no likelihood of solid state transformations due to mixing and/or grinding, and hence it is a good technique for the qualitative identification of pharmaceutical polymorphs. Unfortunately, the mineral oil has a number of large spectral contributions (2952, 2923, 2853, 1458, and 1376 cm^{-1}), which may overlap important absorption bands corresponding to the sample of interest. Typically, this technique is used for qualitative identification assays whenever the alkali halide pellet technique is inappropriate (halide exchange or pressure-induced spectral changes).

3. Diffuse Reflectance

The diffuse reflectance (DR) technique is an important solid state sampling technique for pharmaceutical problem solving and methods development [12]. Sometimes referred to as DRIFTS (diffuse reflectance infrared Fourier transform spectroscopy), this technique is extensively used in the mid- and near-IR spectral regions. The technique involves irradiation of the powdered sample by an infrared beam (Fig. 4). The incident radiation undergoes absorption, reflection, and diffraction by the particles of the sample. Only the incident radiation that undergoes diffuse reflectance contains absorptivity information about the sample. A number of significant advantages exist for diffuse reflectance analysis. Samples may be investigated neat, or diluted within a nonabsorbing matrix such as KBr or KCl (usually at a 1–5% w/w active to nonabsorbing matrix material ratio). A macro and micro sampling cup is usually provided with the diffuse reflectance accessory, and approximately 400 and 10 mg of sample are required for each cup, respectively. The sample is also 100% recoverable, so that other solid state investigations may take place on the same material. In addition, solid state transformations can be monitored by diffuse reflectance IR when interfaced to a variable-temperature, environmentally controlled diffuse reflectance chamber [13]. The DR technique lends itself to polymorph studies since the technique is noninvasive, the polymorph character remains intact due to limited sample

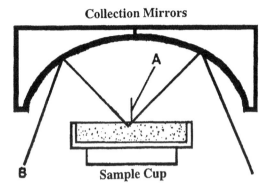

Fig. 4 Schematic representation of the diffuse reflectance sampling accessory. *Key*: A, blocker device to eliminate specular reflectance; B, path of IR beam.

handling, and the technique is quantitative [14]. One disadvantage to diffuse reflectance IR is that it is a particle-size-dependent technique [12]. Development of quantitative polymorph assays require that the particle size of each component be limited to a specific range, including both components of a mixture, and the nonabsorbing matrix if the mixture is not sampled neat. It must also be kept in mind that for a quantitative assay, all calibration, validation, and subsequent samples to be assayed must fall within the particle-size range; otherwise, significant prediction errors may arise. A number of qualitative and quantitative examples of the diffuse reflectance technique as applied to pharmaceutical problem solving and methods development will be highlighted in the Applications section.

4. Microscopy

The first linkage between a microscope and an IR spectrophotometer was reported in 1949 [15]. Today, every manufacturer of IR spectrophotometers offers an optical/IR microscope sampling accessory. The use of optical and IR microscopy is a natural course of action for any solid state investigation. Optical microscopy provides significant information about a sample, such as its crystalline or amorphous nature, particle morphology, and size. Interfacing the microscope to an IR spectrophotometer ultimately provides unequivocal identification of one particular crystallite. Hence, we have the tremendous benefit of IR microscopy for the identification of particulate contamination in bulk or formulated drug products.

The IR microscopy sampling technique is the ultimate sampling technique since only one particle is required for analysis. Due to the restrictions of diffraction effects, typically, the particles of interest must be greater in size than $10 \times 10 \ \mu m$. Once the sample of interest is placed upon an IR optical window, the slide is placed onto the microscope stage and visually inspected. Once the sample of interest is in focus, the field of view is aperatured down to the sample. Depending upon sample morphology, thickness, and transmittance properties, a reflectance and/or transmittance IR spectrum may be acquired by the IR microscope accessory. Obvious advantages for the technique exist, such as nondestructive sampling, reflectance and/or transmittance measurements, minimal sample requirements, and the ability to monitor solid state transformations with the interface of a hot stage, important for polymorphism studies [16].

5. Thermogravimetric/Infrared Analysis

In this IR sampling technique, a thermogravimetric (TG) analyzer is interfaced to an IR spectrophotometer so that the evolved gas from the sample/TG furnace is directed to an IR gas cell. This IR sampling technique lends itself to the identification and quantitation of residual solvent content for a pharmaceutical solid [17], and also to the investigation of pharmaceutical pseudopolymorphs.

Analogous to standard thermogravimetric analysis procedures, the sample of interest for TG/IR analysis is placed into a TG sample cup (approximately 10 mg) and introduced into the TG furnace. The TG apparatus then monitors the weight loss of the sample as a function of temperature (typical heating rate of 10°C per minute). From the TG data, the amount of weight loss over a specific temperature range can imply possible residual solvent content of a sample. Unfortunately, only the percent weight loss can be calculated from the TG experiment. An unequivocal identification of the evolved gas cannot be made. Hence, we see the significant advantage of interfacing the TG apparatus to an IR spectrophotometer.

Although residual solvent content of a sample can be identified and quantified by TG/IR analysis, gas chromatography techniques usually outperform TG/IR experiments (lower detection limits and ease of quantitation). One significant advantage of the TG/IR technique is the investigation of pharmaceutical pseudopolymorphs. Thermal analysis techniques such as TG analysis and differential scanning calorimetry (DSC) typically can determine the presence of pseudopolymorphism based upon correlating TG weight loss and detection of DSC endotherms at the same temperatures. Unfortunately, the identity of the solvent of crystallization cannot be identified by the thermal analysis techniques. Infrared analysis of the evolved gas at the same temperature as the TG weight loss/DSC endotherm can provide identification of the solvent of crystallization.

6. Photoacoustic

The photoacoustic effect was first discovered by Alexander Graham Bell in the early 1880s [18], but it was not applied to Fourier transform infrared (FTIR) spectroscopy until a century later [19,20]. Significant advantages of FTIR photoacoustic spectroscopy (PAS) include the following: (1) Spectra may be acquired on opaque materials (commonly found in pharmaceutical formulations); (2) Minimal sample preparation is necessary; and (3) Depth profiling is possible.

The PAS phenomenon involves the selective absorption of modulated IR radiation by the sample. The selectively absorbed frequencies of IR radiation correspond to the fundamental vibrational frequencies of the sample of interest. Once absorbed, the IR radiation is converted to heat and subsequently escapes from the solid sample and heats a boundary layer of gas. Typically, this conversion from modulated IR radiation to heat involves a small temperature increase at the sample surface ($\sim 10^{-6}$°C). Since the sample is placed into a closed cavity cell that is filled with a coupling gas (usually helium), the increase in temperature produces pressure changes in the surrounding gas (sound waves). Since the IR radiation is modulated, the pressure changes in the coupling gas occur at the frequency of the modulated light, and so does the acoustic wave. This acoustical wave is detected by a very sensitive microphone, and the subsequent electrical signal is Fourier processed and a spectrum produced.

Depth profiling of a solid sample may be performed by varying the interferometer moving-mirror velocity (modulated IR radiation). By increasing the mirror velocity, the sampling depth varies, and surface studies may be performed. Limitations do exist, but the technique has proven to be quite effective for solid samples [21]. In addition, unlike diffuse reflectance sampling techniques, particle size has a minimal effect upon the photoacoustic measurement.

B. Raman Sampling Techniques for Solids

Sampling techniques for Raman spectroscopy are relatively general since the only requirement is that the monochromatic laser beam irradiate the sample of interest and the scattered radiation be focused upon the detector.

1. General Techniques

Raman spectroscopy may be performed on very small samples (e.g., a few nanograms), and sample preparation is simple. Powders do not need pressing into discs or diluting with KBr; they just need to be irradiated by the laser beam. Solid samples are usually examined in stainless steel (or equivalent) sample holders generally requiring around 100 mg of material. Typically, liquid samples are analyzed in quartz or glass cuvettes, which may have mirrored rear surfaces to improve the signal intensity. Glass is a very weak Raman scatterer, and so many liquids can be simply analyzed in a bottle or in, for example, an NMR tube, although fluorescence from some glasses can be problematic. Water is a good solvent for Raman studies since the Raman spectrum of water is essentially one broad, weak band at 3500 cm^{-1}. The complete Stokes Raman spectrum covering shifts in the range 10–3500 cm^{-1} can be obtained, and the intensity of Raman scattering is directly proportional to the concentration of the scattering species, an important factor for quantitative analysis. The Raman effect is relatively weak, however, and hence a material needs to be present at a level of about 1% for accurate assessments, whereas IR can be used to detect materials to a level of approximately 0.01%. Fluorescence can also be problematic in Raman studies but is typically due to the NMR sample tubes being utilized or impurities within the sample of interest. Data massaging techniques can sometimes blank out Raman spectral contributions due to fluorescent materials.

2. Microscopy

Analogous to IR microscopy work, Raman spectra can be acquired on small amounts of material through the use of a Raman microprobe [22]. In a similar fashion to the IR microscope, a sample for Raman microanalysis is first viewed optically by the microscope. The field of view may be aperatured down or the laser focused to the sample, and the subsequent Raman scattering experiment performed. Since a high intensity of monochromatic light from the laser is focused upon a small amount of sample, sample degradation by the laser must be monitored [23]. Otherwise, the Raman microprobe is ideal for investigating polymorphism (single crystals), particulate contamination, and small amounts of samples.

IV. PHARMACEUTICAL APPLICATIONS

A. Mid- and Near-Infrared Spectroscopy

The topics of polymorphism and pseudopolymorphism dominate the majority of publications that deal with utilizing infrared spectroscopy for the physical characterization of pharmaceutical solids. Typically, in each of the publications, IR spectroscopy is only one technique used to characterize the various physical forms. It is important to realize that a multidisciplinary approach must be taken for the complete physical characterization of a pharmaceutical solid. Besides polymorphism, mid- and near-IR have been utilized for identity testing at the bulk and formulated product level, contaminant analysis, and drug–excipient interactions. A number of these applications will be highlighted within the next few sections.

1. Polymorphism and Pseudopolymorphism

Infrared spectroscopy has been widely used for the qualitative and quantitative characterization of polymorphic and pseudopolymorphic compounds of pharmaceutical interest. Since solid state IR can be used to probe the nature of (pseudo)polymorphism on the molecular level, this method is particularly useful in instances where full crystallographic characterization of (pseudo)polymorphism was not found to be possible. Recently, a significant number of publications have appeared that discuss where a multidisciplinary, spectroscopic approach to polymorph characterization has taken place. Qualitative mid-IR studies have taken place on DuP 747 [24], fosinopril sodium [25], diflunisal [26], SC-41930 [14], spironolactone [27], 1,2-dihydro-6-neopentyl-2-oxonicotinic acid [28], indomethacin [29], 4'-methyl-2'-nitroacetanilide [30], and numerous compounds in a series of articles by Kuhnert-Brandstätter and Sollinger [31,32]. Near-infrared has also been used to investigate polymorphic systems such as glycine [33] and SC-25469 [34].

In the case of DuP747 [24], XRD, DSC, and thermomicroscopic studies determined the polymorphic system to be monotropic. Distinct diffuse reflectance IR, Raman, and solid state ^{13}C NMR spectra existed for each physical form. The complementary nature of IR and Raman gave evidence that the polymorphic pair were roughly equivalent in conformation. It was concluded that the polymorphic character of DuP 747 resulted from different modes of packing. Further crystallographic information is required in order to determine the crystal packing and molecular confirmation of this polymorphic system.

Analogous to the DuP 747 study, complete crystallographic information was not possible on the fosinopril sodium polymorphic system [25]. Two known polymorphs (A and B) were studied via a multidisciplinary approach (XRD, IR, NMR, and thermal analysis). Complementary spectral data from IR and solid state ^{13}C NMR revealed that the environment of the acetal sidechain of fosinopril sodium differed in the two forms. In addition, possible *cis-trans* isomerization about the C_6N peptide bond may exist. These conformational differences are postulated as the origin of the observed polymorphism in fosinopril sodium in the absence of the crystallographic data for form B (single crystals not available).

Solid state characterization studies of the previously mentioned polymorphic systems [26–34] all utilize IR as a means to differentiate the various crystal modifications. In some cases, the observation of variations in IR absorption intensities has led to conclusions regarding intramolecular hydrogen bonding [26]. For other systems, fairly complete IR spectral band assignment has allowed for determination of structure for the polymorphic system. In one study [29], DSC-IR was used to identify the polymorphs and determine simultaneously the correlation between thermal events and structural changes.

In each of the aforementioned studies, qualitative IR spectroscopy was used. It is important to realize that IR is also quantitative in nature, and several quantitative IR assays for polymorphism have appeared in the literature. Sulfamethoxazole [35] exists in at least two polymorphic forms, which have been fully characterized. Distinctly different diffuse reflectance mid-IR spectra exist, permitting quantitation of one form within the other. When working with the diffuse reflectance IR technique, two critical factors must be kept in mind when developing a quantitative assay: (1) the production of homogeneous calibration and validation samples, and (2) consistent particle size for all components, including subsequent samples for analysis. During the assay development for sulfamethoxazole, a number of mixing techniques were investigated in an effort to achieve homogeneous samples. Inhomogeneity of calibration and validation samples can lead to inaccurate IR absorption values and subsequent prediction errors. This is also the case with particle size. Variation in the particle size of the nonabsorbing matrix or sample can influence the diffuse reflectance IR spectrum and again lead to prediction errors. After mixing and particle-size factors were optimized, a quantitative diffuse reflectance IR assay was developed in which independent validation samples were predicted within 4% of theoretical values.

Quantitation of one polymorphic form within another by mid-IR diffuse reflectance has also been performed on SC-41930 [14]. The high melting point form (HM) was spiked with low melting point form (LM) between 0 and 25% w/w. The 1670 cm^{-1} absorbance feature in the LM was used for quantitation. Based upon the spectral data, a detection limit of approximately 1% w/w of LM in HM was estimated.

Beside mid-IR, near-IR spectroscopy has been used to quantitate polymorphs at the bulk and dosage product level. For SC-25469 [34], two polymorphic forms were discovered (α and β), and the β-form was selected for use in the solid dosage form. Since the β-form can be transformed to the α-form under pressure by enantiotropy, quantitation of the β-form in the solid dosage formulation was necessary. Standard mixtures of both forms in the formulation matrix were prepared, and spectra were measured in the near-IR via diffuse reflectance. Utilizing a standard, near-IR multiple linear regression, statistical approach, the α- and β-forms could be predicted to within 1% of theoretical. This extension of the diffuse reflectance IR technique shows that quantitation of polymorphic forms at the bulk and/or dosage product level can be performed.

The utilization of IR spectroscopy is very important in the characterization of pseudopolymorphic systems, especially hydrates. It has been used to study the pseudopolymorphic systems SQ-33600 [36], mefloquine hydrochloride [37], ranitidine HCl [38], carbovir [39], and paroxetine hydrochloride [40]. In the case of SQ-33600 [36], humidity-dependent changes in the crystal properties of the disodium salt of this new HMG-CoA reductase inhibitor were characterized by a combination of physical analytical techniques. Three crystalline solid hydrates were identified, each having a definite stability over a range of humidity. Diffuse reflectance IR spectra were acquired on SQ-33600 material exposed to different relative humidity (RH) conditions. A sharp absorption band at 3640 cm^{-1} was indicative of the OH stretching mode associated with either strongly bound or crystalline water (Fig. 5A). The sharpness of the band is evidence of a bound species even at the lowest levels of moisture content. The bound nature of this water contained in low-moisture samples was confirmed by variable-temperature (VT) diffuse reflectance studies. As shown in Fig. 5B, the 3640 cm^{-1} peak progressively decreased in intensity upon thermal

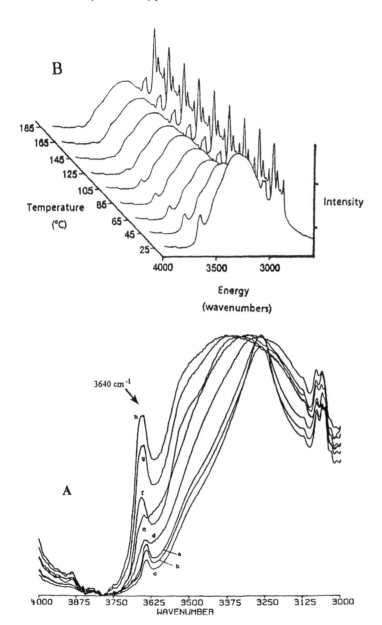

Fig. 5 A, diffuse reflectance IR spectra of SQ-33600 displaying the absorption band due to bound water of hydration after exposure to different relative humidity conditions of (a) initial sample containing 3.5% water, (b) 6%, (c) 15%, (d) 31%, (e) 43%, (f) 52%, (g) 70%, and (h) 84%. B, variable-temperature diffuse reflectance IR spectra of SQ-33600 initially containing 3.5% water.

dehydration and completely disappeared at a temperature of 145°C. No other spectral features changed during the VT experiments, confirming the integrity of the SQ-33600 molecule.

Whenever IR spectroscopy (mid-, near-, or far-IR) is used for the physical characterization of pharmaceutical solids, it is important that the sample retains its integrity upon exposure to the IR source. During the development of a near-IR assay for a pseudopolymorphic system (within the Bristol-Myers Squibb laboratories), the sample was exposed for significant periods of time to the energy from the tungsten source lamp. It became apparent from the IR spectra measured that some form of sample degradation was occurring after repeated sampling of the same material by the spectrophotometer. Upon continuous exposure of the hydrate drug substance to the near-IR source, the hydrate combination band (5137 cm^{-1}) lost intensity over time. It was determined that the sample was dehydrating upon exposure to the IR beam, leading to significant errors in the development of a quantitative near-IR pseudopolymorph assay.

2. Particulate/Contaminant Analysis

Infrared microscopy is well suited for in situ analysis of contaminants found in pharmaceutical processes. Due to the nondestructive nature of the analysis, further experiments such as energy dispersive x-ray analysis may be performed on the same sample once IR investigations are complete. To illustrate the potential of IR microspectroscopy, one application from the Bristol-Myers Squibb laboratories is presented.

A series of foreign particulates was found in several bulk lots and final product lots (tablets) of a developmental drug. Black, red, and brown particles were isolated from the bulk drug material, whereas black particles were observed embedded into the tablets. Only the black particles will be focused upon in this discussion. Since the foreign materials were opaque, IR microspectroscopy data were obtained in the reflectance mode. Figure 6A displays the relatively simple IR spectrum of the isolated black particle. No absorption bands corresponding to aromatic or aliphatic CH groups (2800–3200 cm^{-1}) were present, whereas a few absorption bands were observed between 1150 and 1250 cm^{-1}, corresponding to CF and CC functional groups. Computer-aided spectral library search of the spectrum revealed an IR spectral match with polytetrafluoroethylene. Subsequent energy dispersive x-ray analysis confirmed that fluorine was present in the black particulate sample. Selectivity of the IR microspectroscopy technique is revealed in Fig. 6B. This spectrum represents the reflectance IR measurement of the black particulate that is embedded into the tablet. No sample extraction process was required to obtain the spectrum. A distinct advantage of the IR microspectroscopy technique is the ability to sample only the area of the sample defined by the microscope's aperture. In addition, no destructive sample preparation is required.

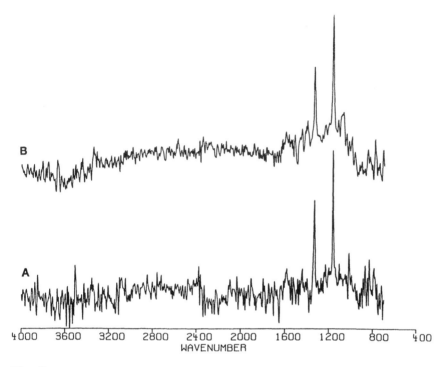

Fig. 6 Infrared spectra (microspectroscopy) of the isolated black particle (A) and a particle embedded into the tablet (B).

3. Other Applications

Near-infrared spectroscopy is quickly becoming a preferred technique for the quantitative identification of an active component within a formulated tablet. In addition, the same spectroscopic measurement can be used to determine water content since the combination band of water displays a fairly large absorption band in the near-IR. In one such study [41] the concentration of ceftazidime pentahydrate and water content in physical mixtures has been determined. Due to the ease of sample preparation, near-IR spectra were collected on 20 samples, and subsequent calibration curves were constructed for active ingredient and water content. An interesting aspect of this study was the determination that the calibration samples must be representative of the production process. When calibration curves were constructed from laboratory samples only, significant prediction errors were noted. When, however, calibration curves were constructed from laboratory and production samples, realistic prediction values were determined (±5%).

Near-infrared has again been used to determine the active ingredient concentration and water content for amiodarone chlorhydrate [42] and ranitidine chlorhydrate [43], but on the intact tablet. These two studies are obvious extensions of the spectroscopic technique to analytical control situations. For the amiodarone chlorhydrate study [42], a specially designed sample cell was constructed so that reflectance near-IR measurements could be made off a tablet surface that was leveled to remove the sugar-coated surface. The near-IR spectra of pure amiodarone chlorhydrate displayed distinct absorption bands for methyl and aromatic CH moieties (1680, 2139 and 2230, 2270 nm, respectively), which allowed quantitation of the active ingredient within a formulation. In addition, a distinct absorption band at 1940 nm corresponded to the hydroxyl group of water, which is present during formulation of the tablet form. Upon construction of calibration curves using a linear regression model, very precise and accurate predicted concentrations of active ingredient were made. The authors hint that this type of technique would make series manipulation possible (on-line production analysis). In the case of ranitidine chlorhydrate [43], an analogous near-IR approach was used to determine the concentration of the active ingredient and water content in crushed tablets. Calibration curves were constructed using only laboratory prepared samples, only commercial samples, and a mixture of laboratory and commercial samples. It was determined that the calibration constructed only with the laboratory samples had the best characteristics. One of the more important conclusions from this investigation is that the optimum calibration range appears to be about 5%. A wider working range does not improve the calibration and tends to prejudice the test results. It is important to note that this conclusion is appropriate for formulation testing when little variation in active ingredient loading within a formulation is expected. This is not the case in bulk drug substance analysis such as the previously discussed quantitative polymorphic assays. Nonetheless, near-IR spectroscopy provided a reliable form of formulation analysis for ranitidine chlorhydrate: only 10–15 samples are required for a calibration curve, and these concentrations can be within an interval of content of 3.5–5%. An additional outcome of this study was that changing of operators of the near-IR equipment did not lead to significant prediction errors. Additionally, studies on the determination of content uniformity of pharmaceutical products, utilizing near-IR, have been performed on aspirin and acetaminophen tablets [44]. Analogous to the previously discussed cases, intact tablets were studied and calibration curves constructed via linear regression models. Upon comparison with chromatographic methods, the spectroscopic approach was equally accurate and less time consuming.

In each of the preceding studies, one major advantage of the near-IR technique is the rapid evaluation of the sample without the need for traditional extraction techniques and subsequent chromatographic or colorimetric analysis. For this reason, near-IR is an ideal form of analysis in a quality control, or on-line, process analysis environment. In a complementary mid- and near-IR study [45], it was determined that near-IR spectroscopy could not differentiate between simvastatin and lovastatin. These two molecules only differ by the presence and absence of an α-methyl group attached to an ester carbonyl, respectively. Instead, mid-IR diffuse reflectance spectroscopy was used to identify and quantify the active components within a formulation. This study also investigated sample preparation variables including cup filling, variance in grinding times, and duplicate spectral measurements. In addition to the previously discussed molecules, enalpril maleate and finastride were also studied.

As previously mentioned, a multidisciplinary approach must be taken for the complete physical characterization of pharmaceutical solids. Another technique in one's arsenal is thermogravimetric-IR (TG/IR) analysis. This technique is ideally suited for the determination of residual solvents in a pharmaceutical solid (bulk drug substance or formulated physical mixture), thermal stability of a product, and the determination and characterization of pseudopolymorphism. An excellent example of utilizing TG/IR for residual solvent analysis in conjunction with pharmaceutical solids has been given by Johnson and Compton [17]. In the investigation of a pharmaceutical solid, three weight losses were measured, at 60, 135, and 175°C. Analysis of the mid-IR spectrum at each temperature revealed that the weight loss at 60°C was associated with water loss, whereas the TG weight loss at 135°C was a combined loss of a mixture of alcohols and water. After IR spectral library searching, isopropanol was identified as the dominant species evolved during this second weight loss. Utilizing the advantage of digital IR spectra, subsequent spectral subtraction techniques ("spectral-stripping") revealed that ethanol was also an evolved component. Decomposition of the sample at 175°C revealed the evolution of acetone, ammonia, carbon dioxide, and water as the major by-products based on the measured mid-IR spectra.

Within the Bristol-Myers Squibb Materials Science laboratory, TG/IR analysis is routinely used for solvate identification for pseudopolymorphic compounds. In one particular example, a submitted bulk drug substance displayed distinctly different XRD and DSC data compared with the research reference standard data. Figure 7 displays the TG weight loss curve and subsequent mid-IR spectra at various time points/temperatures for this questionable sample. Based upon absorption bands at 2979 and 1217 cm^{-1}, an organic species (aliphatic CH, CO moieties, respectively) is present in the evolved gas. Subsequent analysis of the IR spectra collected at the completion of the weight loss (spectral library matching) reveals that a butanol solvate was present. Thus, TG/IR provided an unequivocal characterization as to the origin of pseudopolymorphism in this particular compound. An obvious extension of the TG/IR technique is the study of pharmaceutical dosage forms subjected to various stress conditions to determine if moisture is released, or more importantly, odor analysis for sulfur-containing active drugs.

In an extension of the diffuse reflectance technique, DR has been used to

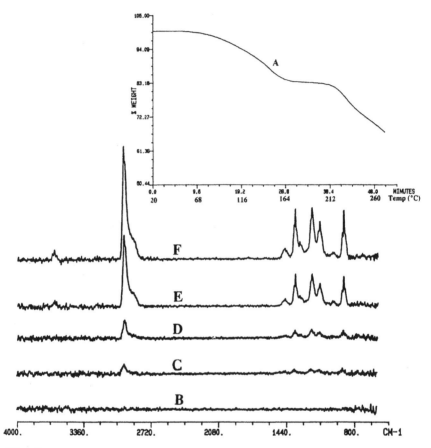

Fig. 7 Thermogravimetric weight loss curve (A) and subsequent IR spectra measured at the designated temperatures: (B) 70°C, (C) 95°C, (D) 120°C, (E) 160°C, and (F) 180°C. A slight lag time exists between the TG weight loss and IR spectral acquisition due to the evolved gas being "carried" into the IR gas cell by the He carrier gas. Each IR spectrum is plotted on the same absolute intensity scale (Abs. units).

study the orientation of various functionalities of a polymer/excipient mixture on microcapsule surfaces [46]. Utilizing mixtures of a block copolymer of d-tartaric acid and 1,8-octanediol with core materials of either talc or kaolin, differences in the molar extinction coefficients calculated from the measured DR spectra allowed determination of the relative orientation of various functionalities within the core material. It was concluded that in the talc mixtures, the OH functionality of the polymer was oriented 60° from normal as compared with 11° for the OH moiety within the kaolin/polymer mixture. For the carbonyl functionality, the talc/polymer showed a 30° orientation off normal as compared with 15° for the kaolin/polymer mixture. The DR spectra provided information on the orientation of a specific functionality of the polymer on the microcapsule surface. This information may be critical in an effort to improve the site specificity of medicinal agents on surfaces.

The extent of homogeneous mixing of pharmaceutical components such as active drug and excipients has been studied by near-IR spectroscopy. In an application note from NIRSystems, Inc. [47], principal component analysis and spectral matching techniques were used to develop a near-IR technique/algorithm for determination of an optimal mixture based upon spectral comparison with a "standard mixture." One advantage of this technique is the use of second-derivative spectroscopy techniques to remove any slight baseline differences due to particle size variations.

Drug–excipient interactions are typically studied via chromatographic techniques. Unfortunately, this form of analysis requires extraction and/or dissolution techniques that may destroy critical physical and chemical information. Hence, the ability to study drug–excipient interactions noninvasively is crucial. Differential scanning calorimetry and diffuse reflectance IR have been used to study the yellow or brown color that develops when aminophylline is mixed with lactose [48]. The DSC thermogram of the aminophylline/lactose mixture is not a direct superposition of the individual component traces, leading to the indication of an incompatibility [49]. Although DSC is able to determine a drug–excipient interaction, the exact nature of the interaction is unknown, hence the need for IR spectroscopy. After complete analysis of the IR spectra of individual components, physical mixtures, and various samples subjected to stress conditions (60°C for three weeks), it was concluded that ethylenediamine is liberated from the aminophylline complex and reacts with lactose through a Schiff base intermediate. This reaction, in turn, results in brown discoloration of the sample. With currently available diffuse reflectance environmental chambers providing control of temperature, pressure, and relative humidity conditions, IR spectroscopy is now a first-line approach to the determination of physicochemical interactions between drugs and excipients.

B. Raman Spectroscopy

The following examples, from the literature and our own work, demonstrate some of the potential applications for Raman spectroscopy in the pharmaceutical sciences.

1. Polymorphism

Raman spectroscopy has been used to characterize polymorphic forms of griseofulvin [50] and sulfathiazole [51]. In both of these studies, the lattice vibrations of the drug crystals (found in the wavenumber region 50–150 cm^{-1}, an area inaccessible using mid-IR spectroscopy) were studied. The Raman spectra showed that, on desolvation, griseofulvin reverted to the lattice structure of the original unsolvated form, and the technique was used to identify the nature of solute–solvent interactions with differentiation between van der Waals and hydrogen bonding. An additional report in the literature includes Raman data on the pseudopolymorphs of griseofulvin and ampicillin [52]. The Raman spectra of griseofulvin and its benzene solvate indicated that the two crystal structures are quite different, whereas the chloroform solvate crystal form is similar in structure to the anhydrous form. For ampicillin, distinct Raman spectra were measured for the two anhydrous forms and the trihydrate. Form I and the trihydrate may be discernable within a mixture, possibly allowing for the quantitation of these forms within another, or within a physical mixture of excipients. Unfortunately, form II does not have any distinct spectral features that allow for unequivocal identification.

More recently, using FT-instrumentation, differences in the Raman spectra of polymorphic forms of cortisone acetate and an experimental pharmaceutical compound, R69, were reported [53]. Of particular value in these studies was the ease of sample preparation, since R69 was reported to be sensitive to grinding. Also, FT-Raman spectroscopy has been used to differentiate polymorphic forms of cimetidine [54], characterized by the "functionality region" of the spectra (1000–1800 cm^{-1}) and, together with FTIR, in the study of spironolactone polymorphs [27]. From our own work [55] we have characterized carbamazepine polymorphs by FT-Raman spectroscopy (Fig. 8). Notably, changes in the stretching frequencies of C=C (around 1624 cm^{-1}) and CN (around 1250 cm^{-1}) bonds and in the deformation of NH (around 1600 cm^{-1}) bands demonstrated altered chemical configurations in the substituent groups of the molecule, while some minor alterations in the parent ring structure were also noted. Variations in lattice vibrations are also seen between the carbamazepine polymorphs (between 50 and 150 cm^{-1}), indicating different packing arrangements of the molecules.

2. Delivery Systems

Interaction between sulfathiazole and povidone was studied by Raman spectroscopy, and the nature of the drug–polymer coprecipitates investigated [51]. The nature of the drug (solvation state) and its bonding to the polymer were assessed with respect to sulfathiazole dissolution rate.

The FT-Raman spectra of a range of drugs (theophylline, indomethacin, diclofenac, and promethazine) in several polymers (sodium alginate, hydroxypropylmethylcellulose, and polyethylene glycol) have been obtained [56,57]. In these studies, the linearity of response of Raman scattering to species concentration was exploited to analyze diclofenac at concentrations of 0.01–6.0% w/w

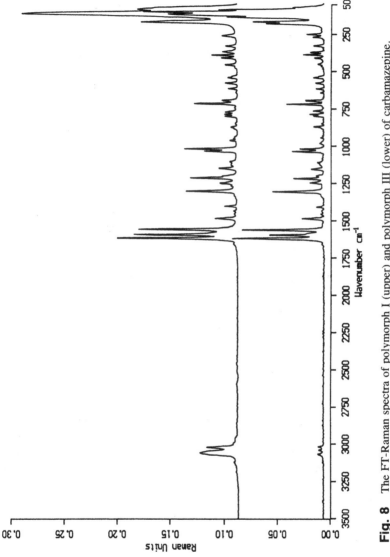

Fig. 8 The FT-Raman spectra of polymorph I (upper) and polymorph III (lower) of carbamazepine.

in sodium alginate. The degradation of poly(sebacic anhydride) in water was also studied [56] since controlled release of a drug can be achieved by leaching or diffusion through the polymer, or by erosion due to dissolution or hydrolysis. Here, the very weak Raman scattering of water permitted a full molecular investigation, which would have been severely limited by infrared spectroscopy. Recently, a series of biomedical poly(ortho esters) has been characterized by FT-Raman spectroscopy [58]. The study demonstrated that semiquantitative relationships between polymer structures and spectral band areas could be obtained, although the authors concluded that in this study FTIR would provide superior molecular information.

3. Mixing

The linearity of response of Raman scattering to species concentration can clearly be used to ensure mixing quality. Raman spectroscopy has been used to determine phenylpropanolamine hydrochloride and acetaminophen in both solid and solution mixtures [59]. Accuracy and precision was shown to be better than 1% for measurement of acetaminophen that was present at 10 times the concentration of phenylpropanolamine hydrochloride, for which the analysis showed accuracy and precision of better than 3%. The study clearly demonstrated the potential of Raman spectroscopy for simultaneous multicomponent determinations on pharmaceutical preparations. A further example of the value of Raman spectroscopy in quality assurance used a fiber optic probe for remote multisite analysis of tablets [60]. With the aim of quality assurance at different production stages, the fiber optic technique provided a limit of detection better than 0.8 mg for a 120-mg tablet containing 10 mg of active substance. Recently, FT-Raman spectroscopy has been used to characterize illicit drugs [61,62], and to identify illicit drugs (e.g., amphetamine sulphate) in street samples cut with carriers (e.g., sorbitol). Again, ease of sampling and the nondestructive nature of the analysis are of advantage compared with currently used forensic techniques such as gas chromatography–mass spectrometry. The application of FT-Raman spectroscopy in identification of materials in a mixture is demonstrated in Fig. 9, where paracetamol (5% w/w) is easily distinguished from a mixture with dicalcium phosphate dihydrate; clearly, this type of study can be performed quantitatively.

4. Crystallinity

Other than the polymorphic studies described earlier, we have used FT-Raman spectroscopy to probe crystal behavior. Succinic acid exists as separate monomers in aqueous solution, which crystallize to form hydrogen-bonded chains in the solid state. The conversion of monomers to chains during cooling has recently been studied [63], with good-quality spectra being obtained for the aqueous system. During the crystallization process ordering of CH_2 features were clearly identified in the spectra. Also, the carbonyl stretching

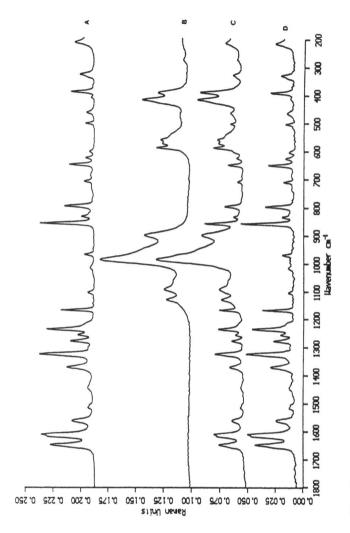

Fig. 9 The FT-Raman spectra of a paracetamol–dicalcium phosphate dihydrate mixture: A, spectrum of pure paracetamol; B, spectrum of pure dicalcium phosphate dihydrate; C, spectrum of a mixture of paracetamol (5% w/w) in dicalcium phosphate dihydrate; D, spectrum C minus spectrum B to identify pure paracetamol (compare with A).

motion of the carboxylic acid group showed a marked shift to lower wavenumber (by about 70 cm^{-1}), which is often noted with hydrogen bonding. Also noted was a structuring of CCO stretching modes on crystallization (band shift to higher wavenumber of approximately 100 cm^{-1}) and the absence of lattice vibrations during initial solute ordering.

The effects of storage conditions (humidity) and drying on the crystallinity of commercial magnesium stearate powder have been investigated by FT-Raman spectroscopy [64,65]. Spectral features attributed to lattice vibrations were lost on drying and hence may be due to water of crystallization being removed from the sample (Fig. 10). The structural alterations induced by drying were also evidenced from an assessment of CH stretching vibrations in the wavenumber range 2600–3200 (Fig. 11). Broadening of bands indicates a less ordered structure, and the peak intensity ratio 2850/2882 cm^{-1} can be used to quantify hydrocarbon-chain flexibility. Thus, quantitative measurements of crystallinity can be made, and the study also demonstrates that spectral features (especially those related to lattice structure) can be obtained typically to about 50 cm^{-1} and, depending on the sample analyzed, even to within 10 cm^{-1} of the laser line (Fig. 10).

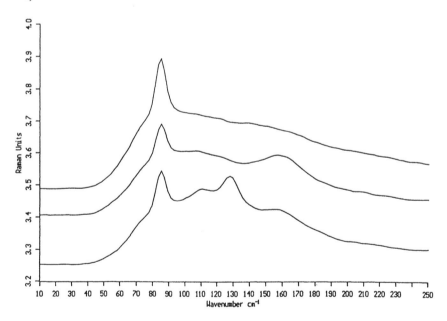

Fig. 10   The FT-Raman spectra over the wavenumber range 10–250 cm^{-1} of magnesium stearate after different drying procedures. Top, heated at 90°C to constant weight under vacuum; middle, heated to 60°C to constant weight under vacuum; bottom, commercially supplied sample.

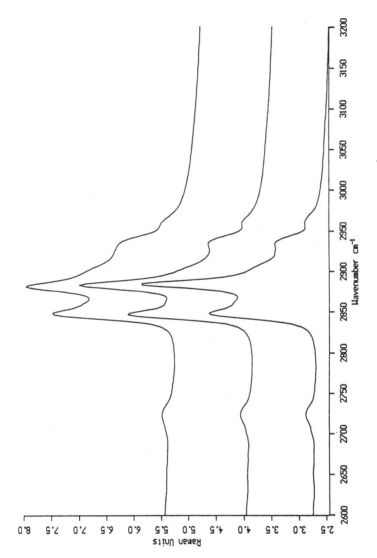

Fig. 11 The FT-Raman spectra over the wavenumber range 2600–3200 cm^{-1} of magnesium stearate after different drying procedures. Top, heated at 90°C to constant weight under vacuum; middle, heated to 60°C to constant weight under vacuum; bottom, commercially supplied sample.

The molecular natures of salts crystallized from salbutamol base have been assessed by FT-Raman spectroscopy [66]. Variations in vibrational frequencies due to electron-withdrawing or -donating substituents were clearly evident; the CCO stretching vibration shifted from 776 cm^{-1} in the free base to 756 cm^{-1} in the benzoate salt. The C=C stretching frequency also shifted from 1610 cm^{-1} to 1603 cm^{-1} with the benzoate ion but showed an increase to 1616 cm^{-1} with sulfate ion. Clearly, the choice of salt affects the molecular nature of the drug, with obvious implications for its physicochemical properties.

5. Other Applications

Raman (and infrared) spectra of drugs such as Dapsone [67], Antineoplaston A10 [68], and benzodiazepines (including diazepam) [69] have been reported and vibrational modes assigned. An interesting application of the technique used multichannel Raman spectrometry for rapid headspace analysis of sealed drug vials [70]. Such a study was feasible because of the very weak Raman spectrum of glass, and spectral quality was sufficient to allow discrimination of 2% oxygen in nitrogen. We have also used FT-Raman spectroscopy for molecular characterization of glucose in aqueous solutions before, during, and after freezing [71]. Again, the advantage of a very weak Raman spectrum from glass and water allowed the analysis of the aqueous system in a liquid nitrogen-cooled glass cell.

V. SUMMARY

Clearly, the potential applications for vibrational spectroscopy techniques in the pharmaceutical sciences are broad, particularly with the advent of Fourier transform instrumentation at competitive prices. Numerous sampling accessories are currently available for IR and Raman analysis of virtually any type of sample. In addition, new sampling devices are rapidly being developed for at-line and on-line applications. In conjunction with the numerous other physical analytical techniques presented within this volume, the physical characterization of a pharmaceutical solid is not complete without vibrational analysis.

REFERENCES

1. C. V. Raman and K. S. Krishnan, *Nature*, *121*, 501 (1928).
2. D. A. Long, *Int. Rev. Phys. Chem.*, 7, 317 (1988).
3. N. B. Colthup, L. H. Daly, and S. E. Wiberley, in *Introduction to Infrared and Raman Spectroscopy*, 3rd ed., Academic Press, New York, 1990.
4. D. A. Long, in *Raman Spectroscopy*, McGraw-Hill, New York, 1977.
5. P. Hendra, C. Jones, and G. Warnes, in *Fourier Transform Raman Spectroscopy*: *Instrumental and Chemical Applications*, Ellis Horwood, New York, 1991.
6. E. A. Cutmore and P. W. Skett, *Spectrochim. Acta*, *48A*, 809 (1993).

7. P. R. Griffiths and J. A. de Haseth, in *Fourier Transform Infrared Spectrometry*, John Wiley & Sons, New York, 1986.

8. J. E. Stewart, in *Infrared Spectroscopy: Experimental Methods and Techniques*, Marcel Dekker, Inc., New York, 1970.

9. V. A. Bell, V. R. Citro, and G. D. Hodge, *Clays and Clay Minerals*, *39*, 290 (1991).

10. S. C. Mutha and W. B. Ludemann, *J. Pharm. Sci.*, *65*, 1400 (1976).

11. J. Hlavay and J. Inczédy, *Spectrochim. Acta*, *41A*, 783 (1985).

12. M. P. Fuller and P. R. Griffiths, *Anal. Chem.*, *50*, 1906 (1978).

13. *The Complete Guide to FT-IR*, Spectra-Tech, Inc. Product Catalog, Stamford, Conn., 1993.

14. D. A. Roston, M. C. Walters, R. R. Rhinebarger, and L. J. Ferro, *J. Pharm. Biomed. Anal.*, *11*, 293 (1993).

15. R. Barer, A. R. H. Cole, and H. W. Thomas, *Nature*, *163*, 198 (1949).

16. J. A. Reffner, J. P. Coates, and R. G. Messerschmidt, *Amer. Lab.*, *19*, 5 (1987).

17. D. J. Johnson and D. A. C. Compton, *Spectroscopy*, *3*(6), 47 (1988).

18. A. G. Bell, *Amer. Assoc. Advan. Sci.*, *29*, 115 (1880).

19. D. W. Vidrine, *Appl. Spectrosc.*, *34*, 314 (1980).

20. M. G. Rockley, *Appl. Spectrosc.*, *34*, 405 (1980).

21. J. F. McClelland, S. Luo, R. W. Jones, and L. M. Seaverson, in *Photoacoustic and Photothermal Phenomena III*, D. Bićanić, ed. Springer-Verlag Berlin, Heidelberg, 1992.

22. T. Hirschfeld, *J. Opt. Soc. Am.*, *63*, 476 (1973).

23. M. Lankers, D. Gottges, A. Materny, K. Schaschek, and W. Kiefer, *Appl. Spectrosc.*, *46*, 1331 (1992).

24. K. Raghavan, A. Dwivedi, G. C. Campbell, Jr., G. Nemeth, and M. A. Hussain, *J. Pharm. Biomed. Anal.*, *12*, 777 (1994).

25. H. G. Brittain, K. R. Morris, D. E. Bugay, A. B. Thakur, and A. T. M. Serajuddin, *J. Pharm. Biomed. Anal.*, *11*, 1063 (1993).

26. M. C. Martiez-Ohárriz, C. Martín, M. M. Goñi, C. Rodríguez-Espinosa, M. C. Tros de Ilarduya-Apaolaza, and M. Sánchez, *J. Pharm. Sci.*, *83*, 174 (1994).

27. G. A. Neville, H. D. Beckstead, and H. F. Shurvell, *J. Pharm. Sci.*, *81*, 1141 (1992).

28. R. S. Chao, and K. C. Vail, *Pharm. Res.*, *4*, 429 (1987).

29. S.-Y. Lin, *J. Pharm. Sci.*, *81*, 572 (1992).

30. R. A. Fletton, R. W. Lancaster, R. K. Harris, A. M. Kenwright, K. J. Packer, D. N. Waters, and A. Yeadon, *J. Chem. Soc. Perkin Trans.*, *II*, 1705 (1986).

31. M. Kuhnert-Brandstätter and H. W. Sollinger, *Mikrochim. Acta [Wein]*, *III*, 233 (1990).

32. M. Kuhnert-Brandstätter and H. W. Sollinger, *Mikrochim. Acta [Wein]*, *III*, 247 (1990).

33. C. E. Miller and D. E. Honigs, *Spectroscopy*, *4*(3), 44 (1989).

34. R. Gimet and A. T. Luong, *J. Pharm. Biomed. Anal.*, *5*, 205 (1987).

35. K. J. Hartauer, E. S. Miller, and J. K. Guillory, *Int. J. Pharm.*, *85*, 163 (1992).

36. K. R. Morris, A. W. Newman, D. E. Bugay, S. A. Ranadive, A. K. Singh, M. Szyper, S. A. Varia, H. G. Brittain, and A. T. M. Serajuddin, *Int. J. Pharm.*, *108*, 195 (1994).

37. A. Kiss, J. Répási, Z. Salamon, Cs. Novák, G. Pokol, and K. Tomor, *J. Pharm. Biomed. Anal.*, *12*, 889 (1994).

38. T. Madan and A. P. Kakkar, *Drug Dev. Ind. Pharm.*, *20*, 1571 (1994).

39. N.-A. T. Nguyen, S. Ghosh, L. A. Gatlin, and D. J. W. Grant, *J. Pharm. Sci.*, *83*, 1116 (1994).

40. I. R. Lynch, P. C. Buxton, and J. M. Roe, *Anal. Proc.*, *25*, 305 (1988).

41. S. Lonardi, R. Viviani, L. Mosconi, M. Bernuzzi, P. Corti, E. Dreassi, C. Murratzu, and G. Corbini, *J. Pharm. Biomed. Anal.*, *7*, 303 (1989).

42. R. Jensen, E. Peuchant, I. Castagne, A. M. Boirac, and G. Roux, *Spectros. Int. J.*, *6*, 63 (1988).

43. P. Corti, E. Dreassi, G. Corbini, S. Lonardi, R. Viviani, L. Mosconi, and M. Bernuzzi, *Pharm. Acta Helv.*, *65*, 28 (1990).

44. NIRSystems, Inc., Application Note, "Determination of active ingredients in solid (pharmaceutical) dosage forms utilizing solid-state standard additions," Silver Spring, Md.

45. J. A. Ryan, S. V. Compton, M. A. Brooks, and D. A. C. Compton, *J. Pharm. Biomed. Anal.*, *9*, 303 (1991).

46. M. L. Shively, G. Lavigne, and A. P. Simonelli, *Drug Dev. Ind. Pharm.*, *17*, 2511 (1991).

47. NIRSystems, Inc., Application Note, "Following the progress of pharmaceutical mixing studies using near-infrared spectroscopy," Silver Spring, Md.

48. K. J. Hartauer and J. K. Guillory, *Drug Dev. Ind. Pharm.*, *17*, 617 (1991).

49. A. Van Dooren and B. Duphar, *Drug Dev. Ind. Pharm.*, *9*, 43 (1983).

50. B. A. Bolton and P. N. Prasad, *J. Pharm. Sci.*, *70*, 789 (1981).

51. B. A. Bolton and P. N. Prasad, *J. Pharm. Sci.*, *73*, 1849 (1984).

52. J. C. Bellows, F. P. Chen, and P. N. Prasad, *Drug Dev. Ind. Pharm.*, *3*, 451 (1977).

53. C. M. Deeley, R. A. Spragg, and T. L. Threlfall, *Spectrochim. Acta*, *47A*, 1217 (1991).

54. A. M. Tudor, M. C. Davies, C. D. Melia, D. C. Lee, R. C. Mitchell, P. J. Hendra, and S. J. Church, *Spectrochim. Acta*, *47A*, 1389 (1991).

55. L. E. McMahon, A. C. Williams, P. York, H. G. M. Edwards, and P. Timmins, in *Proceedings of the XIVth International Conference on Raman Spectroscopy*, N.-T. Yu and X.-Y. Li, eds., John Wiley & Sons, Chichester, p. 908 (1994).

56. M. C. Davies, J. S. Binns, C. D. Melia, and D. Bourgeois, *Spectrochim. Acta*, *46A*, 277 (1990).

57. M. C. Davies, J. S. Binns, C. D. Melia, P. J. Hendra, D. Bourgeois, S. P. Church, and P. J. Stephenson, *Int. J. Pharm.*, *66*, 223 (1990).

58. A. M. Tudor, C. D. Melia, M. C. Davies, S. J. Church, and J. Hellier, *Spectrochim. Acta*, *49A*, 759 (1993).

59. T. H. King, C. K. Mann, and T. J. Vickers, *J. Pharm. Sci.*, *74*, 443 (1985).

60. S. Hameau, M. Jouan, and N. Q. Dao, *Analysis*, *16*, 173 (1988).

61. C. M. Hodges, P. J. Hendra, H. A. Willis, and T. Farley, *J. Raman Spectrosc.*, *20*, 745 (1989).

62. C. M. Hodges and J. Akhavan, *Spectrochim. Acta*, *46A*, 303 (1990).

63. A. B. Brown, P. York, A. C. Williams, and C. Doherty, *Particle Technology Forum. A.I.Ch.E., Proceedings*, *1*, 322 (1994).

64. D. G. Sawh, D. Q. M. Craig, J. M. Newton, A. C. Williams, and G. T. Simpkin, *J. Pharm. Pharmacol.*, *45* (Suppl. 2), 1149 (1993).

65. D. G. Sawh, D. Q. M. Craig, J. M. Newton, A. C. Williams, and M. C. R. Johnson, *Pharm. Res.*, *10* (Suppl.), S-150 (1993).

66. A. B. Brown, P. York, A. C. Williams, H. G. M. Edwards, and H. Worthington, *J. Pharm. Pharmacol.*, *45* (Suppl. 2), 1135 (1993).

67. R. D'Cunha, V. B. Kartha, and S. Gurnani, *Spectrochim. Acta*, *39A*, 331 (1983).

68. D. Michalska, *Spectrochim. Acta*, *49A*, 303 (1993).

69. G. A. Neville and H. F. Shurvell, *J. Raman Spectrosc.*, *21*, 9 (1990).

70. L. P. Powell and A. Campion, *Anal. Chem.*, *58*, 2350 (1986).

71. L. S. Taylor, P. York, A. C. Williams, H. G. M. Edwards, and V. Mehta, *J. Pharm. Pharmacol.*, *45* (Suppl. 2), 1100 (1993).

4

Magnetic Resonance Spectrometry

David E. Bugay
Bristol-Myers Squibb Pharmaceutical Research Institute, New Brunswick,
New Jersey

I. INTRODUCTION

Nuclear magnetic resonance (NMR) spectroscopy in pharmaceutical research has been used primarily in a classical, organic chemistry framework. Typical studies have included (1) the structure elucidation of compounds [1,2], (2) investigating chirality of drug substances [3,4], (3) the determination of cellular metabolism [5,6], and (4) protein studies [7–9], to name but a few. From the development perspective, NMR is traditionally used again for structure elucidation, but also for analytical applications [10]. In each case, solution-phase NMR has been utilized. It seems ironic that although ~90% of the pharmaceutical products on the market exist in the solid form, solid state NMR is in its infancy as applied to pharmaceutical problem solving and methods development.

During the course of developing pharmaceutical compounds today, it is becoming increasingly important to characterize the drug in its dispensed form, which is frequently a solid. It has long been known that drugs may exist in more than one polymorphic form [11]. These forms sometimes display very significant differences in solubility, bioavailability, processability, and physical/chemical stability [12]. The study of the solid state of a pharmaceutical compound must take place not only at the bulk level, but also with the dosage form. Sometimes the extreme conditions of processing the formulation into the dosage form can alter the solid [13] or increase its interaction with excipients [14]. Therefore, solid state analytical techniques are important to characterize pharmaceutical solids.

One technique that is starting to be used for the characterization of pharmaceutical solids is solid state NMR [15]. With recent advances in NMR hardware and software development, the acquisition of high-resolution, multi-nuclear NMR spectra of pharmaceutical solids is now possible. Some of the advantages that exist for NMR in the solution phase exist in the solid state as well. Unfortunately, the same solution-phase disadvantages exist for solid state NMR, in addition to the fact that magnetic interactions are of a much larger magnitude in the solid state than in the solution phase. These disadvantages, however, may be overcome by specific data acquisition techniques that yield highly resolved solid state NMR spectra. It is important to realize that solid state NMR is a nondestructive, multinuclear technique that has the ability to probe the chemical environment of each nucleus. Additionally, it is a quantitative technique that may be used in conjunction with other solid state techniques such as thermal gravimetric analysis, microscopy, infrared spectroscopy, differential scanning calorimetry, and x-ray diffraction techniques (single crystal and powder) for the investigation of pharmaceutical solids.

This chapter provides a basis of theory for solid state NMR and a brief review of experimental aspects. The main focus of the discussion section is on the application of solid state NMR techniques to the study of polymorphism.

Qualitative characterization techniques are outlined as well as the development of quantitative methods for the determination of one polymorphic form in another at the bulk or dosage product level.

II. SOLID STATE NMR THEORY

Conventional utilization of solution-phase NMR data acquisition techniques on solid samples yields broad, featureless spectra (Fig. 1A). The broad nature of the signal is due primarily to dipolar interactions, which do not average out to zero in the solid state, and chemical shift anisotropy (CSA), which again occurs because our compound of interest is in the solid state. Before one describes the two principal reasons for the broad, featureless spectra, it is important to understand the main interactions that a nucleus with a magnetic moment experiences when situated within a magnetic field in the solid state. In addition, manifestations of these interactions in the solid state NMR spectrum need to be discussed.

A. Zeeman Interaction

The principal interaction experienced is the Zeeman interaction (H_z), which describes the interaction between the magnetic moment of the nucleus and the externally applied magnetic field, B_0 (tesla). The nuclear magnetic moment, μ (ampere meter^2) is proportional to the nuclear spin quantum number (I) and the magnetogyric ratio (γ, radian telsa^{-1} second^{-1}):

$$\mu = \frac{\gamma I h}{2\pi} \tag{1}$$

Thus, the Zeeman interaction occurs only with nuclei that possess a spin greater than zero, and it yields $2I + 1$ energy levels of separation $\nu_0 = \gamma B_0/2\pi$. The Hamiltonian is described by

$$H_z = -\gamma \frac{h}{2\pi} \vec{B_0} \vec{I_z} \tag{2}$$

where ν_0 is the corresponding Larmor frequency. The interaction is linear with the applied magnetic field, thus giving the impetus to manufacture higher magnetic field spectrometers, since a larger separation of energy levels leads to a greater population difference, a subsequent increase in the signal-to-noise ratio (S/N), and increased spectral resolution. Since the Zeeman interaction incorporates the magnetogyric ratio, a constant for each particular nucleus, the resonant frequency for each nucleus is different at a specific applied magnetic field (Table 1). The magnitude of the Zeeman interaction for a ^{13}C nucleus in a 5.87 T magnetic field is 62.86 MHz. Small perturbations to the Zeeman effect are produced by other

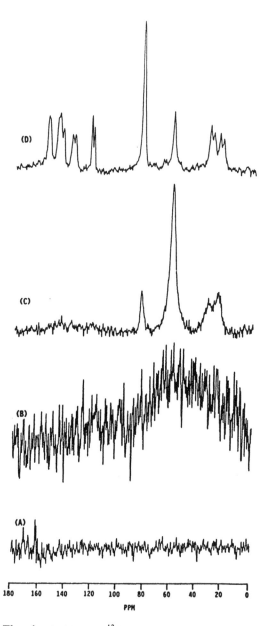

Fig. 1 Solid state ^{13}C NMR spectra of didanosine (VIDEX-ddI) acquired by (A) conventional solution-phase pulse techniques, (B) high-power proton decoupling only, (C) high-power proton decoupling and magic-angle spinning (5 kHz), and (D) high-power proton decoupling combined with magic-angle spinning (MAS) and cross-polarization (CP). In each case, 512 scans were accumulated. (From Ref. 15.)

Table 1 NMR Frequency Listing

Isotope		Spin	Natural abundance	Sensitivity		NMR frequency (MHz) at a field of		
				Rel.[a]	Abs.[b]	2.35 T	5.87 T	11.74 T
1	H	1/2	99.98	1.00	1.00	100.00	250.00	500.00
2	H	1	1.5×10^{-2}	9.65×10^{-3}	1.45×10^{-6}	15.35	38.38	76.75
10	B	3	19.58	1.99×10^{-2}	3.90×10^{-3}	10.75	26.87	53.73
11	B	3/2	80.42	0.17	0.13	32.08	80.21	160.42
13	C	1/2	1.11	1.59×10^{-2}	1.76×10^{-4}	25.14	62.86	125.72
14	N	1	99.63	1.01×10^{-3}	1.01×10^{-3}	7.22	18.06	36.12
15	N	1/2	0.37	1.04×10^{-3}	3.85×10^{-6}	10.13	25.33	50.66
17	O	5/2	3.7×10^{-2}	2.91×10^{-2}	1.08×10^{-5}	13.56	33.89	67.78
19	F	1/2	100.00	0.83	0.83	94.08	235.19	470.39
23	Na	3/2	100.00	9.25×10^{-2}	9.25×10^{-2}	26.45	66.13	132.26
25	Mg	5/2	10.13	2.67×10^{-3}	2.71×10^{-4}	6.12	15.30	30.60
27	Al	5/2	100.00	0.21	0.21	26.06	65.14	130.29
29	Si	1/2	4.7	7.84×10^{-3}	3.69×10^{-4}	19.87	49.66	99.33
31	P	1/2	100.00	6.63×10^{-2}	6.63×10^{-2}	40.48	101.20	202.40
33	S	3/2	0.76	2.26×10^{-3}	1.72×10^{-5}	7.67	19.17	38.35
35	Cl	3/2	75.53	4.70×10^{-3}	3.55×10^{-3}	9.80	24.50	48.99
37	Cl	3/2	24.47	2.71×10^{-3}	6.63×10^{-4}	8.16	20.39	40.78
39	K	3/2	93.1	5.08×10^{-4}	4.73×10^{-4}	4.67	11.67	23.33
41	K	3/2	6.88	8.40×10^{-5}	5.78×10^{-6}	2.56	6.40	12.81
43	Ca	7/2	0.145	6.40×10^{-3}	9.28×10^{-6}	6.73	16.82	33.64
67	Zn	5/2	4.11	2.85×10^{-3}	1.17×10^{-4}	6.25	15.64	31.27
79	Br	3/2	50.54	7.86×10^{-2}	3.97×10^{-2}	25.05	62.63	125.27
81	Br	3/2	49.46	9.85×10^{-2}	4.87×10^{-2}	27.01	67.52	135.03
127	I	5/2	100.00	9.34×10^{-2}	9.34×10^{-2}	20.00	50.02	100.04
195	Pt	1/2	33.8	9.94×10^{-3}	3.36×10^{-3}	21.50	53.75	107.50

[a]Determined for an equal number of nuclei at a constant field.   [b]Product of the relative sensitivity and natural abundance.
Source: Ref. 15.

interactions such as dipole–dipole, quadrupolar, shielding, and spin–spin coupling. Typically, these small perturbations are less than 1% (<600 kHz) of the Zeeman interactions, which range in frequency from 10^6 to 10^9 Hz.

B. Dipole–Dipole Interaction

Dipole–dipole interaction is the direct magnetic coupling of two nuclei through space. This interaction may involve two nuclei of equivalent spin, or nonequivalent spin, and is dependent upon the internuclear distance and dipolar coupling tensor. Additionally, the total interaction, labeled H_D, is the summation of all possible pairwise interactions (homo- and heteronuclear). It is important to note that the interaction is dependent on the magnitude of the magnetic moments (this is reflected in the magnetogyric ratio) and on the angle (θ) that the internuclear vector makes with B_0. Therefore, this interaction is significant for spin $1/2$ nuclei with large magnetic moments, such as ^{1}H and ^{19}F. Also, the interaction decreases rapidly with increasing internuclear distance (r), which generally corresponds to contributions only from directly bonded and nearest-neighbor nuclei. The dipolar interaction for a pair of nonequivalent, isolated spins I and S is described by

$$H_D^{IS} = \frac{\gamma_I \gamma_S}{r^3} \left(\frac{h}{2\pi}\right)^2 \vec{I} \cdot \hat{D} \cdot \vec{S} \qquad (3)$$

Since the dipolar coupling tensor, D, contains a $1 - 3 \cos^2 \theta$ term, this interaction is dependent on the orientation of the molecule. In solution-phase studies where the molecules are rapidly tumbling, the $1 - 3 \cos^2 \theta$ term is integrated over all angles of θ and subsequently disappears. Within solid state NMR, the molecules are fixed with respect to B_0; thus, the $1 - 3 \cos^2 \theta$ term does not approach zero. This leads to broad resonances within the solid state NMR spectrum since dipole–dipole interactions typically range from 0 to 10^5 Hz in magnitude.

In the case of pharmaceutical solids that are dominated by carbon and proton nuclei, the dipole–dipole interactions may be simplified. The carbon and proton nuclei may be perceived as "dilute" and "abundant" based upon their isotopic natural abundance, respectively (Table 1). Homonuclear ^{13}C—^{13}C dipolar interactions essentially do not exist because of the low concentration of ^{13}C nuclei (natural abundance of 1.1%). On the other hand, ^{1}H—^{13}C dipolar interactions contribute significantly to the broad resonances, but this heteronuclear interaction may be removed through simple high-power proton decoupling fields, similar to solution-phase techniques.

C. Chemical Shift Interaction

The three-dimensional magnetic shielding by the surrounding electrons is an additional interaction that the nucleus experiences in either the solution or the solid state. This chemical shift interaction (H_{cs}) is the most sensitive interaction to changes in the immediate environment of the nucleus and provides the most diagnostic information in a measured NMR spectrum. The effect originates from the small magnetic fields that are generated about the nucleus by currents induced in orbital electrons by the applied field. These small perturbations upon the nucleus are reflected in a change in the magnetic field experienced by the nucleus. Therefore, the field at the nucleus is not equal to the externally applied field, and hence the difference is the nuclear shielding, or chemical shift interaction:

$$H_{CS} = \gamma_I \frac{h}{2\pi} \vec{I} \cdot \hat{\sigma} \cdot \vec{B_0} \tag{4}$$

It is important to note the orientation dependence of the shielding constant, σ, and the fact that shielding is proportional to the applied field, whence the need for chemical shift reference materials such as tetramethylsilane.

Solution state NMR spectra yield "average" chemical shift values, which are characteristic of the magnetic environment for a particular nucleus. The average signal is due to the isotropic motion of the molecules in solution. In other words, B_0 "sees" an average orientation of a specific nucleus. For solid state NMR, the chemical shift value is also characteristic of the magnetic environment of a nucleus, but normally the molecules are not free to move. It must be kept in mind that the shielding will be characteristic of the nucleus in a particular orientation of the molecule with respect to B_0. Therefore, a specific functional group oriented perpendicular to the magnetic field will give a sharp signal characteristic of this particular orientation (Fig. 2A). Analogously, if the functionality is oriented parallel to B_0, then a sharp signal characteristic of that orientation will be observed (Fig. 2B). For most polycrystalline pharmaceutical samples, a random distribution of all orientations of the molecule will exist. This distribution produces all possible orientations and is thus observed as a very broad NMR signal (Fig. 2C). The magnitude of the chemical shift anisotropy is typically between 0 and 10^5 Hz.

D. Spin–Spin Couplings and Quadrupolar Interactions

Two additional interactions experienced by the nucleus in the solid state are spin–spin couplings to other nuclei, and quadrupolar interactions, which involve nuclei of spin greater than $1/2$. Spin–spin (H_{sc}), or J, coupling,

$$H_{SC} = \vec{I} \cdot \hat{J} \cdot \vec{S} \tag{5}$$

originates from indirect coupling between two spins by means of their electronic surroundings and are several orders of magnitude smaller (possibly 0–10^4 Hz, typically only several kHz) than dipole interactions. Although J coupling interactions occur in both the solid and the solution state, these couplings are

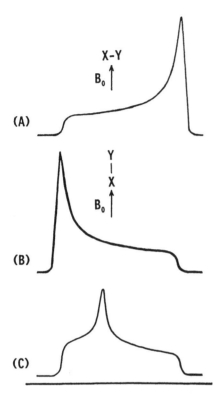

Fig. 2 Schematic representation of the ^{13}C NMR signal of a single crystal containing the functional group AB, oriented (A) perpendicular to the applied field, and (B) parallel to the applied field. The lineshape in (C) represents the NMR signal of a polycrystalline sample with a random distribution of orientations yielding the chemical shift anisotropy pattern displayed. (From Ref. 15.)

typically used in solution-phase work for conformational analysis [16]. Quadrupolar interactions (H_Q)

$$H_Q = \vec{I} \cdot \hat{Q} \cdot \vec{I} \tag{6}$$

arise from dipolar coupling of quadrupolar nuclei, which display a nonspherically symmetrical field gradient, with nearby spin-$\frac{1}{2}$ nuclei. The magnitude of the interaction is dependent upon the relative magnitudes of the quadrupolar and Zeeman interactions and may completely dominate the spectrum ($0-10^9$ Hz), but since pharmaceutical compounds are primarily hydrocarbons, quadrupolar interactions typically do not interfere.

Summarizing the interactions, the isotropic motions of molecules in the solution state yield a discrete average value for the scalar spin–spin coupling and chemical shift interactions. For the dipolar and quadrupolar terms, the average obtained is zero, and the interaction is not observed in solution-phase studies. In sharp contrast, interactions in the solid state are orientation-dependent, subsequently producing a more complicated spectrum, but one that contains much more information. In order to yield highly resolved, "solution-like" spectra of solids, a combination of three techniques is used: dipolar decoupling [17], magic-angle spinning [18,19], and cross-polarization [20]. The remaining portion of this section discusses these three techniques and their subsequent effect upon measured spectra of pharmaceutical compounds that are dominated by carbon nuclei.

E. Dipolar Decoupling

Discussed earlier was the fact that high-powered proton decoupling fields can eliminate the heteronuclear ^{1}H—^{13}C dipolar interactions that may dominate a solid state NMR spectrum. The concept of decoupling is familiar to the solution-phase NMR spectroscopist, but it needs to be expanded for solid state NMR studies. In solution-phase studies, the decoupling eliminates the scalar spin–spin coupling, not the dipole–dipole interactions (this averages out to zero due to isotropic motions of the molecules). Irradiation of the sample at the resonant frequency of the nucleus to be decoupled (B_2 field) causes the z component of the spins to flip rapidly compared with the interaction one wishes to eliminate. Scalar interactions usually require 10 W of decoupling power or less. In pharmaceutical solids work, decoupling is used primarily to remove the heteronuclear dipolar interactions between protons and carbons. The magnitude of the dipolar interaction ($\sim$50 kHz) usually requires decoupling fields of 100 W and subsequently removes both scalar and dipole interactions. Even with the use of high-power decoupling, broad resonances still remain, due principally to chemical shift anisotropy (Fig. 1B).

F. Magic-Angle Spinning

Molecules in the solid state are in fixed orientations with respect to the magnetic field. This produces chemical shift anisotropic powder patterns for each carbon atom since all orientations are possible (Fig. 2). It was shown as early as 1958 that rapid sample rotation of solids narrowed dipolar-broadened signals [18]. Several years later, it was recognized that spinning could remove broadening caused by CSA yet retain the isotropic chemical shift [19].

The concept of magic-angle spinning arises from the understanding of the shielding constant, σ (Eq. 4). This constant is a tensor quantity and, thus, can be related to three principal axes:

$$\sigma_{zz} = \sum_{i=1}^{3} \sigma_i \cos^2 \theta_i \qquad (7)$$

where σ_i is the shielding at the nucleus when B_0 aligns along the ith principal axis, and θ_i is the angle this axis makes with B_0. Under conditions of mechanical spinning, this relationship becomes time-dependent and a $3 \cos^2 \theta - 1$ term arises. By spinning the sample at the so-called "magic-angle" of 54.7°, or 54°44', this term becomes zero and thus removes the spectral broadening due to CSA (Fig. 1C) providing that the sample rotation (kHz) is greater than the magnitude of the CSA (kHz).

Chemical shift anisotropy may range from 0 to 20 kHz, and so our spin rates must exceed this value or spinning sidebands are observed. Figure 3A displays the solid state ^{31}P NMR spectrum of fosinopril sodium (Monopril), a novel ACE inhibitor [21], acquired under proton decoupling and static spinning conditions. At a relatively low spin rate of 2.5 kHz, broadening due to CSA is removed and the center band and sidebands appear (Fig. 3B). As the sample is spun at higher rates (Figs. 3C and D), the sidebands become less intense and move out from the center band. Even at a spin rate of 6 kHz (maximum spin rate for the probe/instrument used), the CSA is not completely removed (Fig. 3E). Although fast enough spin rates may not be achieved to remove CSA totally for specific compounds, slower than optimal rates will still narrow the resonances. Spinning sidebands, at multiples of the spin rate, will complicate the spectrum, but they can be easily identified by varying the spin rate and observing which signals change in frequency. In addition, specific pulse sequences may be used to minimize the spinning sidebands [22]. While increasing the magnetic field strength increases the signal-to-noise ratio (Zeeman interaction), it also increases the CSA, since this interaction is field-dependent (Eq. 4). Utilizing today's high-field spectrometers (>4.7 T), spinning sidebands may exist but can be identified and used to gain additional information or be potentially eliminated if necessary.

The techniques of magic-angle spinning and heteronuclear dipolar decoupling produce solid state NMR spectra that approach the linewidths and appearance of solution-phase NMR spectra. Unfortunately, there is an inherent lack of sensitivity in the general NMR experiment due to the nearly equivalent population of the two spin states for spin-$1/2$ nuclei. In addition, the sensitivity of the experiment is decreased with pharmaceutical compounds, since they are composed primarily of carbon atoms where the ^{13}C observable nuclei have a natural abundance of only 1%. The long relaxation times of specific carbon nuclei also pose a problem since quick, repetitive pulsing cannot occur. The technique of cross-polarization provides a means of both signal enhancement and reduction of long relaxation times.

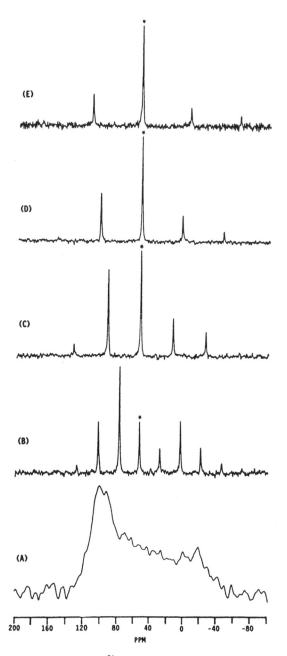

Fig. 3 Solid state ^{31}P NMR spectra of fosinopril sodium acquired under single pulse, high-power proton decoupling and various conditions of magic-angle spinning: (A) static, (B) 2.5 kHz, (C) 4.0 kHz, (D) 5.0 kHz, and (E) 6.0 kHz. The isotropic chemical shift is designated by an asterisk. (From Ref. 15.)

G. Cross-Polarization

The concept of cross-polarization as applied to solid state NMR was implemented by Pines et al. [20]. A basic description of the technique is the enhancement of the magnetization of the rare spin system by transfer of magnetization from the abundant spin system. Typically, the rare spin system is classified as ^{13}C nuclei and the abundant system as ^1H spins. This is especially the case for pharmaceutical solids and the remaining discussion of cross-polarization focuses on these two spin systems only.

Figure 4 describes the cross-polarization (CP) pulse sequence and the behavior of the ^1H and ^{13}C spin magnetizations during the pulse sequence in terms of the rotating frame of reference [23]. Step 1 of the sequence involves rotation of the proton magnetization onto the y' axis by application of a 90° pulse (rotating-frame magnetic field B_{1H}). Subsequent spin-locking occurs along y' by an on-resonance pulse along y' for a specific period of time, t. At this point, a high degree of proton polarization occurs along B_{1H}, which will decay with a specific time referred to as $T_{1\rho H}$. As the ^1H spins are locked along y', an on-resonance pulse, B_{1C}, is applied to the ^{13}C spins. The ^{13}C spins are also "locked" along y' and decay

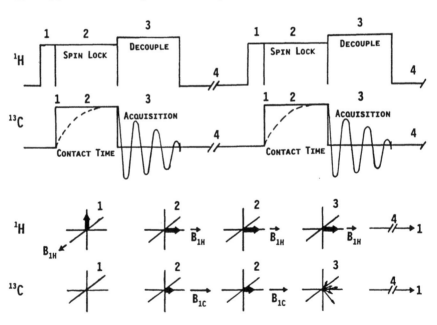

Fig. 4 The top diagram represents the pulse sequence for the cross-polarization experiment; the bottom diagram describes the behavior of the ^1H and ^{13}C spin magnetizations during the sequence. The steps in the two diagrams correspond to each other and are fully explained in the text. (From Ref. 15.)

with the time $T_{1\rho C}$, (Step 2). By correctly choosing the magnitude of the spin-locking fields B_{1H} and B_{1C}, the Hartmann–Hahn [24] condition,

$$\gamma_{1_H} B_{1H} = \gamma_{13_C} B_{1C} \tag{8}$$

will be satisfied, and transfer of magnetization will occur from the ^1H spin reservoir to the ^{13}C spin reservoir. Once this equality is obtained by the correct spin-locking fields, the dilute ^{13}C spins will take on the characteristics of the more favorable ^1H system. In the end, a maximum magnetization enhancement equal to the ratio of the magnetogyric ratios may be achieved (γ_H/γ_C). Once the carbon magnetization has built up during this "contact period" (t), B_{1C} is switched off and the carbon free-induction decay recorded (Step 3). During this data acquisition period, the proton field is maintained on for heteronuclear decoupling of the ^1H—^{13}C dipolar interactions. Step 4 involves a standard delay period in which no pulses occur and the two spin systems are allowed to relax back to their equilibrium states.

The entire CP pulse sequence provides two advantages for the solid state NMR spectroscopist: significant enhancement of the rare spin magnetization and reduction of the delay time between successive pulse sequences since the rare spin system takes on the relaxation character of the abundant spin system. Signal enhancement (magnetization enhancement) for less sensitive nuclei is immediately apparent from the CP process. Less obvious is the fact that the rare spin system signal is not dependent on the recycle time in Step 4 (regrowth of carbon magnetization), but on the transfer process in Step 2 and the relaxation behavior of the proton spin system in Step 4. Since the single spin-lattice relaxation time of the proton system is typically much shorter than the carbon system (1s to 10s versus 100s of seconds), the delay time is much shorter. This corresponds to a greater number of scans per unit time, yielding better S/N.

Throughout the cross-polarization pulse sequence, a number of competing relaxation processes are occurring simultaneously. The recognition and understanding of these relaxation processes are critical in order to apply CP pulse sequences for quantitative solid state NMR data acquisition or ascertaining molecular motions occurring in the solid state.

H. Relaxation

Familiar to most chemists is the notion of spin-lattice relaxation [25]. Labeled as T_1, the spin-lattice relaxation time is defined as the amount of time for the net magnetization (M_z) to return to its equilibrium state (M_0) after a spin transition is induced by a radiofrequency pulse:

$$\frac{\partial M_z}{\partial t} = \frac{-1}{T_1}(M_z - M_0) \tag{9}$$

The "lattice" term originates from the idea that the spin system gives up energy to its surroundings as it tries to re-establish spin equilibrium. The process of transferring spin energy to other modes of energy may be classified as a relaxation mechanism. For T_1 in solids, all of the following mechanisms may contribute: dipole–dipole (T_{1DD}), spin–rotation (T_{1SR}), quadrupolar (T_{1Q}), scalar (T_{1SC}), and chemical shift anisotropy (T_{1CSA}). The simple inversion-recovery pulse sequence may be used to measure T_1 times for solid state samples [26]. In the use of simple high-powered decoupling, single pulse sequences for quantitative NMR studies, the inversion-recovery pulse sequence may be used to determine T_1's of interest. A multiple of five times T_1 may then be incorporated as the recycle time between successive pulses, assuring sufficient time for the magnetization to return back to equilibrium. In this way, the NMR signal observed is truly representative of the number of nuclei producing it.

To understand the cross-polarization process, two other rate processes must be defined: (1) spin-lattice relaxation in the rotating frame $(T_{1\rho})$ and (2) the cross-polarization relaxation time (T_{CH}). The spin-lattice relaxation in the rotating frame characterizes the decay of magnetization in a field B_1, which is normally much smaller than the externally applied field B_0. In Steps 1 and 2 of the CP sequence (Fig. 4), the carbon and proton spin systems are locked by the application of fields B_{1H} and B_{1C}, and each system decays with its characteristic time. Figure 5 represents a thermodynamic model for the relaxation processes during CP. During contact between the two spin systems, magnetization is transferred at the rate T_{CH}. Since competing relaxation processes are occurring, the following conditions must be met to obtain a spectrum by CP: $T_{1C} > T_{1H} \geq T_{1\rho H} >$ CP time $> T_{CH}$. It is apparent that for quantitative NMR studies of solids by CP, the individual relaxation processes must be measured to assure that the signal is truly proportional to the amount of species present.

With an understanding of magnetic interactions in the solid state, inherent relaxation processes, and experimental techniques to overcome these difficulties, Schaefer et al. acquired the first "liquid-like" NMR spectrum of a solid by CP/MAS techniques [27]. Before a review of recent applications of CP/MAS NMR to the pharmaceutical sciences, the brief Experimental section details some important information regarding practical solid state NMR.

III. EXPERIMENTAL ASPECTS

Manufacturers of today's modern NMR spectrometers normally offer a number of different models of instruments that are capable of measuring solid state NMR spectra. Usually, a dedicated solid state NMR instrument is available along with solution-phase models that are capable of solids work with the purchase of an additional solids accessory package. For any pharmaceutical company that is contemplating the purchase of an NMR for solids work, it is this author's opinion

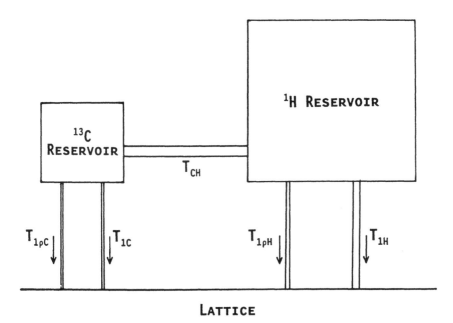

Fig. 5 Thermodynamic model of the cross-polarization sequence and representation of the competing relaxation processes. (From Ref. 15.)

that a dedicated solid state NMR instrument should be purchased. Although this is not the venue to discuss the criteria for purchasing a solid state NMR instrument, it is appropriate to briefly comment on instrumentation, types of samples that may be investigated, and a series of standard compounds that may be used for spectral referencing, MAS setting, and optimization of cross-polarization.

The basic components of the solid state spectrometer are the same as the solution-phase instrument: data system, pulse programmer, observe and decoupler transmitters, magnetic system, and probes. In addition, high-power amplifiers are required for the two transmitters and a pneumatic spinning unit to achieve the necessary spin rates for MAS. Normally, the observe transmitter for ^{13}C work requires broadband amplification of approximately 400 W of power for a 5.87-T, 250-MHz instrument. The amplifier should have triggering capabilities so that only the radiofrequency (rf) pulse is amplified. This will minimize noise contributions to the measured spectrum. So that the Hartmann–Hahn condition may be achieved, the decoupler amplifier must produce an rf signal at one-fourth the power level of the observe channel for carbon work.

Any spectrometer system capable of doing complicated two-dimensional NMR work will have a sufficient data system and pulse programmer. The selection of the magnetic field strength is probably the hardest decision in the purchase process. Sensitivity and resolution improve with increasing field strength, yet spinning sidebands are more prevalent. The intended use of the spectrometer system must be considered in the decision process for the correct magnetic field strength. The probes for solid state NMR work are much different than their solution-phase counterparts. Figure 6 displays a double air bearing, magic-angle spinning, multinuclear probe for use in a Bruker AM-250 NMR spectrometer. This variable-temperature probe has the ability to spin samples at a maximum spin rate of 5.0 kHz between temperatures of 77 and 400 Kelvin (K). The rotors (Fig. 7) that contain the sample are constructed of zirconium oxide with either Kel-F caps for ambient-temperature studies, or boron nitride caps for variable-temperature work. Depending upon the compressibility of the powdered, pharmaceutical sample, anywhere from 200 to 400 mg of sample is required to fill the rotor. Within the Materials Science NMR laboratory at Bristol-Myers Squibb, a large number of nontraditional samples have been investigated by solid state NMR (Table 2). The main limitation is the ability to balance the rotor so that high-speed spinning may take place.

Although a series of standard referencing compounds have not been established for solid state NMR as opposed to solution-phase studies [28], a series of unofficial standard samples are currently being used [29]. Table 3 lists a series of compounds, their intended use, and appropriate references further detailing their intended uses and limitations.

IV. PHARMACEUTICAL APPLICATIONS

The most widely used application of solid state NMR in the pharmaceutical industry is in the area of polymorphism, or pseudopolymorphism, and this will be the focus of this section.

A. Polymorphism and Pseudopolymorphism

The ability of a molecule to crystallize in more than one form is defined as polymorphism, whereas the ability to crystallize in more than one form based on the solvation state of the molecule is termed pseudopolymorphism. It is widely recognized that the identification, characterization, and quantitation of pharmaceutical (pseudo)polymorphs is critical during the drug development process [11,12,36,37]. Different physical forms of a drug substance can display radically different solubilities, which directly affects the dissolution and bioavailability characteristics of the compound. In addition, the chemical stability of one form, as compared with the other, may vary. The physical stability is also crucial. During various processing steps (grinding, mixing, tablet pressing, etc.) the physical form of the drug substance may be compromised, leading to dissolution

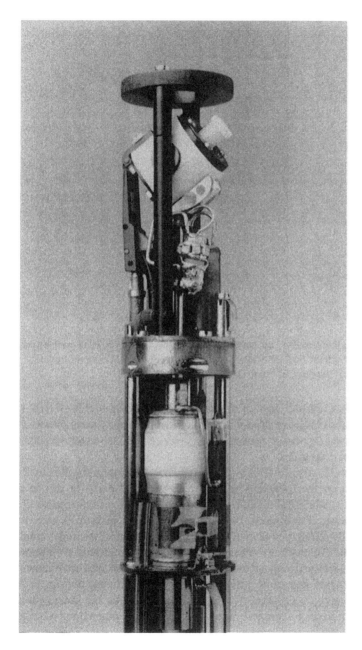

Fig. 6   Double air bearing, magic-angle spinning, multinuclear probe for use in a Bruker AM-250 NMR spectrometer. For illustrative purposes, the capped rotor has only been half inserted into the stator, which is positioned at the magic angle.

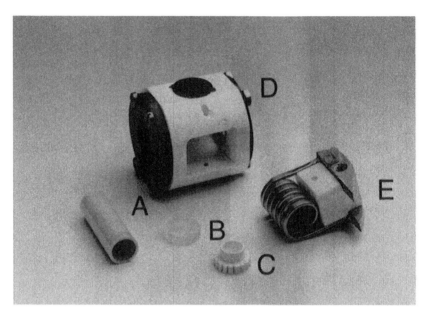

Fig. 7   A zirconium rotor (A), and several types of caps: (B) Kel-F, (C) boron nitride, and the stator (D) in which the doubly tuned transmitter/receiver coil (E) is inserted. Here, the rotor is 22 mm in length and 7 mm in outer diameter.

problems. For these reasons, the full characterization of polymorphic systems is critical to numerous entities (preformulation/physical pharmacy group, chemical process development, regulatory/patent office, and analytical laboratories) within drug development companies.

The use of solid state NMR for the investigation of polymorphism is easily understood based on the following model. If a compound exists in two, true polymorphic forms, labeled as A and B, each crystalline form is conformationally different. This means for instance, that a carbon nucleus in form A may be situated in a slightly different molecular geometry compared with the same carbon nucleus in form B. Although the connectivity of the carbon nucleus is the same in each form, the local environment may be different. Since the local environment may be different, this leads to a different chemical shift interaction for each carbon, and ultimately, a different isotropic chemical shift for the same carbon atom in the two different polymorphic forms. If one is able to obtain pure material for the two forms, analysis and spectral assignment of the solid state NMR spectra of the two forms can lead to the origin of the conformational differences in the two polymorphs. Solid state NMR is thus an important tool in conjunction with thermal analysis, optical microscopy, infrared (IR) spectroscopy, and powder

Table 2 Various Types of Pharmaceutical Samples Investigated by Solid State NMR Spectroscopy

Sample	NMR nucleus	Brief description of studies
Bulk drug	^{13}C, ^{31}P, ^{15}N, ^{25}Mg, ^{23}Na	Solid state structure elucidation, drug–excipient interaction studies (variable temperature), (pseudo)polymorphic characterization at the qualitative and quantitative level, investigation of hydrogen bonding with salt compounds
Excipients	^{13}C, ^{25}Mg, ^{23}Na	Determination of the degree of cross-linking for polymer systems, drug–excipient interaction studies (variable temperature), (pseudo)polymorphic characterization at the qualitative and quantitative level
Capsules	^{13}C, ^{31}P	Identifying residue from dissolution baths, drug–capsule interaction studies
Tablets	^{13}C, ^{31}P	Drug–excipient interaction studies, (pseudo)-polymorphic characterization at the qualitative and quantitative levels

Source: Ref. 15.

and single-crystal x-ray crystallographic techniques for the study of polymorphism.

There are a number of significant advantages in the use of solid state NMR for the study of polymorphism. Compared with diffuse reflectance infrared and x-ray powder diffraction techniques, solid state NMR is a bulk technique that does not have to consider particle-size effects on the intensity of the measured signal. In addition, NMR is an absolute technique under proper data acquisition procedures, meaning that the intensity of the signal is directly proportional to the number of nuclei producing it. By proper assignment of the NMR spectrum, the origin of the polymorphism can be inferred from differences in the resonance position for identical nuclei in each polymorphic form. Finally, the investigation of polymorphism by solid state NMR can be performed at either the bulk drug or dosage form level. This ability provides a significant tool for the investigation of polymorphic conversion under various processing techniques (e.g. various fluidized-bed granulation, dry blending, lyophilization, and tableting conditions).

B. Qualitative Analysis

The majority of applications of solid state NMR used in the investigation of pharmaceutical polymorphs are performed in conjunction with other analytical

Table 3 Reference Samples for Solid State NMR

Compound name	Intended use	Reference
Adamantane ($C_{10}H_{16}$)	External referencing for ^{13}C spectra, optimizing Hartmann–Hahn match, linewidth measurement, sensitivity measurement	[30]
Hexamethylbenzene ($C_{12}H_{18}$)	Optimizing Hartmann–Hahn match	[30]
Glycine ($C_2H_5O_2N$)	External referencing for ^{13}C spectra, sensitivity measurement, magic-angle setting for ^{13}C experimentation	[30]
Ammonium phosphate ($NH_4H_2PO_4$) monobasic	External referencing for ^{31}P spectra	[30]
Potassium bromide (KBr)	Magic-angle setting for ^{13}C experimentation	[31]
ZNP[a], $Zn[S_2P(OC_2H_5)_2]_2$	Magic-angle setting for ^{31}P experimentation	[32]
DMPPO[b], $C_8H_{11}OP$	Magic-angle setting for ^{31}P experimentation	[33]
Samarium Acetate Tetrahydrate ($CH_3CO_2)_3Sm \cdot 4H_2O$	^{13}C CP/MAS chemical-shift thermometer	[34]
TTAA[c]	^{15}N CP/MAS chemical-shift thermometer	[35]

[a]ZNP: zinc(II) bis (O,O'-diethyldithiophosphate).
[b]DMPPO: dimethylphenylphosphine oxide.
[c]TTAA: 1,8-dihydro-5,7,12,14-tetramethyldibenso (b,i)–(1,4,8,11-tetraazacyclotetradeca-4,6,11,13-tetraene-$^{15}N_4$ (tetramethyldibenzo-tetraaza(14)annulene).
Source: Adapted from Ref. 15.

techniques. Byrn et al. have reported differences in the solid state NMR spectra for different polymorphic forms of benoxaprofen and nabilione, and pseudopolymorphic forms of cefazolin [38]. Although single x-ray crystallography was initially used to study the polymorphs, the solid state ^{13}C CP/MAS NMR spectrum of each form was distinctly different. These studies primarily focused on the bulk drug material, although a granulation of benoxaprofen was studied. It was concluded that solid state NMR could be used to differentiate the form present in the granulation even in the presence of excipients. In further studies at Purdue University, the crystalline forms of prednisolone *tert*-butylacetate [39], cefaclor dihydrate [40], and glyburide [41] were studied. The five crystal forms of prednisolone *tert*-butylacetate were again determined by single x-ray crystallography, and solid state NMR was used to determine the effect of crystal packing on the ^{13}C chemical shifts of the different steroid forms. Although conformational changes were observed in the ester sidechain by x-ray crystallography, no major differences were noted in the NMR spectra, indicating that the environment remains relatively unchanged. Significant chemical shift differences were noted for carbonyl atoms involved in hydrogen bonding. This theme is consistent in the NMR study of cefaclor dihydrate. Again, the effects of hydrogen bonding were discernible by solid state NMR. The study of glyburide was principally concerned with the structural conformation of the molecule in solution and solid state. The solution conformation was determined by ^{1}H and ^{13}C NMR, and in the solid by single-crystal x-ray crystallography, IR, and solid state ^{13}C NMR. The solid state NMR results suggested that this method would be useful for comparing solid and solution state conformations of molecules.

In a series of papers by Harris and Fletton, solid state NMR has been used to investigate the structure of polymorphs. The majority of studies involve x-ray and IR techniques and also address the possibility of quantitative measurements of polymorphic mixtures. The three pseudopolymorphic forms of testosterone were examined by IR and ^{13}C CP/MAS NMR [42]. The two forms of molecular spectroscopy were able to differentiate the forms, but NMR had the ability to investigate nonequivalent molecules in a given unit cell. A series of doublet resonances were noted for a series of different carbon atoms. This implied that the specific carbon atom within the molecule may resonate at two different frequencies depending on the crystallographic site of the molecule. In addition to the hydrogen-bonding explanation of the crystallographic splittings, the use of NMR to quantitatively determine the amount of each pseudopolymorph present in a mixture was addressed. In the study of androstanolone [43], high-quality ^{13}C NMR spectra were obtained by CP/MAS techniques and allowed the characterization of the anhydrous and monohydrate forms. Again, crystallographic splittings were noted for the two forms and were related to hydrogen bonding. An identical approach to study pharmaceutical polymorphic structure was used in the investigation of the two polymorphs of 4′-methyl-2′-nitroacetanilide [44]. An additional study of cortisone acetate [45,46] by solid state NMR revealed differences in the NMR spectra for the six crystalline forms. Further multidisciplinary approaches to the physical characterization of pharmaceutical compounds are detailed in separate studies on cyclopenthiazide [47], the excipient lactose [48], frusemide [49], losartan [50], fosinopril sodium [51], and leuko-triene antagonists MK-679, MK-571 [52], and L-660,711 [53]. In each case, solid state NMR was used in conjunction with other techniques such as differential scanning calorimetry (DSC), IR, x-ray diffraction, microscopy, and solubility/dissolution studies to fully characterize the polymorphic systems.

In further studies of polymorphism by solid state NMR, conversion from one solid state form to another by ultraviolet (UV) irradiation [54] and variable-tem-perature techniques [55] are outlined. In the first study, NMR was employed to follow the chemical transformation within the organic crystals of p-formyl-trans-cinnamic acid (p-FCA). A photoreactive β-phase may be crystallized from ethanol, whereas a photostable γ-phase is produced from acetone. After irradiation of the β-phase with UV radiation, and subsequent acquisition of the solid state ^{13}C NMR spectrum, the photoproduct was easily identified by NMR. The second conversion study [55] investigated the four forms of p-amino-benzenesulphonamide sulpha-nilamide (α, β, γ, and δ). The first three forms were fully investigated by solid state ^{13}C NMR and x-ray crystallography techniques. Subsequent variable-temperature studies monitored the interconversion of the α and β forms to the γ form. Coalescence of some NMR signals in the γ form also suggested that phenyl-ring motion occurred within the crystal. Conclusions from the study indicated that solid state NMR had the ability to differentiate pharmaceutical polymorphs, determine asymmetry in the unit cell, and investigate molecular motion within the solid state.

In each of the aforementioned studies, various solid state NMR pulse sequences were utilized for data acquisition. These studies typically used the standard cross-polarization/magic-angle spinning sequence previously described. In some of the investigations, additional pulse sequences were used for observing heteroatoms [51], suppressing spinning sidebands, and spectral editing [42–45,49–51]. Although not complete, a brief description of some important pulse sequences follows with schematics of the pulse sequence and traditional applications.

1. Single Pulse Excitation/Magic-Angle Spinning

The single pulse excitation/magic-angle spinning (SPE/MAS) pulse se-quence is just an adaptation of the solution-phase, gated decoupling NMR experiment [56]. In the SPE/MAS sequence (Fig. 8A), high-power proton decoupling occurs during the signal acquisition period. This decoupling period primarily removes the heteronuclear dipolar coupling that typically broadens solid state NMR spectra. This pulse sequence is normally used to acquire decoupled spectra on heteronuclear atoms (signified as X in the pulse sequence) that possess

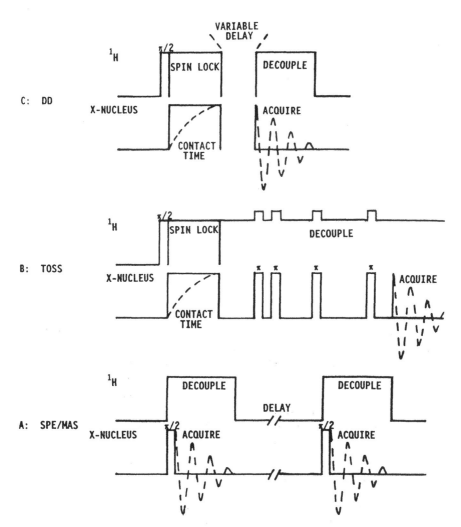

Fig. 8 Schematic diagrams for the following pulse sequences: (A) single pulse excitation/magic-angle spinning, (B) total suppression of sidebands, and (C) delayed decoupling, or dipolar dephasing.

high sensitivity (such as the ^{31}P spectra presented in Fig. 3). Throughout the sequence, MAS is used to minimize broadening due to CSA.

2. *Total Suppression of Sidebands*

As demonstrated in Fig. 3, even with high-speed MAS, spinning sidebands do occur. These sidebands may be confused with actual resonances in the NMR spectrum and be misinterpreted. Dixon et al. have developed a pulse sequence that suppresses spinning sidebands typically observed in NMR spectra acquired with high-field magnets operating at 5 T and above [22]. This pulse sequence has been labeled TOSS, *to*tal *s*uppression of *s*idebands, and it is illustrated in Fig. 8B. After initial cross-polarization, Hahn spin echoes are generated with a series of pulses on the X channel. The timing of these pulses is programmed such that the sidebands are inverted but the isotropic peak is unaffected. Upon addition of spectra with inverted and normal sideband intensities, the sidebands cancel and the isotropic peak remains. An example of the benefits of the pulse sequence is shown in Figs. 9A and B. At a spin rate of 4.0 kHz, the ^{31}P CP/MAS spectrum of fosinopril sodium displays significant spinning sidebands. Upon utilization of the TOSS sequence, the spinning sidebands are nearly removed, simplifying the interpretation of the spectrum. This pulse sequence is very advantageous for the qualitative assignment of resonances in solid state NMR spectra.

3. Delayed Decoupling (Dipolar Dephasing)

Another pulse sequence helpful for the assignment of resonances in a qualitative CP/MAS NMR analysis is the delayed decoupling, or dipolar dephasing, experiment. This pulse sequence (Fig. 8C) is simply the CP/MAS experiment with a variable delay time inserted between the spin-locking period and the start of decoupling/acquisition. Nonprotonated carbons are selectively observed by this pulse sequence [57] due to the inherently weaker ^{13}C—^{1}H dipolar interaction for carbons with nonbonded protons. During the variable delay time (usually programmed between 50 and 100 μs) directly protonated carbons dephase rapidly, and their signal disappears. Subsequently, a simplified spectrum is measured in which only quaternary carbons are observed (Figs. 9C and D).

C. Quantitative Analysis

In a number of the publications describing the use of solid state NMR for polymorphic characterization, the majority of the work has dealt with qualitative studies, with brief references to the possibility of quantitative analysis. An excellent guide to the utilization of magic-angle spinning and cross-polarization techniques for quantitative solid state NMR data acquisition has been outlined by Harris [58]. As mentioned before in the theory section, in order to acquire solid state NMR spectra in which the signal intensities truly reflect the nuclei producing it, data acquisition parameters such as recycle time, pulse widths, cross-polarization time, Hartmann–Hahn match, and decoupling power must be explicitly determined for each chemical system. In Harris's article, the problems of solid state NMR acquisition techniques as applied to quantitative measurements are addressed. Additionally, errors in the setting of the magic angle,

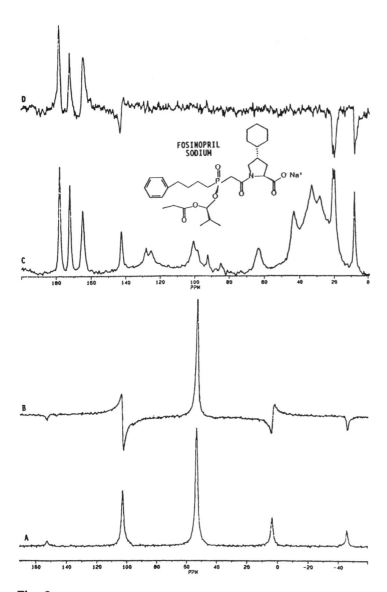

Fig. 9 Examples of simplifying solid state NMR spectra by the TOSS and delayed decoupling pulse sequences. Shown is a comparison of the ^{31}P CP/MAS NMR spectrum of fosinopril sodium utilizing the standard pulse sequence (A) and the TOSS routine (B). Also shown is the full ^{13}C CP/MAS NMR spectrum of fosinopril sodium (C) and the nonprotonated carbon spectrum (D) obtained from the delayed decoupling pulse sequence utilizing a 80 μs delay time. Signals due to the methyl carbon resonances (0–30 ppm) are not completely eliminated due to the rapid methyl group rotation, which reduces the carbon–proton dipolar couplings.

Hartmann–Hahn match, and cross-polarization mixing time are discussed in relation to obtaining quantitative NMR results.

Once qualitative solid state CP/MAS NMR studies reveal that distinct resonances exist for two different (pseudo)polymorphic forms of a drug substance, quantitative method development may begin. From previous discussions it is known that competing rate processes occur in the cross-polarization NMR experiment. So that the measured NMR signal truly represents the number of nuclei producing it, explicit measurements must be made for these various mechanisms, namely T_{CP} and $T_{1\rho H}$. For SPE experiments, T_1 must be measured for the nucleus of interest. Outlined in the following are a number of key pulse sequences that must be utilized in order to develop quantitative solid state NMR assays. As before, a brief description of the sequence is given with a schematic diagram and application note.

1. Contact Time (T_{CH}) Measurement

During the Hartmann–Hahn condition, proton magnetization is being transferred to the carbon spins at the rate of $1/T_{CH}$ *and* to the lattice at the rate of $1/T_{1\rho H}$. In order for CP to occur, $1/T_{CH} < 1/T_{1\rho H}$ ($1/T_{1\rho c}$ is being ignored since it is greater than $1/T_{1\rho H}$. In order for CP to occur, $1/T_{CH} < 1/T_{1pH}$ ($1/T_{1pC}$ is being ignored since it is greater than $1/T_{1pH}$). The ^{13}C magnetization will achieve a maximum value at a specific contact time, which also signifies optimum sensitivity for the CP experiment. Thus, the contact time must by varied and the signal intensity measured via the pulse sequence outlined in Fig. 10A. By plotting the contact time versus the signal intensity, the optimal contact time for each resonance may be determined (Fig. 11). Once the contact time is determined for the distinct resonance observed in the NMR spectrum for each polymorph, this value, or a compromised/average value, may be inserted into the quantitative CP/MAS pulse sequence.

2. Proton T_1 Measurement

In addition to measuring T_{CH} for the polymorphic system in question, the proton T_1 value must be determined since the repetition rate of a CP experiment is dependent upon the recovery of the proton magnetization. Common convention states that a delay time between successive pulses of $1–5 \times T_1$ must be used. Figure 10B outlines the pulse sequence for measuring the proton T_1 through the carbon intensity. One advantage to solids NMR work is that a common proton T_1 value will be measured, since protons communicate through a spin-diffusion process. An example of spectral results obtained from this pulse sequence is displayed in Fig. 12.

Once the proton T_1 and T_{CH} values are determined for a polymorphic system, physical mixtures of the two polymorphs can be generated (calibration samples). Subsequent acquisition of the solid state NMR spectra under quantitative conditions yields signal intensities representative of the amount of each solid state phase

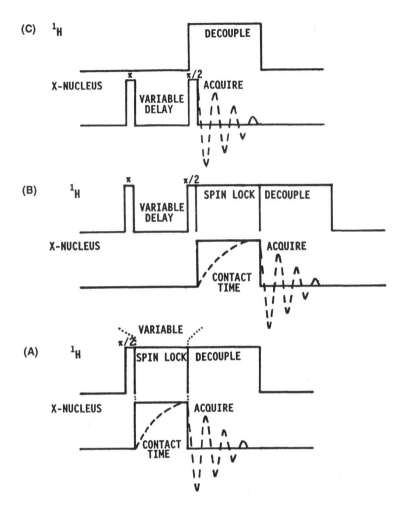

Fig. 10 Schematic diagrams for the following pulse sequences: (A) contact time (T_{CH}) measurement, (B) proton T_1 measurement, and (C) X-nucleus T_1 measurement.

present. Measurement of the relative signal intensities may then lead to determination of the amount of one solid state phase present in an unknown mixture.

3. Heteroatom T_1 Measurements

In some cases, CP is not necessary to obtain a suitable solid state NMR spectrum. In these cases, the SPE/MAS sequence may be used and for quantitative analysis only the X-nucleus T_1 time needs to be determined. The standard inversion-recovery experiment (Fig. 10C) can be used to measure this

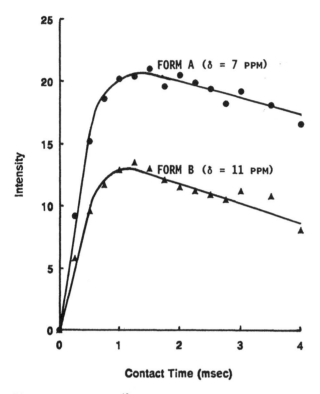

Fig. 11   Plot of the ^{13}C signal intensity as a function of contact time for two distinct methyl resonances of two polymorphic forms of a developmental drug substance.

value, keeping in mind that MAS and high-power decoupling is still necessary. As before, once the X-nucleus T_1 time is determined, $1–5 \times T_1$ may be inserted as the delay time between successive pulses for quantitative data acquisition. Figure 13 shows the ^{31}P T_1 determination for the two polymorphic forms of fosinopril sodium.

Quantitative solid state ^{13}C CP/MAS NMR has been used to determine the relative amounts of carbamazepine anhydrate and carbamazepine dihydrate in mixtures [59]. The ^{13}C NMR spectra for the two forms did not appear different, although sufficient S/N for the spectrum of the anhydrous form required long accumulation times. This was determined to be due to the slow proton relaxation rate for this form. Utilizing the fact that different proton spin-lattice relaxation times exist for the two different pseudopolymorphic forms, a quantitative method was developed. The dihydrate form displayed a relatively short relaxation time, permitting interpulse delay times of only 10 seconds to obtain full-intensity spectra of the dihydrate form while displaying no signal due to the anhydrous

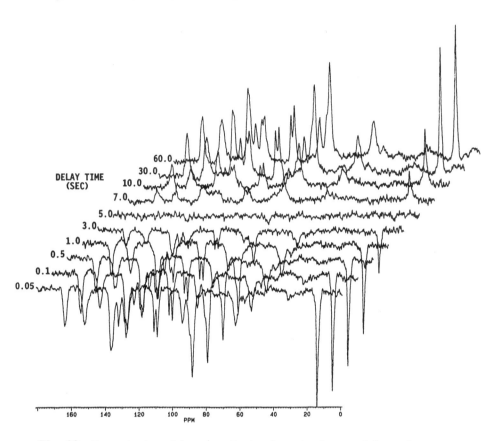

Fig. 12 Determination of the proton T_1 time for a developmental drug substance as measured through the ^{13}C intensity.

form. By utilizing an internal standard (glycine) and the differences in the relaxation rate of the two forms, the peak area of the dihydrate could be measured and related through a calibration curve to the amount of anhydrous and dihydrate content in mixtures of carbamazepine.

D. Other Studies

Finally, a series of papers have been published on the solid state NMR spectra of a number of analgesic drugs. Jagannathan recorded the solid state ^{13}C NMR spectrum of acetaminophen in bulk and dosage forms [60]. From the solution-phase NMR spectrum, assignments of the solid state NMR resonances could be made in addition to explanations for the doublet structure of some resonance: (dipolar coupling). Spectra of the dosage product from two sources indicatec

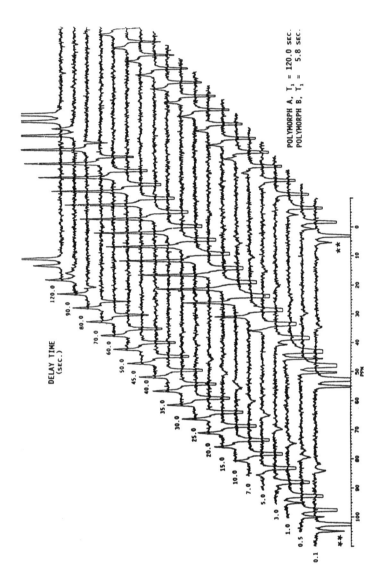

Fig. 13 Determination of ^{31}P T_1 values for fosinopril sodium using the inversion-recovery method. The spectra represent approximately a 50/50% w/w mixture of polymorphic forms A (δ = 52 ppm) and B (δ = 55 ppm). Spinning sidebands are represented by asterisks.

identical drug substance, but different levels of excipients. The topic of drug–excipient interactions was addressed by solid state ^{13}C NMR in the investigation of different commercially available aspirin samples [61,62]. In each commercial aspirin product, the only difference in the measured NMR spectrum was due to variations in the excipients, indicating that there were no interactions between the drug and the excipients under dry blending conditions. After lyophilization of two of the products, one aspirin sample did show a different NMR spectrum, indicating possible interaction during lyophilization, or conversion to a different solid state form during processing.

V. SUMMARY

Within various pharmaceutical laboratories (industrial and academic), the multinuclear technique of solid state NMR has primarily been applied to the study of polymorphism at the qualitative and quantitative levels. Although the technique ideally lends itself to the structure determination of drug compounds in the solid state, it is anticipated that in the future, solid state NMR will become routinely used for method development and problem solving activities in the analytical/materials science/physical pharmacy area of the pharmaceutical sciences. During the past few years, an increasing number of publications have emerged in which solid state NMR has become an invaluable technique. With the continuing development of solid state NMR pulse sequences and hardware improvements (increased sensitivity), solid state NMR will provide a wealth of information for the physical characterization of pharmaceutical solids.

REFERENCES

1. P. A. Mirau, *J. Magn. Reson.*, *96*, 480 (1992).
2. S. K. Branch and U. Holzgrabe, *PZ. Wiss.*, *3*, 211 (1990).
3. T. J. Wozniak, R. J. Bopp, and E. C. Jensen, *J. Pharm. Biomed. Anal.*, *9*, 363 (1991).
4. A. H. Beckelt, *Biochem. Soc. Trans.*, *19*, 443 (1991).
5. D. Regoli, *Trends Pharmacol. Sci.*, *11*, 398 (1990).
6. F. M. H. Jeffrey, A. Rajagopal, C. R. Malloy, and A. D. Sherry, *Trends Biochem. Sci.*, *16*, 5 (1991).
7. V. Stoven, J. Y. Lallemand, D. Abergel, S. Bovaziz, M. A. Delsuc, A. Ekondzi, E. Guittet, S. Laplante, R. LeGoas, T. Malliavin, A. Mikov, C. Reisdorf, M. Rogin, C. van Heijenoort, and Y. Yang, *Biochimie*, *72*, 531 (1990).
8. J. P. M. van Duynhoven, P. J. M. Folkers, C. W. J. M. Prinse, B. J. M. Harmsen, R. N. H. Konings, and C. W. Hilbers, *Biochemistry*, *31*, 1254 (1992).
9. G. M. Clore and A. M. Gronenborn, in *Theoretical Biochemistry and Molecular Biophysics*, D. L. Beveridge and R. Lavery, eds., Adenine Press, Schenectady, N.Y. 1991, pp. 1–16.

10. J. K. Baker and C. W. Myers, *Pharm. Res.*, *8*, 763 (1991).
11. A. Burger and R. Ramberger, *Mikrochim. Acta [Wien]*, *II*, 259 (1979); *Mikrochim. Acta [Wien]*, *II*, 273 (1979).
12. J. Haleblian and W. McCrone, *J. Pharm. Sci.*, *58*, 911 (1969).
13. M. Otsuka, T. Matsumoto, and N. Kaneniwa, *J. Pharm. Pharmacol.*, *41*, 665 (1989).
14. K. A. Connors, G. L. Amidon, and V. J. Stella, in *Chemical Stability of Pharmaceuticals*, John Wiley & Sons, New York, 1986.
15. D. E. Bugay, *Pharm. Res.*, *10*, 317 (1993).
16. M. Karplus, *J. Am. Chem. Soc.*, *85*, 2870 (1963).
17. T. C. Farrar and E. D. Becker, in *Pulse and Fourier Transform NMR*, Academic Press, New York, 1971.
18. E. R. Andrew, A. Bradbury, and R. G. Eades, *Nature*, *183*, 1802 (1959).
19. E. R. Andrew, *Prog. Nucl. Magn. Reson. Spectrosc.*, *8*, 1 (1972).
20. A. Pines, M. G. Gibby, and J. S. Waugh, *J. Chem. Phys.*, *56*, 1776 (1972); *J. Chem. Phys.*, *59*, 569 (1973).
21. A. Salvetti, *Drugs*, *40*, 800 (1990).
22. W. T. Dixon, J. Schaefer, M. D. Sefcik, E. O. Stejskal, and R. A. McKay, *J. Magn. Reson.*, *49*, 341 (1982).
23. C. A. Fyfe, in *Solid State NMR for Chemists*, CFC Press, Guelph, Canada, 1983.
24. S. R. Hartmann and E. L. Hahn, *Phys. Rev.*, *128*, 2042 (1962).
25. R. J. Abraham, J. Fisher, and P. Loftus, in *Introduction to NMR Spectroscopy*, John Wiley & Sons, New York, 1988, pp. 84–86.
26. T. C. Farrar and E. D. Becker, in *Pulse and Fourier Transform NMR*, Academic Press, New York, 1971, pp. 20–22.
27. J. Schaefer, E. O. Stejskal, and R. Buchdahl, *Macromolecules*, *8*, 291 (1975).
28. P. Granger, *Appl. Spectrosc.*, *42*, 1 (1988).
29. W. L. Earl and D. L. VanderHart, *J. Magn. Reson.*, *48*, 35 (1982).
30. *The CP/MAS Accessory Product Description Manual*, Bruker Instruments, Inc., Billerica, Mass., 1987.
31. J. S. Frye and G. E. Maciel, *J. Magn. Reson.*, *48*, 125 (1982).
32. A. Kubo and C. A. McDowell, *J. Magn. Reson.*, *92*, 409 (1991).
33. R. C. Crosby and J. F. Haw, *J. Magn. Reson.*, *82*, 367 (1989).
34. G. C. Campbell, R. C. Crosby, and J. F. Haw, *J. Magn. Reson.*, *69*, 191 (1986).
35. B. Wehrle, F. Aguilar-Parrilla, and H.-H. Limbach, *J. Magn. Reson.*, *87*, 584 (1990).
36. A. Martin, J. Swarbrick, and A. Cammarata, in *Physical Pharmacy*, Lea & Febiger, Philadelphia, 1983, pp. 575–576.
37. S. R. Byrn, in *Solid-State Chemistry of Drugs*, Academic Press, New York, 1982, pp. 79–146.
38. S. R. Byrn, G. Gray, R. R. Pfeiffer, and J. Frye, *J. Pharm. Sci.*, *74*, 565 (1985).
39. S. R. Byrn, P. A. Sutton, B. Tobias, J. Frye, and P. Main, *J. Am. Chem. Soc.*, *110*, 1609 (1988).
40. H. Martinez, S. R. Byrn, and R. R. Pfeiffer, *Pharm. Res.*, *7*, 147 (1990).
41. S. R. Byrn, A. T. McKenzie, M. M. A. Hassan, and A. A. Al-Badr, *J. Pharm. Sci.*, *75*, 596 (1986).

42. R. A. Fletton, R. K. Harris, A. M. Kenwright, R. W. Lancaster, K. J. Packer, and N. Sheppard, *Spectrochim. Acta*, *43A*, 1111 (1987).

43. R. K. Harris, B. J. Say, R. R. Yeung, R. A. Fletton, and R. W. Lancaster, *Spectrochim. Acta*, *45A*, 465 (1989).

44. R. A. Fletton, R. W. Lancaster, R. K. Harris, A. M. Kenwright, K. J. Packer, D. N. Waters, and A. Yeadon, *J. Chem. Soc. Perkin Trans.*, *II.*, 1705 (1986).

45. R. K. Harris, A. M. Kenwright, B. J. Say, R. R. Yeung, R. A. Fletton, R. W. Lancaster, and G. L. Hardgrove, Jr., *Spectrochim. Acta*, *46A*, 927 (1990).

46. E. A. Christopher, R. K. Harris, and R. A. Fletton, *Solid State Nucl. Magn. Reson.*, *1*, 93 (1992).

47. J. J. Gerber, J. G. vanderWatt, and A. P. Lötter, *Int. J. Pharm.*, *73*, 137 (1991).

48. H. G. Brittain, S. J. Bogdanowich, D. E. Bugay, J. DeVincentis, G. Lewen, and A. W. Newman, *Pharm. Res.*, *8*, 963 (1991).

49. C. Doherty and P. York, *Int. J. Pharm.*, *47*, 141 (1988).

50. K. Raghavan, A. Dwivedi, G. C. Campbell, Jr., E. Johnston, D. Levorse, J. McCauley, and M. Hussain, *Pharm. Res.*, *10*, 900 (1993).

51. H. G. Brittain, K. R. Morris, D. E. Bugay, A. B. Thakur, and A. T. M. Serajuddin, *J. Pharm. Biomed. Anal.*, *11*, 1063 (1993).

52. R. G. Ball and M. W. Baum, *J. Org. Chem.*, *57*, 801 (1992).

53. E. B. Vadas, P. Toma, and G. Zografi, *Pharm. Res.*, *8*, 148 (1991).

54. K. D. M. Harris and J. M. Thomas, *J. Solid State Chem.*, *93*, 197 (1991).

55. L. Frydman, A. C. Olivieri, L. E. Diaz, B. Frydman, A. Schmidt, and S. Vega, *Mol. Phys.*, *70*, 563 (1990).

56. R. J. Abraham, J. Fisher, and P. Loftus, in *Introduction to NMR Spectroscopy*, John Wiley & Sons, New York, 1988, pp. 113–116.

57. S. J. Opella and M. H. Frey, *J. Am. Chem. Soc.*, *101*, 5854 (1979).

58. R. K. Harris, *Analyst*, *110*, 649 (1985).

59. R. Suryanarayanan and T. S. Wiedmann, *Pharm. Res.*, *7*, 184 (1990).

60. N. R. Jagannathan, *Current Science*, *56*, 827 (1987).

61. C. Chang, L. E. Díaz, F. Morin, and D. M. Grant, *Magn. Res. Chem.*, *24*, 768 (1986).

62. L. E. Díaz, L. Frydman, A. C. Olivieri, and B. Frydman, *Anal. Lett.*, *20*, 1657 (1987).

5

Particle Morphology: Optical and Electron Microscopies

Ann W. Newman
Bristol-Myers Squibb Pharmaceutical Research Institute, New Brunswick, New Jersey

Harry G. Brittain
Ohmeda, Inc., Murray Hill, New Jersey

I. INTRODUCTION

Evaluation of the morphology of a pharmaceutical solid is of extreme importance, since this property exerts a significant influence over the bulk powder properties of the material. In addition to providing insights into the micromeritic properties of the solid, microscopy can also be used to develop preliminary estimations of the particle-size distribution. A determination can be easily made regarding the relative crystallinity of the material, and it is often possible to deduce crystallographic information as well. Unknown particulates can often be identified solely on the basis of their microscopic characteristics, although it is useful to obtain confirmatory support for these conclusions with the aid of microscopically assisted techniques.

Both optical and electron microscopies have found wisespread use for the characterization of pharmaceutical solids. Optical microscopy is limited in the range of magnification suitable for routine work, having an approximate upper limit of 600×. This magnification limit does not preclude the investigation of most pharmaceutical materials, however, and the use of polarizing optics introduces a power into the technique not available with other methods. Electron microscopy work can be performed at extraordinarily high magnification levels (up to 90,000× on most units), and the images that can be obtained contain a considerable degree of three-dimensional information.

The two methods are complementary in that each can provide information inaccessible to the other. When these techniques are used in conjunction, substantial characterization of a solid material becomes possible. This information can be extremely useful during the early stages of drug development, since only a limited amount of the drug candidate is normally available at that time.

An excellent discussion of the synergistic aspects of optical and electron microscopies has been provided by McCrone [1]. He concludes that electron microscopy yields excellent topographic and shape information, and that it is most useful in forensic situations involving trace evidence characterization and identification. Light microscopy can be used to deliver information on the internal properties of small particles, fibers, and films. When polarizing optics are used in light microscopy, the optical properties of the crystals under investigation can also be determined. In that case, molecular (rather than elemental) information can be obtained on the analyte. In either case, the two methods of microscopy have impressive records for problem solving.

II. THE MORPHOLOGY OF SOLID PARTICLES

A crystal is a polyhedral solid, bounded by a number of planar faces. The arrangement of these faces is termed the *habit* of the crystal, with the crystal faces being identified using Miller indices. The crystal is built up through the repetition of a fundamental building block, known as the *unit cell*. The molecules in the solid form a three-dimensional basic pattern, known as the space lattice. Through the application of simple geometry, it has been shown that only 14 different kinds of simple space lattices are possible. By taking combinations of the various lattices possible for each crystallographic system, it can also be determined that all solids must belong to one of 230 space groups. This information is summarized in any one of the textbooks concerned with optical crystallography [2,3]. A review of crystallography from the pharmaceutical viewpoint is also available [4].

The habit of pharmaceutical compounds has been used for purposes of identification, although the method can only be reliably used when the crystallization solvent used to generate the test crystals is carefully controlled. Since the faces of a crystal must reflect the internal structure of the solid, the angles between any two faces of a crystal will remain the same even if the crystal growth is accelerated or retarded in one direction or another. Toxicologists have made extensive use of microscopy following multiple recrystallization, and they have developed useful methods for compound identification [5].

Crystals are characterized by the repetition of their constituent atoms in a three-dimensional network. There are only a finite number of symmetrical arrangements possible for a crystal lattice, and these may be characterized by the relationships existing among the crystal axes and angles between these. This information has been summarized in Table 1. It is worth noting that the ratio between the lengths of the axes existing in a given crystal must be constant.

Table 1 Characteristics of the Six Crystal Systems

System	Description
Cubic	Three axes of identical length (identified as a_1, a_2, and a_3) intersect at right angles.
Hexagonal	Four axes (three of which are identical in length, and denoted as a_1, a_2, and a_3) lie in a horizontal plane, and they are inclined to one another at 120°. The fourth axis, c, is different in length from the others, and it is perpendicular to the plane formed by the other three.
Tetragonal	Three axes (two of which, denoted as a_1 and a_2, are identical in length) intersect at right angles. The third axis, c, is different in length with respect to a_1 and a_2.
Orthorhombic	Three axes of different lengths (denoted as a, b, and c) intersect at right angles. The choice of the vertical c axis is arbitrary.
Monoclinic	Three axes (denoted as a, b, and c) of unequal length intersect such that a and c lie at an oblique angle, and the b axis is perpendicular to the plane formed by the other two.
Triclinic	Three axes (denoted as a, b, and c) of unequal length intersect at three oblique angles.

In one of the first attempts to produce a systematic procedure for the identification of compounds based upon crystal morphology, Shead proposed to use profile angles as the analytical parameter [6,7]. This method was based on the use of sublimation to obtain thin crystal plates of simple geometrical forms.

When crystals are imperfectly formed, or if the relationship between the faces cannot be discerned, it is still useful to provide a more empirical judgement regarding the overall shape of the particles. A classification scheme, similar to that provided by Amidon [8], is proposed in Table 2. Solids illustrating these definitions are provided in Fig. 1.

In addition to the main types described in Table 2, crystals may form treelike patterns known as dendrites, with such aggregates being termed *dendritic* or *arboraceous*. If an aggregate is composed of tiny crystals radiating from a center, it is termed a *spherulite* or *rosette*.

Deviation of crystal habits from the ideal may often be observed, and such effects are normally the result of alterations in the crystallization conditions. Three factors responsible for the controlling of crystal habits can be listed [9]:

1. Rate of deposition during growth. Rapid deposition on a face encourages formation of a point.
2. Shielding of certain faces. If a face is in contact with the walls of a container, it usually becomes very large.
3. Composition of the mother liquor. The presence of impurities or additives during the crystallization process can often lead to the observation of significant alterations in observed crystal habits and types.

Particle morphology can be examined with optical microscopy and scanning electron microscopy. A discussion of both techniques is presented.

Table 2 Empirical Classification of Observed Crystal Habits

Descriptor	Description
Avicular	Needlelike particle having a similar width and thickness. If the crystals are very thin, the term *fibrous* is used.
Columnar	Rodlike particle, having a width and thickness exceeding that of a needle-type particle. The term *prismatic* may also be used.
Blade	Long, thin, and flat particle, which can also be referred to as being *lath-shaped*.
Plate	Flat particles of similar length and width. These may also be denoted as being *lamellar* or *micaceous*.
Tabular	Also flat particles of similar length and width, but possessing greater thickness than flakes.
Equant	Particles of similar length, width, and thickness.

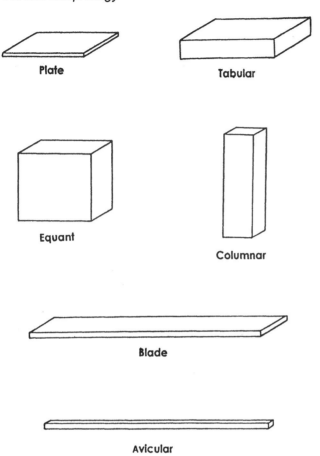

Plate

Tabular

Equant

Columnar

Blade

Avicular

Fig. 1 Classification of particulates according to their observed habits.

III. OPTICAL MICROSCOPY

The literature pertaining to optical microscopy is significantly large, with a number of texts and review articles having been written over the years. A representative selection of sources is provided at the end of this chapter [10–15].

A. The Light Microscope

The compound microscope is essentially a two-stage magnifier. One lens (or combination of lenses), termed the *objective*, forms a real image of an illuminated or self-luminous object, and a second lens (or combination of lenses), denoted as the *eyepiece*, acts as a simple magnifier whose object is the real image formed by the objective. The two lens systems are mounted in a telescoping body tube, with the sample placed on the *stage*. A simplified diagram of a compound microscope is found in Fig. 2. The overall magnification is equal to the magnification of the eyepiece (M_e) times the magnification of the objective (M_o).

The magnification of the eyepiece is readily determined using the conventional formula for a simple magnifier:

$$M_e = \frac{\text{projection distance}}{F_e} \tag{1}$$

where F_e is the focal length of the eyepiece. The magnification of the objective depends on the distance between the real image and the upper focal plane of the objective, or the *optical tube length* of Fig. 2. Since this quantity is not normally measured, an approximate magnification may be calculated by assuming the optical and mechanical tube lengths to be equal and the object to be very close to the lower focal plane of the objective. In that case,

$$M_o = \frac{\text{image distance}}{\text{object distance}} = \frac{\text{mechanical tube length}}{F_o} \tag{2}$$

where F_o is the focal length of the objective. In practice, it is much easier to use a stage micrometer in conjunction with a standard $10\times$ objective to empirically determine the magnification of a given objective.

The power of a microscope to reveal details depends less on the magnification and more on the clarity or sharpness of the image produced by the objective. A simple definition of the *resolving power* of a microscope is the smallest distance between two points in the object such that the two points can be distinguished in the image.

The angle between the most divergent rays that can pass through an objective is termed the *angular aperture* (AA) of the objective. It is customary to express the aperture of an objective in terms of the sine of this angle, and to define the *numerical aperture* (NA):

$$NA = n \sin \frac{AA}{2} \tag{3}$$

where n is the refractive index of the medium between the objective and the object. It is evident that an increase in the refractive index of the medium will increase the numerical aperture and therefore yield more brilliance in the field of observation. For this reason, a variety of oils having refractive indices between 1.4 and 1.6 have been used as immersion fluids. Objectives intended for use with immersion fluids have much shorter focal lengths than do objectives intended for dry use.

The *depth of focus*, or penetrating power, of a lens is the maximum

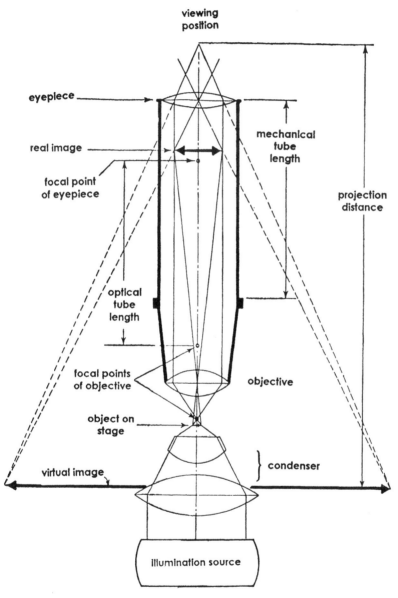

Fig. 2 Elements of the compound microscope.

separation of two planes that are both apparently in sharp focus, although not equidistant from the lens. A large depth of focus is highly desirable, since it will permit the observer to discern various strata in the object without refocusing the system. The depth of focus varies approximately as the square of the focal length, and inversely with the numerical aperture.

B. Polarizing Microscopy

The polarizing microscope is essentially a light microscope equipped with a linear polarizer located below the condenser, and an additional polarizer mounted on top of the eyepiece. A rotating stage is also found to be very useful, as is the ability to add other optical accessories. Polarization optical analysis is based on the action of the analyte crystal on the properties of the transmitted light. This method can yield several directly measured parameters, such as the sign and magnitude of any observed birefringence, knowledge of the refractive indices associated with each crystal direction, what the axis angles are, and what the relations among the optical axes are.

To conduct a polarizing microscope analysis, the light from the source is rendered linearly polarized by the initial polarizer. The second polarizer mounted above the sample (the analyzer) is oriented such that its axis of transmission is orthogonal to that of the initial polarizer. In this condition of "crossed polars," no transmitted light can be perceived by the observer. The passage, or lack thereof, of light though the crystal as a function of the angle between the crystal axes and the direction of polarization is of key importance to the method.

The refractive index of light passing through an isotropic crystal will be identical along each of the crystal axes, and such crystals therefore possess *single refraction*. Crystals within the cubic system will be isotropic along all three crystal axes. Anisotropic substances will exhibit different refractive indices for light polarized with respect to the crystal axes, thus exhibiting *double refraction*. Crystals within the hexagonal and tetragonal systems possess one isotropic direction, and they are termed *uniaxial*. Anisotropic crystals possessing two isotropic axes are termed *biaxial*, and they include all crystals belonging to the orthorhombic, monoclinic, and triclinic systems. Biaxial crystals will exhibit different indices of refraction along each of the crystal axes.

Isotropic samples will have no effect on the polarized light no matter how the crystal is oriented, since all crystal axes are completely equivalent. This effect is known as complete or *isotropic extinction* (Fig. 3a). Noncrystalline, amorphous samples will exhibit the same effect.

When the sample is capable of exhibiting double refraction, the specimen will then appear bright against a dark background. For example, when a uniaxial crystal is placed with the unique c axis horizontal on the stage, it will be alternately dark and bright as the stage is rotated. Furthermore, the crystal will

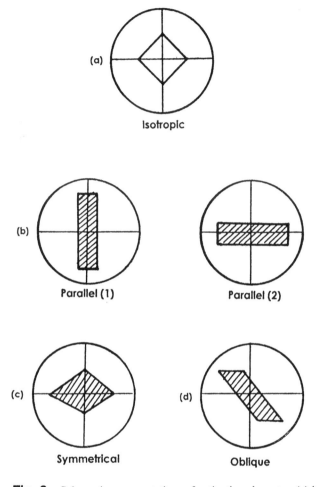

Fig. 3 Schematic representations of extinction character: (a) isotropic, (b) parallel, (c) symmetrical, and (d) oblique extinction.

be completely dark when the c axis is parallel to the transmission plane of the polarizer or analyzer. If the crystal has edges or faces parallel to the c axis, then it will be extinguished when such an edge or face is parallel to one of the polarizer directions. This condition is denoted as *parallel extinction* (Fig. 3b). At all intermediate positions, the crystal will appear light and usually colored. A rhombohedral or pyramidal crystal will be extinguished when the bisector of a silhouette angle is parallel to a polarization direction, and this type of extinction is termed *symmetrical extinction* (Fig. 3c).

When, however, the uniaxial crystal is mounted with the c axis vertical on the stage, it is found that the crystal remains uniformly dark as the stage is rotated (another case of isotropic extinction).

For biaxial crystals, similar results are obtained as with uniaxial crystals. The exception to this rule is that in monoclinic and triclinic systems, the polarization directions need not be parallel to faces or to the bisectors of face angles. If the prominent faces or edges of an extinguished crystal are not parallel to the axes of the initial polarizer, the extinction is said to be *oblique* (Fig. 3d).

Knowledge of the type of extinction permits a determination of the system to which a given crystal belongs. A summary of the extinction characters covering biaxial and uniaxial crystals is provided in Table 3.

If a biaxial or uniaxial crystal is situated in such a position that neither of its crystal axes are aligned with the plane of polarized illumination light, it is useful to consider the effect of the crystal on the light vectors. The linearly polarized light can be resolved into its components lying along the crystal axes, and it is found that these two components do not pass through the crystal with the same velocity. This phenomenon is known as *birefringence*, and the slower direction is said to be retarded with respect to the faster direction. Both uniaxial and biaxial crystals are termed either positive or negative depending on the refractive index difference as defined with respect to the long and short crystal axes.

When these components are recombined upon leaving the crystal, destructive interference will take place. If white light is used to illuminate the sample, then the effect of birefringence will be to produce a *polarization color*. The colors resulting from birefringence will not be that of the ordinary spectrum, but instead they form what is known as *Newton's series*. This color sequence was first observed in the interference fringes of soap bubbles. The degree of birefringence is determined by the transversed pathlength, and consequently Newton's series is divided into *orders* by the red colors that appear regularly as

Table 3 Differentiation of Crystal Systems by the Character of Extinction

System	Character of observed extinction
Cubic	Isotropic or complete extinction.
Hexagonal	Parallel or symmetrical extinction. Can be isotropic if viewed down the *c* axis, but then a six-sided silhouette should be observed.
Tetragonal	Parallel or symmetrical extinction. Can be isotropic, but then a four-sided silhouette should be observed.
Orthorhombic	All crystals will show parallel or symmetrical extinction.
Monoclinic	Some crystals will show parallel or symmetrical extinction, and others oblique extinction.
Triclinic	All crystals will show oblique extinction.

the thickness of the crystal increases. The "first-order" colors (gray, white, yellow, orange, and red) are followed by the brighter colors of the "second-order" (violet, blue, green, yellow-green, yellow, orange, and dark bluish red). Higher degrees of birefringence, correlated with thicker crystals, are denoted as being "third-order" (blue-green, green, yellow-green, and purplish red). The "fourth-order" colors are observed as light greens and pinks, and at the "fifth-order" the birefringence is observed as an intense white. The colors are ordinarily charted on the Michel–Levy scale of birefringence [16].

The particles making up sodium starch glycolate, viewed using ordinary illumination and between crossed polarizers, are shown in Fig. 4. When imaged using simple transmitted light, the starch particles appear as globular and do not exhibit any defined crystal faces. When viewed between crossed polarizers, the same particles are characterized by a "Maltese cross" effect caused by the intrinsic birefringence. Addition of a mica plate to the optical train reveals alternate birefringence colors existing within the four quadrants of each starch granule. This distinctive behavior can be used as a qualitative identity test for any starch material, since all granules (regardless of their source) yield essentially equivalent microscopic images.

C. Thermal Microscopy

Optical microscopy can also be used to study phase conversion processes, being particularly useful in the characterization of melting or freezing phase transitions. A compilation of the thermomicroscopic properties of an extensive ensemble of compounds of pharmaceutical interest has been published by Kuhnert-Brandstätter [17].

Thermal microscopic work is conducted by mounting the sample in a system whose temperature can be accurately controlled and monitored, which is usually termed a *hot stage* or *cold stage* depending on the type of thermal control employed. Most cold-stage systems operate from room temperature down to approximately –50°C, while most hot-stage systems are functional between room temperature and 300–350°C. Special hot stages have been designed for operation at higher temperatures, but to work at the highest temperatures considerable care must be taken not to adversely affect the quality of the microscope optical components.

The temperature at which transitions among polymorphic states occur can be accurately determined, since the optical properties of the crystals will generally undergo a drastic change with the phase transformation. The most useful crystal properties for such determinations are the birefringence, extinction position, optic axial angle, or refractive index [18]. The loss of crystal birefringence is probably the easiest parameter to use, since the transition will generally terminate in an isotropic condition. It should be recognized that the phase transition is a

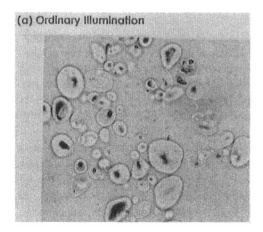

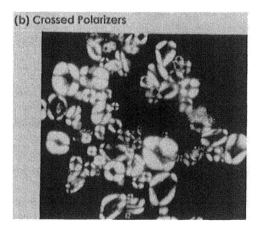

Fig. 4   Sodium starch glycolate viewed (a) using ordinary illumination and (b) between crossed polarizers.

kinetically controlled process; thus, it is important that slow heating rates be used during the monitoring.

Based on the reversibility of their phase transformation behavior, polymorphs can easily be classified as being either *enantiotropic* (interchange reversibly with temperature) or *monotropic* (irreversible phase transformation). Enantiotropic polymorphs are each characterized by phase stability over well-defined temperature ranges. In the monotropic system, one polymorph will be stable at all temperatures, and the other is only metastable. Ostwald formulated the rule of successive reactions, which states that the phase that will crystallize out of a melt will be the state that can be reached with the minimum loss of free energy [19]. In systems capable of exhibiting polymorphism, this will of necessity be the metastable phase.

The dehydration, desolvation, or decomposition temperature of a compound can also be evaluated from an examination of discontinuities in the optical properties. These processes are most effectively monitored if the sample is immersed in an oil, since the evolution of gases normally accompanies these transformations. The evolution of gases is evident in the generation of gaseous bubbles from the solids under study.

Undoubtedly the most important use of thermal microscopy is in the accurate determination of melting points and melting point ranges. Since the melting point of a solid is strongly affected by the purity of the compound, such measurements can be used to deduce the relative purities of different lots of the same compounds. At the melting point, the crystal optical properties change drastically, with all the axially dependent properties of the solid being replaced by the analogous isotropic properties of the melted material. With proper accuracy in the temperature-measuring system, melting points should be accurate to within ±0.1°C.

Molecular weight determinations can be made on the basis of the lowering of the freezing point of a suitable solvent system [20]. In this methodology, a solid solution (approximately 10% by weight) is prepared by melting the solute in the solvent, and allowing the melt to freeze. The melting point of the solid solution is then determined, and the molecular weight calculated from the freezing point depression data in the conventional manner. The melting point is taken as the disappearance of the last crystals in the solvent. A standard series of solvent systems are recommended for such work, with the most appropriate solvent being one whose melting point is fairly close to that of the analyte. A selection of recommended solvent systems is found in Table 4. The melting point and molar depression constant must be determined for each lot of solvent used, since these properties will depend critically on the purity of the material.

D. Chemical Microscopy

Probably the most extensive use of particle morphology and microscopy has been in the area of chemical microscopy. With this approach, derivatives of the analyte species are prepared, crystallized, and identified through the morphological characteristics of these derivatives [21]. Most of these applications have been superseded by modern methods of analysis, but the microscopic method can still be used by skilled practitioners for the study of trace quantities of analyte. The literature developed during the heyday of chemical microscopy is too large to be reviewed here, but advances in the field are still chronicled in the Annual Reviews issue of *Analytical Chemistry* [22]. A substantial review of the optical characteristics of organic compounds is available [23].

Table 4 Solvent Systems Useful in the
Microscopic Determination of Molecular Weight

System	Melting point (°C)
Anthraquinone	285
Dimethylglyoxime	246
Saccharine	228
Borneol	204
Hexachloroethane	186
Camphor	178
Salicylic acid	159
Bornyl chloride	130
Quinone	116
Tetrabromoethane	93
Naphthalene	80
Cyclopentadecanone	65
Camphene	50
Phenol	41

One of the most systematic investigations into the use of organic chemical microscopy was made by Dunbar and his associates, who used a series of specific reagents to develop specific tests for various functional groups. Methods were developed for amines [24,25], carboxylic acids, anhydrides, and acid chlorides [26,27], aldehydes and ketones [28], hydroxy compounds [29], amino acids [30], and cations [31].

In another extensive series of studies, Clarke and coworkers developed sensitive microchemical tests for the determination of alkaloids [32–34], anesthetics [35], antihistamines [36], antimalarials [37], and analgesics [38]. One of the useful techniques introduced by Clarke was that of the hanging-microdrop [32], which permitted identification tests to be made on submicrogram quantities of analyte. Results obtained on the cinchona alkaloids are shown in Table 5 to illustrate the methodology.

IV. SCANNING ELECTRON MICROSCOPY

Scanning electron microscopy is commonly used to study the particle morphology of pharmaceutical materials. Its use is somewhat limited because the information obtained is visual and descriptive, but usually not quantitative. When the scanning electron microscope is used in conjunction with other techniques, however, it becomes a powerful characterization tool for pharmaceutical materials.

A conventional scanning electron microscope (SEM) can be thought of as

Table 5 Results of Microchemical Tests Performed on Cinchona Alkaloids

Alkaloid	Reagent	Crystal habit	Sensitivity (μ)
Cinchonidine	Sodium carbonate	Small rosettes	0.1
	Platinum chloride	Small rods	0.025
Cinchonine	Sodium carbonate	Rosettes	0.025
	Platinum iodide	Curved needles	0.05
Quinidine	Potassium iodide	Triangular crystals	0.25
	Sodium carbonate	Dense rosettes	0.1
Quinine	Sodium phosphate	Needles	0.1
	Platinum iodide	Curved serrated needles	0.05

Source: Ref. 32.

an inverted light microscope [39,40]. It consists of an electron gun for electron beam generation (illumination), a column with lenses for beam focusing, a sample chamber, and a detector (Fig. 5). The conventional scanning electron microscope exhibits a resolution of approximately 100 Å [39], and its large depth of field yields three-dimensional images. Samples can be large and thick, but conductive coatings are required for nonconductive samples. High vacuum is required for the electron source and sample chamber. The environmental scanning electron microscope (ESEM), on the other hand, will provide high-resolution images without conductive coatings and high vacuums in the sample chamber.

Instrumentation, image formation, accessories, and applications of conventional scanning electron microscopy will be discussed. Information about the ESEM will also be presented.

A. Electron Beam Generation and Focusing

The illumination system begins with the electron gun [41]; its parts are outlined in Fig. 6. A thermionic emission electron gun contains a filament, such as tungsten [40], lanthanum hexaboride (LaB_6) [42], or cerium hexaboride (CeB_6) [43], which emits electrons when heated by an electric current. To achieve maximum beam insensity, the electrons should be ejected at the tip of the filament in a narrow path called crossover, as shown in Fig. 7 [40]. The electrons emitted from the filament are driven down the microscope column by a voltage difference in the electron gun housing. The voltage difference results from the potentials on the Wehnelt cylinder housing surrounding the filament (cathode) and an anode plate. The electrons are

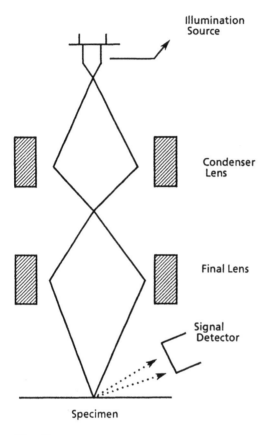

Fig. 5 Diagram of a conventional scanning electron microscope.

attracted to the positive charge on the anode plate and continue down the column. The beam crossover should occur in front of the cathode and must be aligned with the hole in the anode for optimum beam intensity. The hole in the anode plate acts as a crude aperture to block peripheral electrons.

As the beam travels down the column, a number of electromagnetic lenses are used to guide the beam to the sample [44]. The condenser lenses are part of the illumination system and are used to deliver electrons from the electron gun crossover to the sample. The condenser lenses determine the beam current reaching the sample. The objective, or final, lens determines the final spot size of the beam. A set of scanning coils are also present in the instrument column to scan the beam in a raster pattern over an area of the sample. At each point, data is collected and the points are combined to form the image. More detail on the data collection is given in the image formation section.

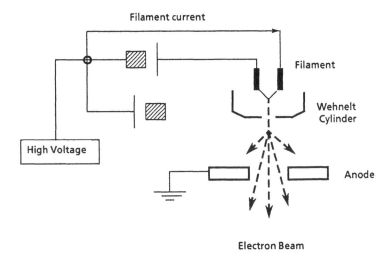

Filament current

Filament

Wehnelt
Cylinder

High Voltage

Anode

Electron Beam

Fig. 6 Diagram of an electron gun configuration.

B. Interactions of the Electron Beam

The interaction of the electron beam with the sample is generally onion-shaped and results in a number of different processes, as shown in Fig. 8. The formation of secondary electrons and backscattered electrons is the basis for the imaging capabilities of the SEM.

On the molecular level, a secondary electron is produced when the beam interacts with an inner electron in the nucleus and ejects it with a discrete amount of energy. When the ejected electron is weakly bound and has an energy below 59 eV, microscopists refer to it as a secondary electron. Due to their low energy, these secondary electrons escape from the surface of the sample and are sensitive to the topography of the sample being examined. Because they are formed by the primary beam before it spreads in the sample, the secondary electrons exhibit high spatial resolution relative to the other processes. Secondary electrons are commonly chosen for imaging due to their topographic sensitivity and high spatial resolution. Secondary electrons do not convey information about the elemental composition of the sample.

Backscattered electrons, however, do give some elemental information about the sample because they are more energetic than secondary electrons and escape from farther within the sample [45,46]. On the molecular level, the electron beam can interact with the nucleus of an atom and be scattered with minimal loss of energy. These incident electrons may be scattered more than once and then ejected from the sample as backscattered electrons. The backscattered electrons originate from a greater depth within the sample and are

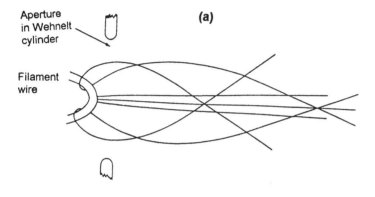

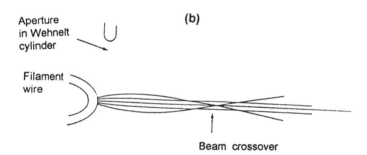

Fig. 7   Diagram of electron beam: (a) no beam crossover; (b) beam crossover. (Adapted from Ref. 40.)

sensitive to the mean atomic number of the sample. Elements with high atomic numbers contain a greater positive nuclear charge and are more likely to produce backscattered electrons. Because the probability of backscattering increases with increasing atomic number, the image contains chemical information. Most pharmaceutical compounds would not be imaged using backscattered electrons due to the low atomic number elements, such as carbon and hydrogen, present in these materials. More detailed explanations and mathematical treatments of beam interactions can be found elsewhere [47–49].

C. Image Formation

The secondary electrons emitted from the sample are collected in the sample chamber. The most common type of secondary electron detector is the scintilla-

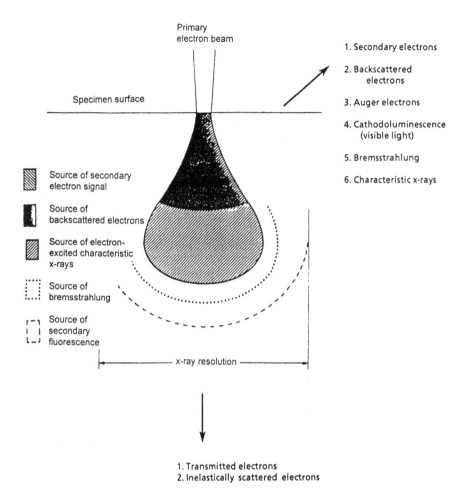

Fig. 8 Schematic of electron beam interaction with a sample and the electron beam interaction volumes for electron–specimen interactions.

tor/light pipe/photomultiplier system, presented in Fig. 9 [41,44]. It consists of a collector screen, scintillator, light pipe, and photomultiplier tube (PMT).

The secondary electrons emitted from the sample are attracted to the detector by the collector screen. Once near the detector, the secondary electrons are accelerated into the scintillator by a positive potential maintained on the scintillator. Visible light is produced by the reaction of the secondary electrons with the scintillator material. The emitted light is detected by a photomultiplier tube, which is optically coupled to the scintillator via a light pipe. The PMT signal is then transferred to the grid of a cathode ray tube (CRT). Data collection

Inside Sample Chamber   Outside Sample Chamber

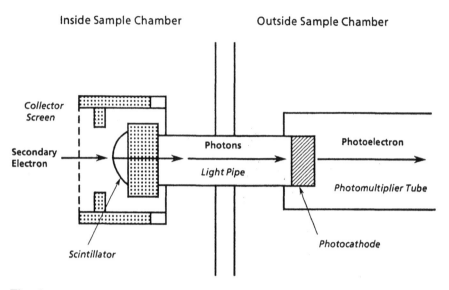

Fig. 9 Block diagram of a scintillator/light pipe/photomultiplier system. (Adapted from Ref. 44.)

on the CRT is synchronized with the scanning of the electron beam to collect data on a point-by-point basis. The brightness of any point on the CRT will be related to the strength of the signal from that point on the sample. In this way, an image of the specimen surface is built up on the CRT screen as the beam moves over the sample. The image is observed on a monitor, and it can be photographed or saved digitally to a computer file, depending on the equipment available.

D. Sample Preparation and Sample Chamber Accessories

Sample preparation for imaging of samples is relatively easy. The sample, such as a powder or tablet, is mounted on a sample mount using double-sided tabs or carbon paint. Most pharmaceutical materials are insulators and, therefore, require vapor deposition of a thin (approximately 200 Å) metal coating using a sputter coater. Common metals used for coating include gold or gold/palladium. The coating reduces charge buildup on the specimen under the electron beam, resulting in clear, stable images.

Sample preparation for more specialized work can require more intensive procedures and accessories [49]. Stages have been made for the SEM to accommodate a variety of experiments [50]. Heating, cooling, and mechanical manipulation would be useful for most pharmaceutical materials, but other disciplines have also used magnetic fields, electric fields, or etching capabilities in their stages [48]. Dynamic experimentation is possible in the SEM, but it is limited by the vacuum requirements of the conventional SEM instrument.

E. Environmental Scanning Electron Microscopy

The introduction of the environmental scanning electron microscope (ESEM) [51–54] has extended the use of electron microscopy, and dynamic electron microscopy experiments are now possible. The benefits of the ESEM include those found with the conventional SEM, such as depth of field, resolution, range of magnification, and signal detection modes. Based, however, on a system of vacuum gradients and pressure-limiting apertures, the sample chamber of the ESEM is designed to accommodate pressures up to 30 torr while maintaining the electron gun and optics at high vacuum. The electron beam travels through the ESEM column and apertures with minimal restriction, and the beam collides with the specimen as described for the conventional SEM. A modified secondary-imaging detector was developed for the environmental SEM to withstand higher pressure ranges [52].

Conductive sample coatings are not needed because the gas molecules in the chamber replenish electrons on the sample surface to prevent charging. Direct observation of either wet or dry specimens is possible based on the continuously variable specimen environment. The instrument accommodates a micromanipulator, heatable stage, and gaseous environment. Energy dispersive x-ray (EDX) units can also be added to the sample chamber for elemental analysis. Samples can be analyzed in their natural state, at elevated relative humidities, elevated temperatures, and in various gas environments (including 100% relative humidity).

The ESEM can be used for a variety of applications in the pharmaceutical field, besides routine imaging and elemental analysis. Polymer coatings can be treated at various pressures, humidities, and temperatures to investigate coverage and stability over a wide range of conditions. Hydrated materials can be imaged without fear of dehydrating the sample. Because the sample chamber can be maintained at high relative humidities, biological samples, such as tissues, can be readily and realistically studied. As the ESEM becomes more widely used in the pharmaceutical industry, many other applications will be reported.

F. Elemental Analysis

Elemental analysis can also be performed on SEM samples using x-ray spectrometer attachments [55]. The techniques are known as energy dispersive x-ray (EDX) analysis and wavelength dispersive x-ray (WDX) analysis and require installation of a detector in the sample chamber.

Both techniques are based on the interactions of the electron beam and the elemental atoms in the sample. As shown in Fig. 10a, the incident beam ejects an electron from an inner atomic shell. The ion is now in a higher energy or

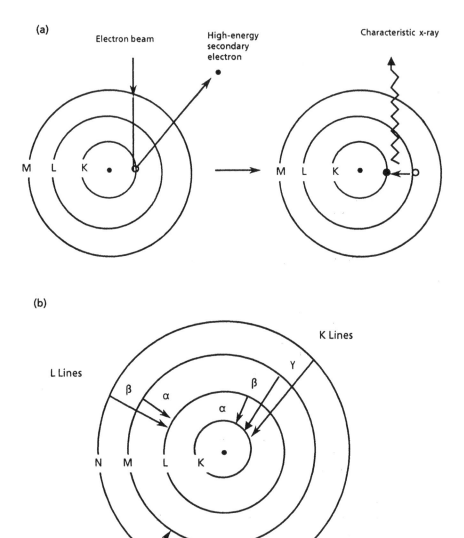

Fig. 10   (a) Diagram of x-ray generation; (b) diagram of x-ray emission nomenclature.

excited state. In order to return to the more stable *ground state*, the atom gives up energy by allowing an electron from an outer shell to drop into the vacancy in the inner shell. The energy released can be in the form of x-rays, and the amount of energy released is related to the difference in energy between the two atomic shells involved in the transition. Because the energy differences between the shells of different elements are unique, the x-rays detected are indicative of a particular element.

The nomenclature for the emitted x-rays is related to the shells involved in the transition, as illustrated in Fig. 10b. For example, if the initial electron is ejected from the K shell and the second electron drops from the adjacent L shell, a K-α x-ray is emitted. If the electron drops from the M shell, the emitted x-ray is a K-β x-ray. The most common lines observed are the *KLM* lines.

The main difference between EDX and WDX is the detector used for analysis. The EDX detectors collect the full spectrum of elements present, allowing immediate elemental identification. A solid state lithium-drifted silicon detector is used for analysis [44]. The x-ray photon emitted from the sample is converted to a charge pulse in the detector, which is then converted to a voltage pulse by a charge-sensitive preamplifier, and then sorted by voltage using a multichannel analyzer (MCA). Signals from an EDX detector are measured in keV, and they range from 0.75 keV up to the energy of the incident beam, such as 20 keV. Most EDX instruments can detect elements including and above boron. The WDX detectors, on the other hand, allow analysis of one element at a time, resulting in better sensitivity. This detector is based on diffraction of x-rays using Bragg's law [41]. A portion of the emitted x-rays strike an analyzing crystal, and only x-rays diffracted according to Bragg's law will be detected by a proportional counter. For this analysis, different detector crystals are used for different elements. For both the EDX and WDX techniques, the analysis is limited to the sample in the field of view of the instrument.

Elemental analysis can also give information about where the element is located in a sample, by using line scans and elemental mapping [56]. Specific energy ranges corresponding to certain elements can be used to monitor the presence of the element in the sample. For a line scan, the element is detected along a specific "line" across the sample. When the element is present, a peak will appear in the line scan, whereas the line will return to the baseline when the element is not present. For an elemental map, the element will be detected in the field of view. When the element is present, the area will appear bright and the element is "mapped" for the sample. When the element is not present, the sample will appear dark.

G. Applications

One of the most common uses of the SEM is to examine particle morphology. The particle morphology of a material can be related to any number of parameters, including flow properties, dispersion, and excipient functionality. A sampling of how SEM has been used in the pharmaceutics field will be presented.

The morphology of excipients plays a major role in formulating products. A study relating morphology and functionality of 14 direct compression tablet excipients effectively used SEM to illustrate this point [57]. One example is the fibrous nature of the disintegrant crosscarmelose (Ac-Di-Sol), which allows intraparticulate wicking of water into the tablet matrices. By breaking the fibers into short lengths, improved flow and blending properties were also achieved. Another disintegrant, crospovidone (crosslinked PVP), was composed of small particles fused into large porous aggregates. The water absorbed into the porous structure provided the wicking of water into the matrix.

The surface characteristics of excipients have also been studied and related to the dispersion and dissolution of poorly soluble drugs [58]. It was found that excipients with rough surfaces (such as Emcompress, with a porous surface) trap the drug particles in the indentations, which can then be blocked by fine excipient particles and decrease dissolution. Smooth surfaces (such as spherical sugar beads), however, produced high dissolution efficiency of the poorly soluble drugs.

Imaging has also been used to study the morphology of bulk drugs. Various crystal forms and morphologies can be obtained from different crystallizations [59,60], manufacturing processes [61,62], and storage conditions [63,64]. In the study of a leukotriene D_4 receptor antagonist, SEM was used to compare crystalline material with lyophilized (amorphous) material [63]. It was also found that water vapor sorbed at room temperature converted the crystalline form of the material to a noncrystalline form resembling the amorphous material. Using polarizing optical microscopy, it was determined, based upon the birefringence patterns observed, that the amorphous material was a liquid crystalline phase.

The particle morphology of bulk drugs can also be a factor when using other analytical techniques. Solid state quantitation methods, such as x-ray powder diffraction and infrared spectroscopy, require homogeneous standards when measuring the amounts of different crystal forms present [65,66]. During development of a quantitative method for the presence of a dihydrate crystal form in monohydrate bulk material, problems with mixing were attributed to the different particle morphologies of the two different forms [67]. The dihydrate material was found to exist as long needles, which did not mix easily with the irregular platelike structure of the monohydrate material, shown in Fig. 11. A slurry technique was developed to uniformly mix the two phases, and the SEM photomicrographs revealed that both phases were present homogeneously in the samples when prepared by this technique.

Compression of pharmaceutical materials has also been investigated with scanning electron microscopy. Simple powder mixtures have been used to illustrate the processes that occur during compression [68]. The excipients chosen

Fig. 11 SEM photomicrographs of (a) monohydrate, (b) dihydrate, and (c) 14.75% dihydrate/monohydrate after acetone slurry preparation.

exhibited plastic and elastic deformation, as well as brittleness. Other tablets studied after compaction (aspirin, anhydrous calcium gluceptate, and metoclopramide hydrochloride) showed crystal growth on the surface of the compacts within 60 minutes of compression [69]. Aspirin showed an increase in size and definition of the crystals with time. Examination of dimetrally fractured compacts exhibited recrystallization in crystal interfaces within the compact as well.

Other applications of the SEM involve film coatings [70–75] and surface morphology of microcapsules and beads [76–78]. One study has used imaging to evaluate various types of film coatings applied to tablets using both aqueous and organic-solvent methods [70]. Continuity and imperfections of the film surface were evaluated, and cross sections of the pellets were also examined to determine the presence of a distinct boundary between the film coating and the core. The physicomechanical properties of the films were found to be highly dependent on the processing techniques used. Another study correlated film thickness and dissolution for acetaminophen beads [71]. Using SEM it was found that a low coating level of 4% resulted in a discontinuous film over the beads with visible holes providing channels for drug release. High coating levels of 16% resulted in a continuous film, which correlated to the drug release being controlled by diffusion through the film barrier and zero-order kinetics.

Size measurements have also been collected for film thicknesses and bead sizes on the electron microscope [79–81]. One example is the study of the relationship between film thickness and the size and mass of a bead in a fluidized-bed unit [79]. The bead diameter and film thickness were determined by SEM and correlated to dissolution data. It was found that larger beads received thicker coatings and exhibited slower release rates than the smaller beads.

Scanning electron microscopy in conjunction with elemental analysis, such as EDX, can also be a powerful problem solving tool. Recently, EDX mapping has been used to gather information about the microscopic internal structure and spatial distribution of drug during the drug release process [82]. The study suggested that film-coated pellets do behave as traditional diffusion-controlled systems with a gradual depletion of drug from the matrix of the core. Another example is the investigation of tablet dissolution problems. Tablets with acceptable and unacceptable dissolution were analyzed by mapping the chlorine atom in the active drug substance (Fig. 12). It was found that tablets with good dissolution exhibited an even dispersion of drug in the tablet, whereas the tablets with poor dissolution showed large concentrated areas of active drug. The dissolution problem was explained by the dispersion of the drug in the tablet, and stricter particle-size specifications were enacted to solve the problem.

These are a few of the many examples of the uses of scanning electron microscopy. The use of this technique with other physical characterization methods results in a powerful pharmaceutical tool.

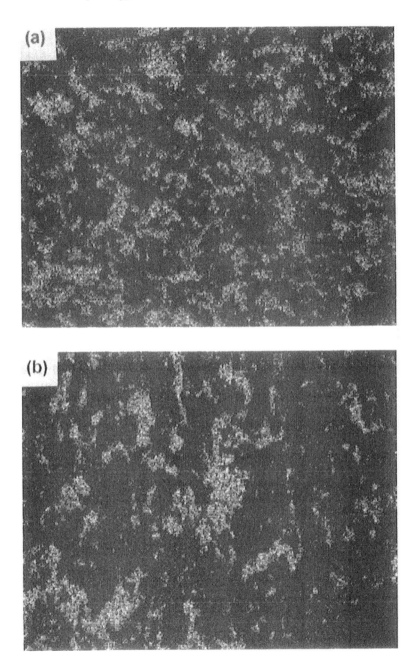

Fig. 12 EDX chlorine elemental maps showing (a) even dispersion of drug in tablet and (b) large concentrated areas of active drug.

ACKNOWLEDGMENT

The authors would like to thank Mr. Joseph DeVincentis for the SEM micrographs presented in Fig. 11.

REFERENCES

1. W. C. McCrone, *Scanning Microscopy*, 7, 1 (1993).
2. C. W. Bunn, *Chemical Crystallography*, 2nd ed., Oxford University Press, New York, 1961.
3. E. E. Wahlstrom, *Optical Crystallography*, 4th ed., John Wiley & Sons, New York, 1969.
4. J. A. Biles, *J. Pharm. Sci.*, *51*, 499, 601 (1962).
5. F. Amelink, *Rapid Microchemical Identification Methods in Pharmacy and Toxicology*, Interscience, New York, 1962.
6. A. C. Shead, *Ind. Eng. Chem. Anal. Ed.*, *9*, 496 (1937); *ibid.*, *10*, 662 (1938).
7. A. C. Shead, *Mikrochem. Acta*, 432 (1957); *ibid.*, 657 (1959).
8. G. E. Amidon, *Pharm. Forum*, *18*, 4089 (1992).
9. T. R. P. Gibb, *Optical Methods of Chemical Analysis*, McGraw-Hill, New York, 1942, p. 239.
10. J. Belling, *The Use of The Microscope*, McGraw-Hill, New York, 1930.
11. G. Needham, *The Practical Use of The Microscope*, Charles C. Thomas Pub., Springfield, Ill., 1958.
12. E. M. Chamot and C. W. Mason, *Handbook of Chemical Microscopy*, Vol 1, 3rd ed., John Wiley & Sons, New York, 1958.
13. A. F. Hallimond, *The Polarizing Microscope*, 3rd ed., Vickers, New York, 1970.
14. N. H. Hartshorne and A. Stuart, *Crystals and the Polarizing Microscope*, 4th ed., Edward Arnold Pub., London, 1970.
15. W. C. McCrone, L. B. McCrone, and J. G. Delly, *Polarized Light Microscopy*, Ann Arbor Science Pub., Ann Arbor, Mich., 1978.
16. E. M. Chamot and C. W. Mason, *Handbook of Chemical Microscopy*, Vol. 1, 3rd ed., John Wiley & Sons, New York, 1958, p. 489.
17. M. Kuhnert-Brandstätter, *Thermomicroscopy in the Analysis of Pharmaceuticals*, Pergamon Press, Oxford, 1971.
18. H.-H. Emons, H. Keune, and H.-H. Seyfarth, "Chemical Microscopy," in *Comprehensive Analytical Chemistry*, Vol. 16, Elsevier, Amsterdam, 1982.
19. W. Ostwald, *Z. Physik. Chem.*, *22*, 306 (1897).
20. W. C. McCrone, *Fusion Methods in Chemical Microscopy*, Interscience, New York, 1957.
21. E. M. Chamot and C. W. Mason, *Handbook of Chemical Microscopy*, Vol. 2, 2nd ed., John Wiley & Sons, New York, 1940.
22. For the most recent review, see P. M. Cooke, *Anal. Chem.*, *64*, 219R (1992).
23. A. N. Winchell, *The Optical Properties of Organic Compounds*, 2nd ed., Academic Press, New York, 1954.
24. R. E. Dunbar and J. Knuteson, *Microchem. J.*, *1*, 17 (1957).
25. R. E. Dunbar and F. Ferrin, *Microchem. J.*, *4*, 167 (1960).

26. R. E. Dunbar and C. C. Moore, *Microchem. J.*, *3*, 491 (1959).
27. R. E. Dunbar and B. W. Farnum, *Microchem. J.*, *5*, 5 (1961).
28. R. E. Dunbar and A. E. Aaland, *Microchem. J.*, *2*, 113 (1958).
29. R. E. Dunbar and F. Ferrin, *Microchem. J.*, *3*, 65 (1959).
30. R. E. Dunbar and F. Ferrin, *Microchem. J.*, *4*, 59 (1960).
31. R. E. Dunbar and F. Ferrin, *Microchem. J.*, *5*, 145 (1961).
32. E. G. C. Clarke and M. Williams, *J. Pharm. Pharmacol.*, *7*, 255 (1955).
33. E. G. C. Clarke, *J. Pharm. Pharmacol.*, *9*, 187 (1957).
34. E. G. C. Clarke, *J. Pharm. Pharmacol.*, *11*, 629 (1959).
35. E. G. C. Clarke, *J. Pharm. Pharmacol.*, *8*, 202 (1956).
36. E. G. C. Clarke, *J. Pharm. Pharmacol.*, *9*, 752 (1957).
37. E. G. C. Clarke, *J. Pharm. Pharmacol.*, *10*, 194 (1958).
38. E. G. C. Clarke, *J. Pharm. Pharmacol.*, *10*, 642 (1958).
39. A. C. Reimschuessel, *J. Chem. Ed.*, *49*(8), A413–458 (1972).
40. A. W. Agar, R. H. Alderson, and D. Chescoe, *Principles and Practice of Electron Microscopy*, North Holland Publishing: Amsterdam, 1974.
41. O. C. Wells, *Scanning Electron Microscopy*, McGraw-Hill, New York, 1974.
42. *The Lanthanum Hexaboride Electron Emission Source and the Vacuum Environment*, Amray Technical Bulletin 112–277, Amray, Bedford, Mass., 1986, pp. 11–18.
43. T. Studt, *R&D Magazine*, January, 56–57 (1992).
44. J. I. Goldstein, D. E. Newbury, P. Echlin, D. C. Joy, C. Fiori, and E. Lifshin, *Scanning Electron Microscopy and X-Ray Microanalysis*, Plenum Press, New York, 1981.
45. G. E. Lloyd, *Mineral. Mag.*, *51*, 3–19 (1987).
46. *Backscattered Electron Detectors and Applications*, Amray Technical Bulletin 122–685, Amray, Bedford, Mass., 1986, pp. 89–106.
47. D. E. Newbury, in *Principles of Analytical Electron Microscopy*, D. C. Joy, A. D. Romig, Jr., and J. I. Goldstein, eds., Plenum Press, New York, 1986, pp. 1–28.
48. D. E. Newbury, D. C. Joy, P. Echlin, C. E. Fiori, and J. I. Goldstein, *Advanced Scanning Electron Microscopy and X-Ray Microanalysis*, Plenum Press, New York, 1986.
49. E. E. Hunter, *Practical Electron Microscopy*, Praeger Scientific, New York, 1984.
50. T. G. Rochow and E. G. Rochow, *An Introduction to Microscopy by Means of Light, Electrons, X-rays, or Ultrasound*, Plenum Press, New York, 1978.
51. R. Harniman, *Res. and Dev.*, September, 112–116 (1988).
52. G. D. Danilatos, *J. Microsc.*, *160*, 9–19 (1990).
53. A. N. Farley and J. S. Shah, *J. Microsc.*, *158*, 379–388 (1990).
54. A. N. Farley and J. S. Shah, *J. Microsc.*, *158*, 389–401 (1990).
55. *Energy Dispersive X-Ray Analysis*, Kevex Instruments, Santa Monica, Calif., 1989.
56. *X-Ray Mapping and the Use of Back-Scattered Electrons in the SEM*, Amray Technical Bulletin 106–274, Amray, Bedford, Mass., 1976, pp. 11–13.
57. R. F. Shangraw, J. W. Wallace, and F. M. Bowers, *Pharm. Tech.*, October, 44–60 (1981).
58. E. Sallam, H. Ibrahim, M. Takieddin, M. Abu Shamat, and T. Baghal, *Drug Dev. Ind. Pharm.*, *14*(9), 1277–1302 (1988).

59. H.-K. Chan, S. Venkataram, D. J. W. Grant, and Y.-E. Rahman, *J. Pharm. Sci.*, *80*, 677–685 (1991).
60. J. J. Gerber, J. G. vanderWatt, and A. P. Lotter, *Int. J. Pharm.*, *73*, 137–145 (1991).
61. H. P. Huang, K. S. Murthy, and I. Ghebre-Sellassie, *Drug Dev. Ind. Pharm.*, *17*(17), 2291–2318 (1991).
62. Y Kawashima, N. Toshiyuki, H. Takeuchi, T. Hino, and Y. Itoh, *J. Pharm. Sci.*, *80*, 472–478 (1991).
63. E. B. Vardas, P. Toma, and G. Zografi, *Pharm. Res.*, *8*, 148–155 (1991).
64. Y. Matsuda, M. Otsuka, M. Onoe, and E. Tatsumi, *J. Pharm. Pharmacol.*, *44*, 627–633 (1992).
65. R. Suryanarayanan, *Pharm. Res.*, *6*, 1017–1024.
66. K. J. Hartauer, E. S. Miller, J. K. Guillory, *Int. J. Pharm.*, *85*, 163–174 (1992).
67. D. E. Bugay, A. W. Newman, P. Findlay, and J. DeVincentis, manuscript in preparation.
68. H. Hess, *Pharm. Tech.*, June, 54–68 (1987).
69. A. G. Mitchell and G. R. B. Down, *Int. J. Pharm.*, *22*, 337–344 (1984).
70. A. Mehta and D. M. Jones, *Pharm. Tech.*, June (1985) pp. 52–60.
71. G. Zhang, J. B. Schwartz, and R. L. Schnaare, *Pharm. Res.*, *8*(3), 331–335 (1991).
72. R. Bodmeier and O. Paeratakul, *Pharm. Res.*, *8*(3), 355–359 (1991).
73. L. E. Appel and G. M. Zentner, *Pharm. Res.*, *8*(5), 600–604 (1991).
74. T. Eldem, P. Speiser, and A. Hincal, *Pharm. Res.*, *8*(1), 47–54 (1991).
75. S. Y. Lin and Y. H. Kao, *Pharm. Res.*, *8*(7), 919–924 (1991).
76. P. R. Pal and T. K. Pal, *Indian Drugs*, *24*(9), 430–437 (1987).
77. C. M. Adeyeye and J. C. Price, *Pharm. Res.*, *8*(11), 1377–1383 (1991).
78. S. P. Sanghvi and J. G. Nairn, *J. Pharm. Sci.*, *80*(4), 394–398 (1991).
79. R. Wesdyk, Y. M. Joshi, N. B. Jain, K. Morris, and A. Newman, *Int. J. Pharm.*, *65*, 69–76 (1990).
80. P. Sheen, P. J. Sabol, G. J. Alcorn, and K. M. Feld, *Drug Dev. Ind. Pharm.*, *18*(8), 851–860 (1992).
81. A. W. Jenkins and A. T. Florence, *J. Pharm. Pharmacol.*, *25* (Suppl.), 57P–61P (1973).
82. C. D. Melia, I. R. Wilding, and K. A. Khan, *Pharm. Tech.*, March, 56–63 (1992).

6

Particle Size Distribution

Cynthia S. Randall
SmithKline Beecham Pharmaceuticals, King of Prussia, Pennsylvania

I. INTRODUCTION

The determination and control of particle size is often a necessity in pharmaceutical analysis and formulation. The size range and distribution of particles in a given product can influence its safety, efficacy, stability, viability of dosage form, and manufacturing processes. This is particularly true in the case of solid dosage forms, where particle size of poorly water-soluble drugs is known to influence dissolution behavior and, hence, bioavailability. It is also true for manufacturing low strengths of highly potent drugs, since the particle size distribution can affect content uniformity. Moreover, particle size of active ingredients as well as excipients can affect tablet characteristics such as porosity and flowability. It is thus not surprising that regulatory agencies are showing increased concern about particle size distribution of drug substances, and they have emphasized the need for particle size specifications in active ingredients.

II. DEFINITIONS

Particle shape plays an important role in particle size determination. The simplest definition of particle size diameter is based on a sphere, which has a unique diameter. In reality, however, many particles are not well represented by this model. Figure 1 illustrates the variety of shapes that may be found in particle samples [1]. As the size of a particle increases, so does its tendency to have an irregular shape [2], complicating statistical analysis. Particle shape coefficients have been derived for different geometries [3], and various "equivalent diame-

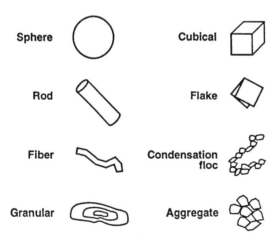

Fig. 1   Illustration of different particle forms that can occur. Note the variety of shapes. (From Ref. 1.)

ters" have been developed to relate the size of such particles to that of a sphere with the same diameter, surface area, or volume; some of these are listed in Table 1. Note that in certain cases, the diameter being measured is a function of the measuring technique, and so direct comparisons of data obtained using different sizing methods may not always be meaningful. For example, sizes determined by a Coulter counter, which are expressed as d_{vn}, may not be comparable with those obtained with a microscope, which have different definitions specific to this instrument.

In describing the "average" diameter of a sample, three parameters can be used: the mean, the median, and the mode. The mean is the sum of all diameters divided by the total number of particles. It is sensitive to extreme values and thus is often not particularly useful. The median is the value above and below which 50% of the particles are found. It is less influenced by outlying values and is preferable to the mean as a single measure of particle size. The mode represents the size occurring most frequently; it is used less frequently than the mean or the median. In a perfectly symmetrical distribution, the mean, median, and mode values are the same.

Since most pharmaceutical substances contain a range of sizes, the size distribution of the particles must also be determined. A frequency distribution curve can be constructed by plotting the number of particles in a given size range against the total size range. The ideal resulting curve is a "normal," or Gaussian, distribution, with the standard deviation measuring the distribution around the mean; see Fig. 2. The standard deviation σ gives an indication of the uniformity of particle size within a sample. In the data of Fig. 2, there is a 68% probability that a particle in the sample will be within ± 1 σ and a 95% chance that it will fall within ± 2 σ.

Table 1 Various Equivalent Diameters for Particle Size Description

Type of mean	Size parameter	Frequency	Mathematical expression	Mean diameter name
Arithmetic	Length	Number	$\Sigma\, nd / \Sigma\, n$	Length-number mean, d_{ln}
Arithmetic	Surface	Number	$\sqrt{\Sigma\, nd^2 / \Sigma\, n}$	Surface-number mean, d_{sn}
Arithmetic	Volume	Number	$\sqrt[3]{\Sigma\, nd^3 / \Sigma\, n}$	Volume-number mean, d_{vn}
Arithmetic	Length	Length	$\Sigma\, nd^2 / \Sigma\, nd$	Surface-length or length-weighted mean, d_{sl}
Arithmetic	Length	Surface	$\Sigma\, nd^3 / \Sigma\, nd^2$	Volume-surface or surface-weighted mean, d_{vs}
Arithmetic	Length	Weight	$\Sigma\, nd^4 / \Sigma\, nd^3$	Weight-moment or volume-weighted mean, d_{wm}

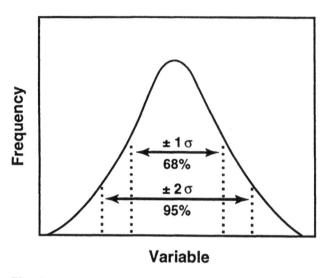

Fig. 2   Normal, or Gaussian, size-frequency distribution curve. Percentage of particles lying within ±1 and ±2 standard deviations about the arithmetic mean diameter are indicated.

Most pharmaceutical powders, however, are milled and as a consequence tend to give skewed distributions, as illustrated by Fig. 3a. Such a distribution can be normalized by plotting frequency versus logarithm of particle size, and a size distribution conforming to this treatment is referred to as a *log-normal* distribution. Figure 3b illustrates a common graphical representation of a log-normal distribution. In these systems, the logarithm of the particle size equivalent to 50% of the distribution is defined as the geometric mean diameter, d_g, which is also equal to the median diameter. The slope of the resulting line represents σ_g, the standard deviation. It is equal to the ratio of the 84.13% undersize divided by the 50% size, or the ratio of the 50% size divided by the 15.87% undersize.

As can be seen in Fig. 3b, it is important to specify whether data are represented as a number distribution (obtained by a counting technique such as microscopy) or as a weight distribution (obtained by methods such as sieving), since the results will not be the same. Hatch and Choate [4] have developed equations for converting one type of diameter to another; the relationships between them are summarized in Table 2. Note that caution should be exercised in using the Hatch–Choate conversions if the distributions do not closely fit the log-normal model. While this distribution is the most frequently used to describe pharmaceutical systems, other distribution functions have also been developed [2,5,6].

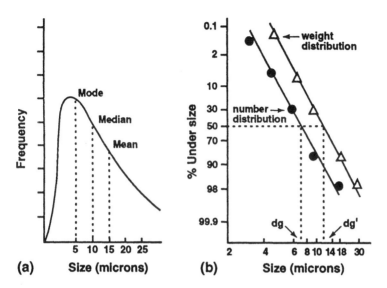

Fig. 3 (a) A skewed size-frequency distribution typical of pharmaceutical powders. (b) Graphical representation of a log-normal distribution. Note the difference between number and weight distributions.

III. SAMPLING

For a meaningful particle size, representative samples of the bulk powder must be analyzed. Obtaining a representative sample is not always easy, due to the tendency of the particles to undergo size segregation. Allen [7] recommends that samples be taken from a moving stream of powder, and that samples be taken from the entire stream over a series of time intervals. Depending on the homogeneity of the powder, different sampling strategies can be employed. A wide variety of equipment is available for sampling, but some procedures are significantly prone to operator variation [1,7]. One of the most reliable devices is the spinning riffler,

Table 2 Hatch–Choate Equations for Conversion of Diameters

Diameter	Number distribution	Weight distribution
Length-number mean	$\log d_{ln} = \log d_g + 1.151 \log^2 \sigma_g$	$\log d_{ln} = \log d_g' - 5.757 \log^2 \sigma_g$
Surface-number mean	$\log d_{sn} = \log d_g + 2.303 \log 2\sigma_g$	$\log d_{sn} \log d_g' - 4.606 \log^2 \sigma_g$
Volume-number mean	$\log d_{vn} = \log d_g + 3.454 \log^2 \sigma_g$	$\log d_{vn} = \log d_g' - 3.454 \log^2 \sigma_g$
Volume-surface mean	$\log d_{vs} = \log d_g + 5.757 \log^2 \sigma_g$	$\log d_{vs} = \log d_g' - 1.151 \log^2 \sigma.$
Weight-moment mean	$\log d_{wm} = \log d_g + 8.059 \log^2 \sigma_g$	$\log d_{wm} = \log d_g' + 1.151 \, \text{l}($

shown in Fig. 4. Whatever the sampling method, it should be confirmed that samples taken represent the particle size distribution of the bulk powder.

IV. METHODS

A variety of techniques are available for sizing particles of pharmaceutical interest. The goal of this chapter is to provide an overview of common techniques currently in use for sizing of powders, and to illustrate their applications. The discussion will focus on techniques used to characterize powders above one micron (μm); however, it should be emphasized that in some cases the same methods may also be applicable to submicron particles.

A. Microscopy

Despite the emergence of other methods for particle sizing, microscopy remains a powerful tool. It has the advantage of providing a direct visual representation of the particles being measured. In addition to size, microscopy can provide details about shape, crystal habit, and possible inhomogeneity within the sample. It requires an extremely small amount of sample, needs no calibration by other methods, and the equipment is relatively inexpensive to acquire and maintain. Even when other techniques are used to size particles, microscopy studies are generally recommended to support their findings. The effective size range for analyzing particles by optical microscopy is about 0.25 to 100 μm.

A diagram of a conventional optical microscope is shown in Fig. 5 [8]. Proper application of a microscope will depend on its resolving power,

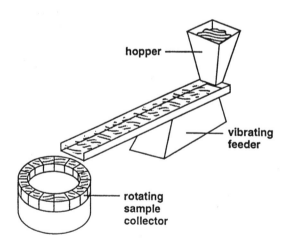

Fig. 4   Spinning riffler used to collect powder samples.

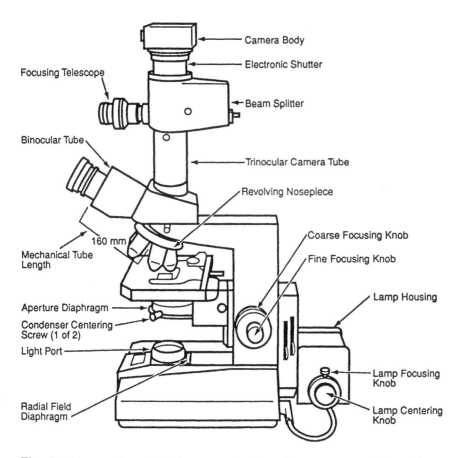

Fig. 5 Diagram of an optical microscope, showing various components. (Adapted from Ref. 8.)

magnification, and depth of field. For light of wavelength λ, resolution (R) is related to the numerical aperture (NA) of the lens by

$$R = \frac{\lambda}{2\,NA} \qquad (1)$$

For very high quality microscopes (NA = 1.4), R is on the order of 0.2 μm; that is, particles separated by less than this distance cannot be distinguished from each other. Magnification is the product of the eyepiece and objective numbers; maximum magnification is 1000 times the numerical aperture. Depth of field is also related to the aperture, decreasing when the latter increases. A large depth of field is useful when evaluating particles of many different sizes; however, this utility will offset by lower resolution. All of these factors must be kept in mind when using a particular microscope. Also, proper illumination and focusing are critical for optimal results [1]. A number of additional recommendations have been proposed for microscopic sizing of pharmaceutical substances [9].

Mention should also be made of electron microscopes. Schematic diagrams comparing optical and electron microscopes are shown in Fig. 6 [1]. In the latter instrument, the light source of the optical microscope is replaced by a beam of electrons. As the collimated beam passes through the specimen, electrons are absorbed or scattered to form an image. Electron microscopes have significantly better resolution than the optical microscope and can achieve much higher magnification, on the order of 100,000 times. This makes them the microscope of choice for examining particles in the submicron range. The scanning electron microscope (SEM) also offers the advantage of information about the topography of the particles. This equipment, however, is expensive; an SEM can cost well over $100,000. Also, sample preparation for SEM analysis is generally more complex and time consuming than for the optical microscope.

Before carrying out particle sizing on a sample, the microscope's ocular scale must be calibrated. This is normally done with a stage micrometer, which has a linear graduated scale. The micrometer is aligned with the eyepiece ocular to determine the length per ocular scale division. The ocular can then be used to read the diameters of particles on a slide. When sufficient particles are sized, the length-number mean can be calculated:

$$d_{\text{ln}} = \frac{\Sigma \, d_x \, \Delta n}{\Sigma \, \Delta n} \tag{2}$$

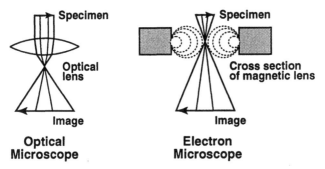

Fig. 6 Comparison of image projection in optical and electron microscopes. In the former, an optical lens focuses and enlarges a beam of light. In the electron microscope, a magnetic coil is used to bend electron rays for focusing and enlargement of the specimen. (Adapted from Ref. 1.)

A similar procedure for sizing can be done with a globe-and-circle graticule. Two common graticules are shown in Fig. 7. The Patterson–Cawood graticule has disks graduated in an arithmetic series, while the Porton graticule disks are graduated in a series based on $\sqrt{2}$ hr. The latter is considered more widely applicable, especially when sizing irregular, i.e., nonspherical particles [10]. In contrast to the stage micrometer, measurement of particles with a graticule gives particle area, from which the surface-number mean can be determined:

$$d_{sn} = \sqrt{\frac{\Sigma\, d_x^2\, \Delta n}{\Sigma\, \Delta n}} \tag{3}$$

To size particles, graticule circles of known area on the graticule are superimposed until one is found that best fits the particle; the surface area is then used to calculate particle size.

Another diameter of interest is the geometric mean diameter, d_g. This is particularly useful if one wishes to compare the microscopic diameter with the one estimated by sieving, d_g' (see Fig. 3b). For microscopy data, d_g can be obtained using the relationship

$$\log d_g = \frac{\Sigma(n\, \log d\,)}{\Sigma n} \tag{4}$$

The geometric mean diameter can then be found from the antilogarithm of log d_g [11].

Various methods are employed to size particles by optical microscopy. For spherical particles, the diameter suffices, but for nonspherical particles, alterna-

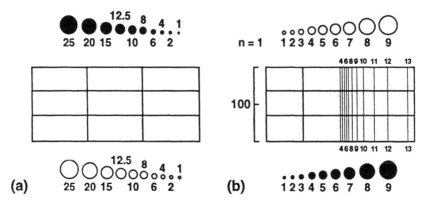

Fig. 7 Common microscope eyepiece graticules used in particle sizing: (a) Patterson–Cawood; (b) Porton.

tive descriptions are used (Fig. 8). Feret's diameter (d_F) is the distance between two tangents on opposite sides of the particle, perpendicular to the fixed direction of the scan. Martin's diameter (d_M) is the length of line, parallel to a fixed direction, bisecting the particle image. The projected area diameter (d_A) represents the diameter of a circle whose area equals the projected area of the particle, i.e.,

$$A = \frac{\Pi}{4 \, d_A^2} \tag{5}$$

and falls between d_M and d_F. The projected diameter gives the best estimate of the particle's cross-sectional area. This parameter, however, is not particularly useful in the case of irregular particles such as needles or fibers. Note that all of these diameters are two-dimensional and do not include a shape factor. As illustrated in Fig. 1, a variety of particle shapes can exist. For this reason, different descriptors of particle shape have recently been proposed for characterization of pharmaceutical substances [9].

Reporting of data from microscopic analysis of pharmaceuticals should include size range (in microns), average particle diameter (d), number of particles in range (n), percent particles in range, and cumulative percent of particles in range [11]. With respect to the number of particles counted, The American Society of Testing and Materials has recommended that the number of particles measured in a given size range should be at least 25 [12]. Obviously, to increase the accuracy even more particles need to be counted [10]. This brings up a significant disadvantage of microscopy: it is labor-intensive, tedious, and subject to human judgment in analyzing particles.

Various methods have been used to reduce operator fatigue; one that is frequently employed is to photograph the field of view and measure the particles on an enlarged print. Photomicrography can improve contrast between particles and their background, as well as provide a permanent record of the sample [1]. Useful results, however, require careful focusing of the object, choice of

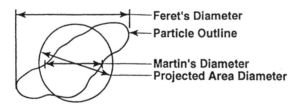

Fig. 8   Various diameters employed in particle sizing by microscopy.

appropriate film, and optimization of exposure time [9]. Different statistical approaches have also been applied to sampling to help improve reliability while minimizing the total number of measurements [1,10].

The limitations of optical microscopy for particle counting have been significantly overcome by interfacing the microscope with the image analyzer (IA). Image analysis describes a computer-based system that processes and analyzes digitized images. A schematic diagram showing the typical components of an IA system is given in Fig. 9. The image is produced by a scanner, e.g., a video camera. The interior of the scanner contains a strip of photosensitive film at the front, and a focused electron beam filament at the rear. As light strikes the film, a charge builds up that is proportional to the intensity of the image. The charge travels to the back of the film, where it is scanned by the electron beam. As the beam is scanned, electrical pulses are given off, which are then converted to arrays of digital values.

Each digital value in an array is referred to as a picture element or pixel, and each pixel is assigned a numerical value related to its brightness, known as its gray level. The pixel is also assigned coordinates related to its position within the image (i.e., x,y coordinates), and it is stored for processing. To analyze a particular object, a gray level "threshold" is first set to define the background. The spatial resolution of the IA system is defined by the number of pixels within the digital image. The greater the resolution, the more closely the digitized image matches the real-time image. Prior to processing, a number of mathematical filters can be applied to enhance image detail and correct shading distortions.

A powerful advantage of IA is the large enhancement of the optical microscope's image. With the human eye, the response to incoming light is logarithmic [13] and is directly related to the ratio of [object brightness minus background brightness] divided by [background brightness], referred to as the

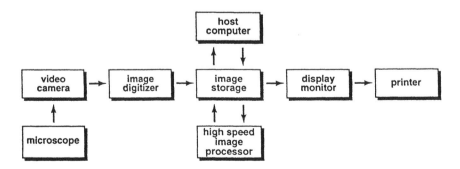

Fig. 9 Schematic diagram of an image analyzer.

image contrast. Improved contrast is required as the image becomes less bright, but the eye can only distinguish roughly 30 gray levels. An IA system typically has 256 gray levels.

Processing of the image involves segmentation and discrimination. In the former, the gray-level image is converted to a binary image, wherein all of the particles are assigned (segmented) as either white or black; the remaining background is given the opposite color. A threshold value for brightness is set to facilitate discrimination between the particles of interest and the background. This is commonly done by using a gray-scale histogram, i.e., a plot of the number of pixels in each gray level. Use of a threshold allows some consistency and objectivity in the determination of particles, which may not be possible with microscopy alone.

Image analysis can be used to determine a variety of morphometric parameters including area, Feret's diameter, Martin's diameter, aspect ratio (ratio of minimum to maximum Feret diameter), perimeter, length, width, and form factor (the ratio of area/[perimeter]^2), which can be related to specific particle shapes. Some of these functions are illustrated in Fig. 10 [14]. In addition, quantitative methods have been developed to measure particle shape [2,15,16].

With respect to particle size, measurements are commonly expressed in terms of the maximum Feret diameter using either the sizing algorithms of intersecting chords or rotating coordinates ("calipering"). An additional type of measurement involves summing up all the pixels within the digitized image to obtain the area, which can then be converted to its equivalent circular diameter. This method is considered more precise than the others, but it cannot be applied to all shapes of particles [17].

Zingerman et al. [18] have discussed several concerns about use of IA for routine particle size determination. Microscope magnification must be optimized using standards of known diameter to ensure accuracy; higher magnification does not guarantee better accuracy. Sample preparation is also important; i.e., samples should be well dispersed to minimize agglomeration, and image contrast should be as high as possible. Finally, one should be aware that IA sizing algorithms may generate errors when applied to irregular particles.

In all microscopic methods, sample preparation is key. Powder particles are normally dispersed in a mounting medium on a glass slide. Allen [7] has recommended that the particles not be mixed using glass rods or metal spatulas, as this may lead to fracturing; a small camel-hair brush is preferable. A variety of mounting fluids with different viscosities and refractive indices are available; a more viscous fluid may be preferred to minimize Brownian motion of the particles. Care must be taken, however, that the refractive indices of sample and fluid do not coincide, as this will make the particles invisible. Selection of the appropriate mounting medium will also depend on the solubility of the analyte [9]. After the sample is well dispersed in the fluid, a cover slip is placed on top

Feature	Display	Functions

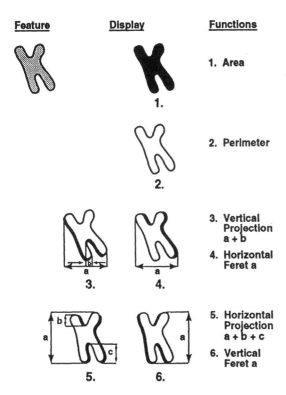

1. Area

2. Perimeter

3. Vertical Projection a + b

4. Horizontal Feret a

5. Horizontal Projection a + b + c

6. Vertical Feret a

Fig. 10 Various functions that can be measured with image analysis. (From Ref. 14.)

of the mixture. Yamate and Stockham [10] have recommended that the eraser end of a pencil be gently applied to the cover slip using a circular motion. This action helps to deagglomerate particles without breaking them. In the case of electron microscopy, specialized techniques must be used that involve evaporation of the sample under vacuum [1].

Obtaining representative fields of view is also important, since size segregation can occur during sample preparation. Small particles tend to stream toward the edges of the mounting fluid, and so fields in this area should be examined along with those at the center. And each particle to be sized must be carefully focused to determine its exact outer boundaries. Other aspects involving microscope observation and photomicrography are discussed in detail by Smith [8].

B. Sieving

Sieves provide mechanical barriers allowing separation of particles on the basis of size down to 38 μm (400 mesh); subsieve material falls in the range from about 1 to 38 μm. A single sieve can be used to separate particles larger than a specified size from those which are smaller. This is done by placing a known amount of sample in the sieve, shaking for a defined time, and then collecting and weighing the sample amount that has passed through the sieve. In the commonly used U.S. Standard sieve series, sieves are classified according to the number of openings woven into wire cloth as shown in Table 3. The square sieve openings in this series increase in linear dimension by the square root of 2. The practical limit for sieve sizing is considered to be 38 μm, although progress has been made in developing small-mesh sieves and electrodeposited screens to analyze particles below this size. Sieving is more generally applicable to larger particles, i.e., above 75 μm, where microscopic characterization becomes increasingly difficult [9].

Table 3   Openings of Standard Sieves, U.S. Series

Assigned number	Sieve opening
2	9.5 mm
3.5	5.6 mm
4	4.75 mm
8	2.36 mm
10	2.00 mm
14	1.40 mm
16	1.18 mm
18	1.00 mm
20	850 μm
25	710 μm
30	600 μm
35	500 μm
40	425 μm
45	355 μm
50	300 μm
60	250 μm
70	212 μm
80	180 μm
100	150 μm
120	125 μm
200	75 μm
230	63 μm
270	53 μm
325	45 μm
400	38 μm

In practice, one often utilizes a series, or "nest," of sieves mounted with the coarsest sieve on top, followed by successively finer ones; after shaking a known amount of sample for a specified time, the amount of sample retained on each sieve is weighed, allowing the percentage of material in a given size range to be determined. To produce an acceptable particle size distribution, the total weight should be within 0.5% of the original mass, and a minimum of five sieves must be used [11]. For reporting of pharmaceutical particle size data, a recommendation has been made to include (1) sieve size, (2) mass collected on each sieve, (3) percentage of sample collected on each sieve, (4) cumulative percentage of sample retained on each sieve, and (5) percentage of sample passing each sieve [11].

Assuming a log-normal distribution, a plot of particle size versus the cumulative percentage of undersize particles can be used to determine the geometric mean weight diameter d_g', i.e., the size corresponding to the 50% value. A typical example of data treatment is shown in Fig. 11 [19]. Note that if the plot produces a straight line, the material can be considered a log-normal Gaussian distribution. A curved line or multiple line segments in the plot, however, indicate a polymodal distribution [11]. Another parameter of interest is σ_g, which can be calculated from the slope of the linear portion of the plot, as discussed in Section II. The value of σ_g is related to sample dispersity; a monodisperse system has a value of unity, while values greater than one correlate

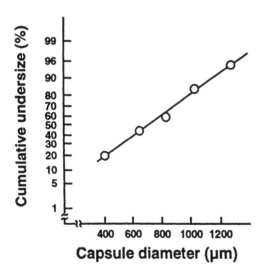

Fig. 11 Sieving data presented as a cumulative percentage undersize plot. The data were used to determine the geometric mean diameter of ethylcellulose microcapsules. (Adapted from Ref. 19.)

with polydispersity. A value approaching unity is usually desirable when milling or micronizing pharmaceutical powders to preserve product homogeneity [11].

Sieving is a rapid, convenient means of sizing larger particles such as granules or coarse excipients in the dry state. But while sieving is relatively simple to perform, several considerations should be kept in mind when evaluating or comparing results. These include characteristics of the sample: amount, particle shape, moisture content, and electrostatic attraction can all influence the ability of the particles to pass through mesh openings. Fine particles, i.e., those below 38 μm, can be particularly problematic as they tend to agglomerate; this tendency may be reduced by decreasing the amount of sample loaded.

Alternatively, wet sieving, which involves washing particles through the sieve with a liquid, can be carried out. Desai et al. have described a wet sieving procedure for chlorhydrol, a highly agglomerated material, which yields satisfactory results [20]. Another potential problem in sieve stacking is "near-size" particles [21]. These are coarse particles that are close in size to a mesh opening; they may or may not actually pass through the openings during the sieving procedure. This situation contributes to variability when comparing data from different sieve stacks.

The sieving procedure itself, particularly the manner and time of shaking, must also be consistent. Note that the U.S. Pharmacopeia (USP) specifies shaking times as well as sample weights depending on particle appearance (i.e., coarse versus fine) [22]. Additional recommendations regarding test sieves, sample size, sieving procedure, and determination of endpoint time have been proposed for sieving of pharmaceuticals [23]. Use of more than one operator often contributes to variation in results. Because hand sieving is inherently difficult to standardize, use of a mechanical sieve shaker may be preferable.

As an alternative to mechanical shaking of sieves, air-jet sieving is sometimes used. In this process, air flows are used to control particle movement through the mesh openings. The success of this technique, however, depends on the ability of the particles to be well dispersed in the air stream, which is difficult if the particles tend to agglomerate [23]. Regardless of the method of agitation, the sieves used should meet applicable standards and be calibrated. Calibration can be done with glass bead references, or by direct microscopic observation of the openings. The sieves should also be regularly examined for evidence of wear, and subjected to rigorous cleaning before a new sample is introduced.

The particle diameter obtained by sieving represents the minimum square aperture through which the particle can pass. It should be noted, however, that shape can also influence sieve particle separation; in general, particles will pass openings through based on their cross-sectional diameter rather than length. Sieving is also influenced by particle shape in a time-dependent manner [24]. As particle length increases, the passage time increases. This is because the elongated particle requires more time to change its orientation and pass through the rectangular mesh openings. Various modifications of the conventional sieving unit have been proposed to separate spherical particles from flat or elongated ones [25,26].

If sieving conditions are properly controlled, this method has useful applications in solid dosage form development. For example, Malmqvist and Nystrom used a sieve classification scheme to determine the presence of agglomerates formed during mixing of fine particles with coarser carrier particles [27]. With prolonged mixing, the large agglomerates disappear and provide an indication of ordered mixing. The same scheme was used to assess the effect of scale-up on mixing time, and it showed that mixing time was substantially reduced with increasing batch size [28]. Agglomeration is also of interest in fluid-bed coating of pellets. Johansson et al. have used sieving to identify agglomerates under various processing conditions, and they have shown a relationship between film yield and agglomeration [29].

C. Electrical Sensing Zone Methods

Electrical zone sensing, sometimes referred to as the Coulter principle, is the basis of operation underlying instruments such as the Coulter counter and the Elzone. In such systems, sample particles are first suspended in an electrically conducting medium. The suspension containing the particles of interest flows through a small orifice or aperture with an immersed electrode on either side (Fig. 12). The base resistance to the current between the electrodes is determined by the aperture size and electrolyte strength. As each sample particle enters the aperture, it displaces a volume of electrolyte solution equal to its own immersed volume, momentarily changing the resistance and creating an electrical pulse; i.e., the particle is "counted." Using an electrolyte solution of sodium chloride, the change in resistance ΔR is related to particle diameter d by

$$\Delta R = \frac{53d^3}{D^4} \tag{6}$$

where D is the aperture diameter [30].

The height of each pulse is thus proportional to particle volume; i.e.,

$$V = kt_L A I \tag{7}$$

where k is a calibration constant determined experimentally, t_L is the threshold volume, I is the aperture current, and A is the amplification factor. The sequence of the electronically amplified pulses (ca. 50,000) can be counted to produce a number versus particle volume distribution (or equivalent spherical distribution).

Electrical zone sensing counters such as the Coulter counter (Coulter Electronics, Hialeah, Fla.) and the Elzone (Particle Data, Inc., Elmhurst, Ill.) can analyze particles in the range of about 1200 to 0.4 μm. To cover this range,

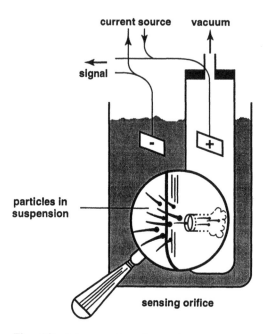

Fig. 12   Diagram of an electrical zone sensing apparatus.

several aperture sizes are needed, which span from 10 μm up to over 100 μm. Larger apertures allow data to be collected more rapidly, but they will count a smaller number of particles because sample concentration must be reduced. For a given electrical condition and aperture size, the instrument response will be linear with particle volume provided the particle to aperture diameter ratio is between 0.04 and 0.4 [31].

Sample preparation for electrical zone sensing involves suspending sample particles in a suitable electrolyte, i.e., one that has a resistivity in the range of 1 to 100 kΩ. Sodium chloride solutions are often used to suspend samples known to have poor solubility in water. Particles that do have appreciable solubility in water can be treated in several ways: Their aqueous solubility can be reduced by adding salts to promote a common ion effect, by changing the solution pH, or by using a nonaqueous solvent. Because most nonaqueous solvents have high electrical resistivities, the range of apertures that can be applied is more limited than with aqueous electrolytes, and the lower limit on particle size detection is inferior to what can be achieved with the latter [31].

The best candidates for electrical zone sensing analysis are those powders which can be easily dispersed in the electrolyte and do not readily agglomerate; thus, high particle loading in the dispersion should be avoided. The occurrence of agglomeration may often lead to nonreproducible results [32]. Sonication of the dispersion prior to analysis can be helpful in alleviating this problem. Karuhn [33] has discussed other strategies for deagglomeration. Another problem can occur if sample concentration is too high; two particles may occupy the same sensing zone simultaneously. This event is referred to as "coincidence," and it leads to a loss of particle counts. Coincidence can be minimized by using a more dilute sample concentration, or by reducing the aperture. In general, these instruments are best suited for measuring narrow distributions of particles; particles that are too large can block the aperture, while very small particles will not be detected. Particle shape can also affect results when the particles deviate significantly from a sphere, e.g., if the particles are flakes.

Electrical zone sensing is a well-established method for particle sizing. Numerous applications since the 1950s have been documented by one of the manufacturers [34]. These applications include a number of pharmaceutical systems, such as measurement of crystal growth in suspensions [35], monitoring of emulsion stability [36], and counting of particulates in parenteral formulations [37]. With respect to solid dosage forms, the technique has been used to assess milling of powders [38] and changes in excipients during tableting processes [39]. One of the most important applications involves characterization of drug dissolution behavior. Nystrom et al. have applied a Coulter counter to measure changes in size distribution of the poorly water-soluble compounds griseofulvin and hydrocortisone acetate [40]. By monitoring changes in the size distribution at frequent intervals, weights and surface areas of the undissolved particles could be calculated as functions of time. A limitation of this method is that the changes in the particle size distribution must be fairly large, and it will only be useful for compounds with aqueous solubility above 5 ppm. For compounds with lower solubility, surfactants such as polysorbate 80 can be used to enhance dispersion [41].

D. Light Scattering

A more detailed treatment of the theory underlying light scattering can be found in various references [42,43]. With respect to applications, a common type of commercial light scattering instrument is based on the theory of Fraunhofer diffraction. This principle is also referred to as static light scattering (SLS) or low-angle forward light scattering, and it applies to particles with dimensions larger than the wavelength of the incoming light. Developed in the 1970s, current examples of such systems are the Malvern Mastersizer (Malvern Instruments, Malvern, U.K.), and the Microtrac Full Range Analyzer (Leeds & Northrup, North Wales, Pa.). As shown in Fig. 13, a laser light passes through the particles (or droplets, in the case of a liquid sample), diffracting the intensity in an angular distribution (Iw). Mathematically, this can be described by the Airy equation,

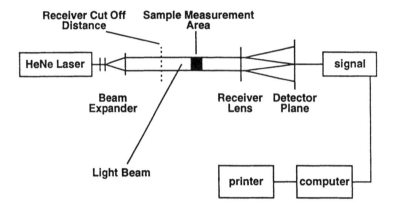

Fig. 13   Schematic diagram of a static light scattering (SLS) system.

$$I(w) = Ek^2A^4\left[\frac{J_1(kAw)}{kAw}\right]^2 \tag{8}$$

where E is the flux per unit area of the incident beam, $k = 2\pi/\lambda$, A is the particle radius, $w = \sin\phi$, with ϕ being the angle relative to the direction of the incident beam, and J_1 is a first-order Bessel function of the first kind.

A Fourier transform lens focuses the diffraction pattern on the detector, where it appears as light and dark concentric rings. The angle of diffraction decreases with increasing particle size, and the distribution of the diffracted light can be related to the distribution of particle sizes. This is done by repeatedly reading the individual patterns at the detector over time ("sweeps"), and summing the data. Normalization of the ring-detector with respect to total scattered light intensity allows calculation of volume distribution particle size. Fraunhofer diffraction–based instruments are applicable in the particle size range of 1.2 μm up to 1800 μm, depending on the lens used, although the standard working range for most instruments is roughly 2 to 300 μm [44]. At the high end of the size range, results should be interpreted with caution and compared with those obtained by another method, such as sieving [45].

Powder samples can be introduced into the laser in the dry form or, more commonly, in suspension. It is important to choose a suspension medium where the sample has low solubility, and to adequately disperse the sample. Addition of surfactants may be needed to facilitate dispersion and prevent flocculation, while sonication is often necessary to reduce aggregation. Sonication, however, must be applied judiciously to avoid breaking individual particles. Microscopy is helpful in establishing a suitable sonication time. Drug loading is also important; higher loading of samples has been shown to increase agglomeration.

For particles under 5 μm, which are close to the wavelength of the light, another type of light scattering may be more useful. This type is known variously as photon correlation spectroscopy (PCS), dynamic light scattering, or light beating spectroscopy, and it is based on a combination of Fraunhofer diffraction with the more complex Mie theory. An experimental set up of such an instrument is shown in Fig. 14. In this technique, the small Doppler frequency shifts of the particles being scattered are related to Brownian motion of the particles, i.e., diffusion due to collisons with molecules in the surrounding medium. Autocorrelation of the photomultiplier signal leads to an exponentially decaying curve. The time constant of the curve is related to the diffusion coefficient of the particles (D), which in turn can be used to determine particle size from the Stokes–Einstein equation:

$$D = \frac{kT}{6\pi\eta R} \tag{9}$$

where k is Boltzmann's constant, T is the temperature, η is the viscosity of the solvent, and R is the hydrodynamic radius of the particle.

It should be emphasized that this equation holds only for a sphere. When it is applied to irregularly shaped particles, significant errors can result. Photon correlation spectroscopy is best suited for narrow distributions of particles, although various mathematical treatments have been developed for analyzing broad distributions [46]. Photon correlation spectroscopy is applicable for

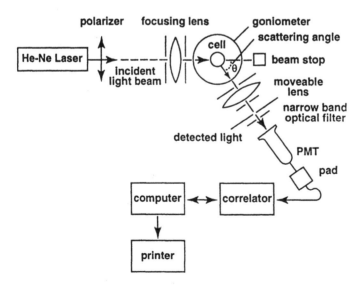

Fig. 14 Schematic diagram of a dynamic light scattering system.

particles ranging from 5 μm down to 0.001 μm, and it has been widely used to characterize a number of pharmaceutical samples, including aerosols [47], colloids [48], emulsions [49], and liposomes [50].

Like microscopy and electrozone sensing, light-scattering methods can be applied to measure a wide range of particles. Additionally, they are absolute methods requiring no calibration and are rapid to perform. Light scattering also offers the ability to evaluate polydispersity of samples in addition to their size. These methods are not affected by sample concentration, provided that the samples are sufficiently dilute to minimize interparticle interactions and multiple scattering effects. One should be aware, however, that size values obtained by light scattering do not necessarily coincide with those obtained by other techniques. For example, in a study of grinding by ball-milling, the feed size measured by SLS differed from coarse as well as fine size fractions measured by sieving by as much as 10% [51]. And, in a comparison of the Coulter counter to SLS for sizing of boron powders, SLS results were found to oversize those of the Coulter counter by as much as 30% [52]. In general, light scattering tends to be biased in favor of particles of larger diameter. Another limiting factor in light-scattering methods is particle shape. Determinations done on irregular and elongated particles are often inaccurate due to orientation effects produced by the particles [53].

In the case of particles in the range of 1–10 μm, it has been argued that SLS gives inaccurate results due to a breakdown of the Fraunhofer assumption; use of the particle's optical parameters (refractive index, absorptivity) allows calculation of more accurate results [54]. An analogous correction has been developed for PCS measurements below 5 μm [55]. With regard to which technique should be used to size particles in the overlapping range of 1–5 μm, a comparative study conducted with latex spheres suggests that PCS has higher resolution than SLS [56]. On the other hand, PCS measurements can be more time consuming than SLS if one wants to obtain accurate peak widths for distributions. Furthermore, PCS requires great care in sample preparation and analysis in order to minimize the effects of dust.

V. APPLICATIONS

A. Importance of Particle Size in Dissolution

The relationship between surface area and dissolution rate is described by the Noyes–Whitney equation,

$$\frac{dC}{dt} = \frac{D}{h} \times S \times (C_s - C) \tag{10}$$

where dC/dt is the dissolution rate at time t, S is the surface area, $C_s - C$ is the concentration gradient, and D is the diffusion coefficient over the boundary layer thickness h [57]. In dissolution rate-limited absorption, we can assume that D and h are relatively constant and that C is $<<C_s$. Therefore,

$$\frac{dC}{dt} = k \times S \times C_s \tag{11}$$

Reduction of particle size increases the total specific surface area exposed to the solvent, allowing a greater number of particles to dissolve more rapidly. Furthermore, smaller particles have a small diffusion boundary layer, allowing faster transport of dissolved material from the particle surface [58]. These effects become extremely important when dealing with poorly water-soluble drugs, where dissolution is the rate-limiting step in absorption. There are numerous examples where reduction of particle size in such drugs leads to a faster dissolution rate [59–61]. In some cases, these in vitro results have been shown to correlate with improved absorption in vivo [62–64].

As an example of how particle size can effect dissolution and bioavailability, Yakou et al. [64] studied the dissolution of two batches of phenytoin; one contained particles in the range of 177–350 μm, while the other batch had particles in the range of 74–350 μm. The initial dissolution rate of the latter was almost twice that of the batch with larger particles. The results of a crossover study in humans using these same batches is shown in Fig. 15. The batch containing the smaller particles gives a significantly higher bioavailability in terms of the area under the curve (AUC).

When considering reduction of particle size, one should be aware of potential instability enhancement. It is well known that grinding can reduce crystallinity [65]; the resulting noncrystalline material may be more hygroscopic

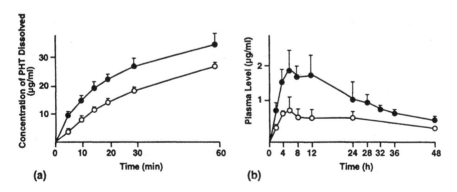

Fig. 15 Effect of phenytoin (PHT) particle size on (a) dissolution in simulated gastric fluid, pH 1.2, 37°C; and (b) bioavailability in human volunteers after oral administration of 300 mg PHT. Dark circles: lot with particle size range 74–350 μm; open circles: lot with particle size range 177–350 μm. (From Ref. 64.)

[65], or susceptible to chemical decomposition [66]. Also, grinding may facilitate unwanted polymorphic conversion [67]. And due to cohesiveness, the smaller particles have a greater tendency to aggregate and adhere to container surfaces. The aggregation tendency may be decreased by addition of a detergent or by sieving. The distribution of particle sizes should also be considered. Simulation and experimental studies suggest that an increased width of distribution can slow dissolution [68].

An increased rate of dissolution resulting from particle size reduction has also been observed for several excipients. For example, griseofulvin systems containing ethylcellulose of size fraction 710–850 μm released the drug almost 25% faster than the same systems containing ethylcellulose of size fraction 1000–2000 μm. The authors interpret these results in terms of more excipient particles being available to interact with the drug in the fraction of lower size [69].

Less frequently, increasing drug particle size can lead to an improved rate of dissolution. In the case of theophylline tablets, those containing 50 or 30 μm particles gave a faster dissolution rate than those of 10 μm [70]. Electron microscopy studies showed the presence of agglomerates in the tablet with the smallest particles, which may explain its slower dissolution rate. In general, agglomerated particles are undesirable; they are expected to reduce surface area, leading to slower dissolution and decreased bioavailability.

B. Effects of Particle Size on Manufacturing Processes

Particle size is often an important processing variable in pharmaceutical manufacturing, and efforts have been made to predict particle size distributions resulting from milling [71,72] and mixing [73]. Other processes of interest that can be influenced by particle size include granulation, spray-drying, compression, and capsule filling. In some cases the particle size of the drug substance itself can be a factor. Such is the case for indomethacin when formulated as controlled release pellets; batches with micronized, i.e., smaller, particles gave a greater yield of product and higher drug load than the batches containing the larger, "as is" particles [74]. Among the micronized batches, differences in particle size distribution continued to influence product yield; particles with a narrow distribution gave better yields than those with a wide distribution, presumably because the latter contained a significant number of larger particles.

Drug particle size can also be a factor in mixing uniformity. When particles of different sizes are mixed, the smaller particles tend to travel or "percolate" downward. This tendency is a particular problem with low-dose drugs, where a small amount of drug is to be mixed with large amounts of diluents. In this situation, it is more difficult to achieve uniform mixing, and undesirable deviations in drug content in the solid dosage form can arise. Various approaches have been used to overcome size aggregation [75,76]. One strategy is to use ordered mixtures, where a carrier material such as lactose is covered with a monolayer of fine particles of drug. Micronization of the drug particles helps to maximize the amount of drug absorbed on the carrier. It has been recommended that the carrier particle size be as large as possible and of a narrow size range to improve homogeneity [77].

Concerns about content uniformity of low-dose drugs have led to the implementation of particle size specifications. If the mean particle size and coefficient of variance of a tablet are known, the minimum tablet dose capable of passing the USP content uniformity test can be determined [78]. Since size reduction increases the total number of particles available for mixing, it would seem reasonable that making the particles as fine as possible would alleviate content uniformity problems. For example, reducing the mean particle size of medazepam from 156 to 82 μm improved tablet uniformity [79]. Further reducing the particle size to 32 μm, however, significantly decreased uniformity. This is because smaller particles tend to be more cohesive and promote agglomeration, which makes mixing more difficult. These particles tend to exhibit a positively skewed or even bimodal distribution. The effect of this inhomogeneity on ethinyloestradiol content in the resulting tablets has been systematically studied by Sallam and Orr [80].

Considerable interest has been focused on the relationship of granule size to tablet properties. A recent paper presents results from evaluation of paraceto-mol tablets prepared with different size fractions [81]. There appears to be a critical granule size range for properties such as flow rate, drug content, hardness, and disintegration time. Figure 16 illustrates this effect with respect to tablet hardness. Granule particle size can also affect content uniformity; in a study involving granulation of lactose and chlorpheniramine maleate solutions, a higher percent deviation from the mean drug content was found with smaller granules [82]. This is ascribed to fines of lactose particles adhering to the filter bag during granulation. Lactose particle size has also been shown to be a significant influence on the spheronization of extrudate, which is used to make spherical pellets. Substitution of coarse lactose for fine lactose produces unsatisfactory results due to agglomeration of the granules [83]. Because of the sensitivity of granule size to formulation ingredients such as lactose, computer optimization techniques have been developed to identify the formula consistent with the desired size range [84].

Flow behavior of powders is also of interest in direct compression. It is generally accepted that the flow rate initially increases with particle size, achieves a maximum in the range of 100–400 μm, and then decreases [85]. An excipient that has been well characterized is lactose, which undergoes particle fragmentation when compacted. For α-lactose monohydrate, it has been shown that the

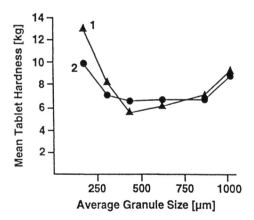

Fig. 16 Effect of paracetamol granule size on tablet hardness. Curve 1: paracetamol granules with no added surfactant; curve 2: paracetamol granules with 0.2% w/v sodium lauryl sulfate. (From Ref. 81.)

tendency to fragment upon compaction decreases with increasing particle size [86,87]. In a study of compaction on different types of lactose, consolidation decreased with decreasing particle size, yielding tablets of higher porosity [88].

The physical characteristics of various celluloses, including microcrystal-line celluloses (Avicel and Emocel), have also been examined. In contrast to lactose, these substances exhibit plastic deformation on compression. In this case, no clear correlation could be established between cellulose particle size and mechanical properties of the compressed tablets [89]. It has been suggested, however, that cellulose containing large, roughly spherical agglomerates may have superior tableting properties to Avicel and Emocel due to the former's increased porosity. Also, Khan and Pilpel [90] have reported that increasing the particle diameter of Avicel results in lower tensile strength. The rationale for this is that larger particles have fewer sites for formation of interparticle bonding.

Particle size of the drug substance itself can also affect direct compression tablet properties. An interesting example of this is the antidiabetic drug tolbutamide [91]. Here, agglomerated crystals of the drug (15–30 μm) were actually preferred over the bulk drug (300 μm) for formulation, due to their superior tensile strength. Apparently, the smaller particle size allows greater interparticle contact points per unit area, facilitating compression.

VI. CONCLUSIONS

All of the methods described have applications to pharmaceutical systems. Choice of the "best" method for particle sizing will depend on several factors. These include the range of particle sizes to be examined, the time of analysis, and the cost of equipment. The chemical and physical properties of the samples themselves are also important, such as solubility (to prepare suitable suspensions for electrical zone sensing and light-scattering studies), refractive index (useful information for microscopy and light scattering), and stability to grinding and sonication. In the case of electron microscopy, the sample's ability to withstand heat and vacuum conditions should also be assessed. Finally, an initial, qualitative knowledge of the particle size range and shape(s) present in the sample is helpful. This information can be most easily obtained with an optical microscope. The microscopy observations can then be used to determine the best method(s) for more quantitative work.

ACKNOWLEDGMENTS

The expert assistance of Cathy Del Tito and Wendy Crowell in preparing the illustrations is gratefully acknowledged.

REFERENCES

1. L. Silverman, C. Billings, and M. First, *Particle Size in Industrial Hygiene* Academic Press, New York, 1971.
2. T. Itoh and Y. Wanibe, *Powder Metallurgy, 34*, 126 (1991).
3. H. Heywood, *Chem. Ind., 56*, 149 (1937).
4. T. Hatch and S. Choate, *J. Franklin Institute, 207*, 369 (1929).
5. K. Ogawa, *Part. Part. Syst. Charact., 7*, 127 (1990).
6. M. Ochiai and R. Ozao, *Thermochim. Acta, 198*, 279 (1992).
7. T. Allen, *Particle Size Measurement*, 4th ed., Chapman and Hall, New York, 1990.
8. R. F. Smith, *Microscopy and Photomicrography: A Working Manual*, CRC Press, Boca Raton, Fla., 1990.
9. G. E. Amidon, *Pharmacopeial Forum, 18*, 4089 (1992).
10. G. Yamate and J. D. Stockham, in *Particle Size Analysis*, J. D. Stockham and E. G. Fochtman, eds., Ann Arbor Science, Ann Arbor, Mich., 1978, pp. 23–33.
11. H. Brittain, *Pharmacopeial Forum, 19*, 4640 (1993).
12. American Society for Testing and Materials (ASTM), "Recommended Practice for Analyses by Microscopical Methods," ASTM: Part 23 and 30, Philadelphia, Pa., 1973.
13. S. Inoue, *Am. Biotech. Lab., 4*, 8 (1989).
14. J. Graf, in *Particle Size Analysis*, J. D. Stockham and E. G. Fochtman, eds., Ann Arbor Science, Ann Arbor, Mich., 1978, pp. 35–44.
15. M. Bergeron, P. Laurin, and R. Tawashi, *Drug Dev. Ind. Pharm., 12*, 915 (1986).
16. A. J. Maeder and N. N. Clark, *Powder Technology, 68*, 137 (1991).
17. T. A. Barber, M. D. Lannis, and J. G. Williams, *J. Parenteral Sci. Technol., 43*, 27 (1989).

18. J. P. Zingerman, S. C. Mehta, J. M. Salter, and G. W. Radebaugh, *Int. J. Pharm.*, *88*, 303 (1992).
19. R. Senjkovic and I. Jalsenjak, *J. Microencapsulation*, *1*, 241 (1984).
20. R. D. Desai, B. Toth, and R. Somkaite, *Drug and Cosmetic Industry*, December, 50 (1972).
21. S. D. Cowley and S. J. Morris, *Powder Bulk Engineering*, February, 34 (1991).
22. *USP XXII*, United States Pharmacopeial Convention, Rockville, Md., 1989, p. 1602.
23. K. Marshall, *Pharmacopeial Forum*, *19*, 4637 (1993).
24. S. Rajpal, L. Hua, C. R. Chang, and J. K. Beddow, *Part. Charact. Technol. 2*, 193 (1984).
25. M. Furuuchi and K. Gotoh, *Powder Technol.*, *73*, 1 (1992).
26. M. Whiteman and K. Ridgway, *Drug. Dev. Ind. Pharm.*, *12*, 1995 (1986).
27. K. Malmqvist and C. Nystrom, *Acta Pharm. Suec.*, *21*, 9 (1984).
28. K. Malmqvist and C. Nystrom, *Acta Pharm. Suec.*, *21*, 30 (1984).
29. M. E. Johansson, A. Ringberg, and M. Nicklasson, *J. Microencapsulation*, *4*, 217 (1987).
30. J. T. Carstensen, *Theory of Pharmaceutical Systems*, Vol. 11, Academic Press, New York, 1973, pp. 23–24.
31. J. Graf, in *Particle Size Analysis*, J. D. Stockham and E. G. Fochtman, eds., Ann Arbor Science, Ann Arbor, Mich., 1977, pp. 65–75.
32. J. Spence, *Anal. Proc. (London)*, *18*, 509 (1981).
33. R. F. Karuhn, in *Particle Size Distribution II*, T. Provder, ed., American Chemical Society, Washington, D.C., 1991, p. 386.
34. Anonymous, "Coulter Counter—Industrial Bibliography," Coulter Electronics Ltd., Bedfordshire, England, May 1986.
35. J. E. Carless and A. A. Foster, *J. Pharm. Pharmacol.*, *18*, 697 (1966).
36. A. Takamura, F. Ishii, S. Noro, M. Tanifuji, and S. Nakajima, *J. Pharm. Sci.*, *73*, 91 (1984).
37. T. F. Stokes, E. D. Sumner, and T. E. Needham, *Am. J. Hosp. Pharm.*, *32*, 821 (1975).
38. M. I. Barnett and E. E. Sims, *Int. J. Pharm. Tech. Prod. Mfr.*, *4*, 59 (1983).
39. E. Shotton and G. S. Leonard, *J. Pharm. Pharmacol.*, *24*, 798 (1972).
40. C. Nystrom, J. Mazur, M. I. Barnett, and M. Glazer, *J. Pharm. Pharmacol.*, *37*, 217 (1985).
41. C. Nystrom and M. Bisrat, *J. Pharm. Pharmacol.*, *38*, 420 (1986).
42. M. Kerker, *The Scattering of Light and Other Electromagnetic Radiation*, Academic Press, New York, 1969.
43. H. C. Van De Hulst, *Light Scattering by Small Particles*, John Wiley & Sons, New York, 1957, pp. 383–393.
44. H. N. Frock, in *Particle Size Distribution: Assessment and Characterization*, T. Provder, ed., American Chemical Society, Washington, D.C., 1987, pp. 146–160.
45. A. M. Juppo, J. Yliruusi, L. Kervinen, and P. Strom, *Int. J. Pharm.*, *88*, 141 (1992).
46. B. B. Weiner and W. W. Tscharnuter, in *Particle Size Distribution: Assessment and Characterization*, T. Provder, ed., American Chemical Society, Washington, D.C., 1987, pp. 48–61.

47. M. M. Clay, D. Pavia, S. P. Newman, and S. W. Clarke, *Thorax*, *38*, 755 (1983).
48. J. J. Torrado, L. Ilum, and S. S. Davis, *Int. J. Pharm.*, *51*, 85 (1989).
49. D. Attwood, C. Mallon, and C. J. Taylor, *Int. J. Pharm.*, *88*, 427 (1992).
50. M. L. Shively, *J. Colloid. Interface Sci.*, *155*, 66 (1993).
51. O. Orumwense, *Powder Technol.*, *73*, 101 (1992).
52. J. A. Davies and D. S. Collins, *Part. Part. Syst. Charact.*, *5*, 116 (1988).
53. H. Kanerva, J. Kiesvaara, E. Muttonen, and J. Yliruusi, *Pharm. Ind.*, *55*, 849 (1993).
54. U. Tuzun and F. A. Farhadpour, *Part. Charact.*, *3*, 151 (1986).
55. P. Bowen, J. A. Dirksen, R. Humphrey-Baker, and L. Jelinek, *Powder Technol.*, *74*, 67 (1993).
56. E. Gulari, A. Annapragada, E. Gulari, and B. Jawad, in *Particle Size Distribution: Assessment and Characterization*, T. Provder, ed., American Chemical Society, Washington, D.C., 1987, pp. 133–145.
57. A. A. Noyes and W. R. Whitney, *J. Am. Chem. Soc.*, *19*, 930 (1897).
58. M. Bisrat and C. Nystrom, *Int. J. Pharm.*, *47*, 223 (1988).
59. J. A. K. Lauwo, *Drug Dev. Ind. Pharm.*, *11*, 1565 (1985).
60. A. Narukar, P.-C. Sheen, E. L. Hurwitz, and M. A. Augustine, *Drug Dev. Ind. Pharm.*, *13*, 319 (1987).
61. R. Voigt, K. Etiz, and M. Bornschein, *Pharmazie*, *42*, 22 (1987).
62. G. T. McInnes, M. J. Asbury, L. E. Ramsay, and J. R. Shelton, *J. Clin. Pharmacol.*, *22*, 410 (1982).
63. S. Yakou, Y. Yajima, T. Sonobe, M. Sugihara, and Y. Fukuyama, *Chem. Pharm. Bull.*, *30*, 319 (1982).
64. S. Yakou, S. Yamazaki, T. Sonobe, M. Sugihara, K. Fukomuro, and T. Nagai, *Chem. Pharm. Bull.*, *34*, 4400 (1986).
65. M. Otsuka and N. Kaneniwa, *Chem. Pharm. Bull.*, *32*, 1071 (1984).
66. G. Ovcharova and R. Nacheva, *Antibiotiki*, *29*, 166 (1984).
67. K. C. Lee and J. A. Hersey, *J. Pharm. Pharmacol.*, *29*, 249 (1977).
68. R. J. Hintz and K. C. Johnson, *Int. J. Pharm.*, *51*, 9 (1989).
69. H. Ibrahim, E. Sallam, M. Takieddin, and M. A. Shabat, *Drug Dev. Ind. Pharm.*, *14*, 1249 (1988).
70. J. L. Montel, G. Mulak, J. Cotty, F. Chanoine, and V. Rovei, *Drug Dev. Ind. Pharm.*, *9*, 399 (1983).
71. J. J. Motzi and N. R. Anderson, *Drug Dev. Ind. Pharm.*, *10*, 915 (1984).
72. A. Devaswithin, B. Pitchumani, and S. R. DeSilva, *Ind. Eng. Chem. Res.*, *27*, 723 (1988).
73. F. K. Y. Lai and J. A. Hersey, *Int. J. Pharm.*, *36*, 157 (1987).
74. S. P. Li, K. M. Feld, and C. R. Kowarski, *Drug Dev. Ind. Pharm.*, *15*, 1137 (1989).
75. H. Egermann, *Pharm. Acta Helv.*, *60*, 322 (1985).
76. M. Alonso, M. Satoh, and K. Miyanami, *Powder Technol.*, *68*, 145 (1991).
77. E. A. Sallam, A. Badwan, and M. Takieddin, *Drug Dev. Ind. Pharm.*, *12*, 1731 (1986).
78. S. H. Yalkowsky and S. Bolton, *Pharm. Res.*, *7*, 962 (1990.
79. H. Egermann, *Paperback APV*, *3*, 29 (1981).

80. E. Sallam and N. Orr, *Drug Dev. Ind. Pharm.*, *12*, 2015 (1986).
81. M. N. Femi-Oyewo and A. O. Adefeso, *Pharmazie*, *48*, 120 (1993).
82. L. S. C. Wan, P. W. S. Heng, and G. Muhari, *Int. J. Pharm.*, *88*, 159 (1992).
83. K. E. Fielden, J. M. Newton, and R. C. Rowe, *Int. J. Pharm.*, *81*, 205 (1992).
84. O. Shirakura, M. Yamada, M. Hashimoto, S. Ishimaru, K. Takayama, and T. Nagai, *Drug Dev. Ind. Pharm.*, *17*, 471 (1991).
85. A. R. Fassihi and I. Kanfer, *Drug Dev. Ind. Pharm.*, *12*, 1947 (1986).
86. G. Alderborn and C. Nystrom, *Powder Technol.*, *44*, 37 (1985).
87. H. Vromans, A. H. de Boer, G. K. Bolhuis, C. F. Lerk, and K. D. Kussendrager, *Drug Dev. Ind. Pharm.*, *12*, 1715 (1986).
88. K. A. Riepma, K. Zuurman, G. K. Bolhuis, A. H. de Boer, and C. F. Lerk, *Int. J. Pharm.*, *85*, 121 (1992).
89. T. Pesonen and P. Paronen, *Drug Dev. Ind. Pharm.*, *16*, 31 (1990).
90. N. F. Khan and N. Pilpel, *Powder Technol.*, *48*, 145 (1986).
91. A. Sano, T. Kuriki, Y. Kawashima, H. Takeeuchi, T. Hino, and T. Niwa, *Chem. Pharm. Bull.*, *40*, 1573 (1992).

7

X-Ray Powder Diffractometry

Raj Suryanarayanan
College of Pharmacy, University of Minnesota, Minneapolis, Minnesota

I. INTRODUCTION

X-rays are electromagnetic radiation lying between ultraviolet and gamma rays in the electromagnetic spectrum. The wavelength of x-rays is expressed in angstrom units (Å); 1 Å is equal to 10^{-8} cm.

Diffraction is a scattering phenomenon. When x-rays are incident on crystalline solids, they are scattered in all directions. In some of these directions, the scattered beams are completely in phase and reinforce one another to form the diffracted beams [1,2]. Bragg's law describes the conditions under which this would occur. It is assumed that a perfectly parallel and monochromatic x-ray beam, of wavelength λ, is incident on a crystalline sample at an angle θ. Diffraction will occur if

$$n\lambda = 2d \sin \theta \tag{1}$$

where d = distance between the planes in the crystal, expressed in angstrom units, and n = order of reflection (an integer).

X-ray powder patterns can be obtained using either a camera or a powder diffractometer. Currently, diffractometers find widespread use in the analysis of pharmaceutical solids. The technique is usually nondestructive in nature. The theory and operation of powder diffractometers is outside the purview of this chapter, but these topics have received excellent coverage elsewhere [1,2]. Instead, the discussion will be restricted to the applications of x-ray powder diffractometry (XPD) in the analysis of pharmaceutical solids. The U.S. Pharmacopeia (USP) provides a brief but comprehensive introduction to x-ray diffractometry [3].

II. QUALITATIVE ANALYSIS

X-ray powder diffractometry is widely used for the identification of solid phases. The x-ray powder pattern of every crystalline form of a compound is unique, making this technique particularly suited for the identification of different polymorphic forms of a compound. The technique can also be used to identify the solvated and the unsolvated (anhydrous) forms of a compound, provided the crystal lattices of the two are different [4]. The technique has limited utility in the identification of noncrystalline (amorphous) materials. Their powder patterns consist of one or more broad diffuse maxima.

A. Reference Diffraction Patterns

The International Centre for Diffraction Data (ICDD, Newtown Square, Pa.) maintains a collection of single-phase x-ray powder patterns [5]. There are separate listings of inorganic and organic compounds. The x-ray diffraction data of ibuprofen, as a representative example, is given here (Fig. 1).

The card pattern contains the Powder Diffraction File (PDF) number (region 1), quality mark of the data (region 2), the chemical formula and the specimen name (region 3), the experimental conditions under which the powder pattern was obtained and the source of the data (region 4), physical data, which includes crystallographic system, space group, lattice parameters, and interaxial angles (region 5), general comments, Crystal Data cell (if different from that reported in region 5), Pearson Symbol Code (PSC), Merck Index number, etc. (region

1 ← **32-1723**   * → 2

$C_{13}H_{18}O_2$			

3 ← Ibuprofen

Rad. CuKα_1 λ 1.5406   Filter Ni   d-sp
Cut off   Int. Diffractometer I/I_{cor}.

4 ← Ref. Gong, P., Polytechnic Institute of New York, Brooklyn, New York, USA, *JCPDS Grant-in-Aid Report*, (1981)

Sys. Monoclinic   S.G. P2$_1$/c   (14)
a 14.667   b 7.899   c 10.731   A 1.8568   C 1.3585

5 ← α β 99.46 γ Z 4   mp

Ref. Ibid.

D$_x$ 1.117   D$_m$ SS/FOM   F30=33.0 (.0114,80)

Color White
Reported by McConnell, *Cryst. Struct. Commun.*, 3

6 ← 73 (1974) as: a=14.667, b=7.886, c=10.730, β=99.36,

Space Group=P2$_1$/c, Z=4. Silicon used as internal standard. PSC: mP132. Merck Index, 9th Ed., 4796. To replace 30-1757.

d Å	Int	hkl	d Å	Int	hkl
14.41	85	100	2.886	4	$\bar{3}13$
7.24	15	200	2.792	2	213
6.93	9	$\bar{1}10$	2.665	2	$\bar{4}21$
6.33	15	011	2.644	3	004
6.02	10	$\bar{1}11$	2.533	6	$\bar{1}31$
5.34	100	210	2.504	4	$\bar{4}22$
5.01	45	$\bar{2}11$	2.437	3	$\bar{3}23$
4.73	10	102	2.409	2	114
4.65	15	$\bar{2}02$	2.379	4	502
4.55	30	211	2.280	1	512
4.40	75	012	2.193	2	232
4.12	2	$\bar{3}10$			
4.06	4	112			
3.973	40	202			
3.897	4	$\bar{3}02$			
3.811	2	120			
3.664	4	311			
3.617	4	400			
3.546	9	212			
3.466	4	$\bar{2}20$			
3.373	1	$\bar{2}21$			
3.290	4	410			
3.219	15	221			
3.053	3	320			
3.015	3	411			

→ 7

Fig. 1   The card pattern of ibuprofen. (Reproduced with permission of the copyrigh owner, ICDD, Newtown Square, Pa., from Ref. 6.)

6), and a table of interplanar spacings, relative intensities and the Miller indices (region 7).

There are also several indicators of the quality of the data. The highest quality data is "*" marked. To qualify for this mark, the chemistry of the compound must be well characterized. The intensities of the x-ray lines must be measured objectively and instrumentally, and there must be no unindexed, space group extinct or impurity lines. Lines with d-spacings ≤ 2.50 Å must retain at least three significant digits after the decimal point. To qualify for the "i" mark, there can be a maximum of two unindexed, space group extinct, or impurity lines provided none of these belong to the strongest eight. Again, there must be no serious systematic errors, and lines with d-spacings ≤ 2.00 Å must retain at least three significant digits after the decimal point. If the data is of low precision or if the data is due to a poorly characterized or multiphase system, an "O" mark is assigned. Patterns that do not meet the criteria for *, i, or O are left blank. When the powder pattern is calculated from structural parameters, the pattern is marked "C." Extensive details about the quality mark guidelines can be found in ICDD publications [7].

B. Sample Preparation

In order to obtain reliable and reproducible results, it is necessary to give careful consideration to sample preparation. The critical issues here are (1) the crystallite size, (2) preferred orientation of the powder sample in the holder, (3) the method by which the sample is filled in the holder, and (4) coplanarity of the sample and holder surface [1]. The issues relating to crystallite size will be discussed in Section IV.D.1. When a powder sample is packed in an x-ray holder, the distribution of crystal orientations can be nonrandom, and this condition is referred to as preferred orientation [2]. Preferred orientation of the particles must be minimized. The way the sample is filled in the holder affects the orientation of the crystallites. The "standard" holders are rectangular aluminum or glass plates, containing a rectangular window in which the powder is packed. The powder loading can be done from the front (top), edge, or back [1]. Studies with mineral powders suggest that preferred orientation problems can be minimized by filling the sample from the side [8]. X-ray sample holders made of quartz or silica, with very low background noise, are now commercially available. These holders can be used to enhance sensitivity and also to analyze small amounts of sample.

C. Phase Identification

Six crystalline solid phases of fluprednisolone and an amorphous phase were characterized using XPD, infrared (IR) spectroscopy, and differential scanning calorimetry [9]. Three of these six crystalline phases were anhydrous (forms I,

II, and III), two were monohydrates (termed α-monohydrate and β-monohydrate) and one was a *tert*-butylamine disolvate. The differences in the powder patterns of the phases were readily evident (Table 1). This study demonstrates the unique ability of x-ray diffractometry for the identification of (1) anhydrous compound existing in both crystalline and amorphous states, (2) different polymorphic forms of the anhydrate, (3) the existence of solvates where the solvent of crystallization is water (hydrate) or an organic solvent (in this case, *tert*-butylamine), and (4) polymorphism in the hydrate.

Depending on the water vapor pressure, cephalexin can exist as an anhydrate ($C_{16}H_{17}N_3O_4S$), a monohydrate ($C_{16}H_{17}N_3O_4S \cdot H_2O$) or a dihydrate ($C_{16}H_{17}N_3O_4S \cdot 2H_2O$) at 25°C [10]. The monohydrate and the dihydrate were characterized by the pronounced differences in their powder x-ray diffraction patterns. Thus, x-ray diffractometry can be used to characterize several hydrated states of a compound.

When stored under increasing relative humidities (RH), cromolyn sodium absorbed water, resulting in a continuous series of interstitial solid solutions [11]. The amount of water absorbed was proportional to the relative humidity of the environment and could be up to about nine molecules of water per molecule of drug. Such an unusual system was characterized by combining XPD with single-crystal x-ray studies. The unit cell parameters of cromolyn sodium were obtained from single-crystal x-ray studies, and this permitted the authors to index the powder pattern. The *b* axis spacing was found to increase dramatically as a function of the relative humidity up to 20% RH (Table 2). Above 40% RH, the unit cell dimensions were nearly constant.

In addition to the ICDD, publications dealing solely with the powder patterns of drugs appear occasionally [12–15]. In 1971, Sadik et al. pointed out that the identification test for kaolin (in NF XIII) was a test for the presence of aluminum, and therefore both kaolin and bentonite gave positive results [16]. Since the two compounds have different crystal structures, their x-ray diffraction patterns are different, and therefore XPD was recommended for identification of these compounds. In the current edition of USP, the identification of bentonite is based on its powder x-ray pattern [3].

X-ray powder diffractometry can be used for the identification not only of the active ingredient, but also of the excipients in a formulation. For example, computer-based fingerprint data are being developed in order to determine whether or not a manufactured product was an FDA-approved formulation [17]. The analytical techniques used in this study included Fourier transform infrared (FTIR) spectroscopy, thermogravimetric analysis, and XPD. X-ray powder diffractometry has also been used to detect and identify impurities in pharmaceutical formulations. The identification of barium sulfate crystals, which occurred as contaminants in an injectable solution, was achieved through polarizing light and scanning electron microscopy, Raman spectroscopy, and

Table 1   The Powder X-Ray Diffraction Patterns of Various Crystalline Phases of Fluprednisolone

Form I		Form II		Form III	
d-spacing (Å)	I/I_0	d-spacing (Å)	I/I_0	d-spacing (Å)	I/I_0
10.3940	11.67	8.8378	20.00	8.9624	38.76
9.0173	28.46	7.8584	18.18	8.4181	17.98
8.3784	39.05	7.1608	55.45	6.3203	39.55
7.1320	9.49	6.4114	48.18	6.1035	44.94
5.9604	33.94	6.0619	98.18	5.5004	100.00
5.4835	100.00	5.6041	100.00	5.0348	91.01
5.1510	12.04	5.2111	56.36	4.7036	19.10
4.7410	19.85	5.0348	38.18	4.5026	14.83
4.4577	20.80	4.4801	57.27	4.2168	21.91
4.2168	19.49	4.0367	34.54	4.0550	10.11
3.7666	10.22	3.9783	20.91	3.9309	4.49
3.5448	9.49	3.5587	24.54	3.6746	19.10
3.4112	11.68	3.4502	16.36	3.5309	12.92
3.0557	7.66	3.2348	18.18	3.4241	15.17
				3.2406	20.22
				3.0455	21.35

α-monohydrate		β-monohydrate		*tert*-butylamine disolvate	
d-spacing (Å)	I/I_0	d-spacing (Å)	I/I_0	d-spacing (Å)	I/I_0
9.9270	58.12	7.8584	50.76	15.3580	100.00
7.4305	31.11	7.1897	14.36	7.6220	7.69
6.4816	99.49	6.4022	35.90	6.8042	6.25
6.2317	68.38	5.6041	100.00	6.0209	16.34
5.6041	72.65	5.3358	49.74	5.2789	36.06
4.8834	100.00	4.9511	51.28	4.6962	8.65
4.6068	50.43	4.5950	4.10	4.3287	10.09
3.9481	24.78	4.3496	13.33	4.1486	8.65
3.7354	35.04	4.2069	10.25	4.0008	3.85
3.5378	31.62	4.0008	14.87	3.7903	5.96
3.2406	27.35	3.8885	21.54	3.5309	4.61
3.0660	14.53	3.7509	15.38		
		3.5657	30.76		
		3.3857	16.41		
		3.2406	14.36		
		3.1078	4.10		

Table 2 Lattice Parameters of Cromolyn Sodium as a Function of Relative Humidity

Percent relative humidity	Cell edge (Å)		
	a	*b*	*c*
0	11.5	14.0	3.58
10	11.8	15.0	3.62
20	11.9	15.5	3.64
30	11.9	15.7	3.66
40	11.9	15.7	3.67
51	11.9	15.8	3.68
76	11.9	15.8	3.70
88	11.9	15.8	3.70

Reproduced with permission of the copyright owner, American Pharmaceutical Association, from Ref. 11.

XPD [18]. Similarly, mineral impurities in talc were analyzed by polarizing light microscopy, differential thermal analysis, and XPD [19]. It must be recognized, however, that small amounts of crystalline impurities (usually <0.5% w/w) may not be detected by XPD. In case of noncrystalline impurities, much higher concentrations may be nondetectable.

D. Special Instrumentation

X-ray diffraction studies are usually carried out at room temperature under ambient conditions. It is possible, however, to perform variable-temperature XPD, wherein powder patterns are obtained while the sample is heated or cooled. Such studies are invaluable for identifying thermally induced or subambient phase transitions. Variable-temperature XPD was used to study the solid state properties of lactose [20]. Fawcett et al. have developed an instrument that permits simultaneous XPD and differential scanning calorimetry on the same sample [21]. The instrument was used to characterize a compound that was capable of existing in two polymorphic forms, whose melting points were 146°C (form II) and 150°C (form I). Form II was heated, and x-ray powder patterns were obtained at room temperature, at 145°C (form II had just started to melt), and at 148°C (Fig. 2; one characteristic peak each of form I and form II are identified). The x-ray pattern obtained at 148°C revealed melting of form II but partial recrystallization of form I. When the sample was cooled to 110°C and reheated to 146°C, only crystalline form I was observed. Through these experiments, the authors established that melting of form II was accompanied by recrystallization of form I.

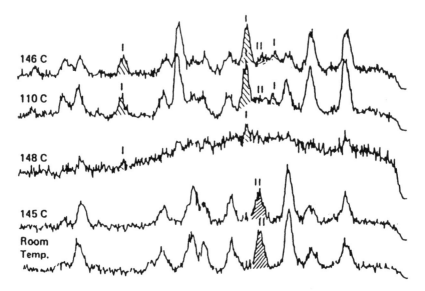

Fig. 2 The x-ray powder diffraction patterns obtained when the polymorphic form II of a pharmaceutical compound was heated to different temperatures. More details are provided in the text. (Reproduced with permission of the copyright owner, Plenum Press, from Ref. 21.)

E. Potential Problems

Most materials of pharmaceutical interest are organic compounds. Jenkins has comprehensively discussed the problems associated with the x-ray diffractometric analysis of organic materials [22]: (1) the unit cells of organic compounds are often large, resulting in low angle lines (large d-spacing). The accuracy of d-spacing data tends to decrease as the d-spacing increases, particularly when the d-spacing value is higher than 8 Å. (2) The unit cells are often of low symmetry, resulting in complex powder patterns. (3) Since organic compounds usually have low mass attenuation coefficients (the mass attenuation coefficient is discussed in Section IV.A.1), the x-rays can penetrate deep into the specimen, leading to large beam transparency errors [23]. (4) Finally, organic compounds are especially prone to preferred orientation. These problems have led to the development of the long-wavelength vacuum x-ray powder diffractometer [24].

III. DEGREE OF CRYSTALLINITY

Solids may be either crystalline or noncrystalline. The crystalline state is characterized by a perfectly ordered lattice, and the noncrystalline (amorphous) state is characterized by a disordered lattice. These represent two extremes of lattice order, and intermediate states are possible. The term *degree of crystallinity* is useful in attempts to quantify these intermediate states of lattice order.

X-ray powder diffractometry is widely used to determine the degree of crystallinity of pharmaceuticals. X-ray diffractometric methods were originally developed for determining the degree of crystallinity of polymers. Many polymers exhibit properties associated with both crystalline (e.g., evolution of latent heat on cooling from the melt) and noncrystalline (e.g., diffuse x-ray pattern) materials. This behavior can be explained by the two-state model, according to which polymeric materials consist of small but perfect crystalline regions (crystallites) that are embedded within a continuous matrix [25]. The x-ray methods implicitly assume the two-state model of crystallinity.

A. Theory

A rigorous procedure for determining the degree of crystallinity of polymers was developed by Ruland [26]. This method has found only limited application for determining the degree of crystallinity of pharmaceuticals [27,28]. The procedure developed by Hermans and Weidinger [29] is much simpler and is based on three assumptions [30]. First, it must be possible to demarcate and measure the crystalline intensity (I_c) and amorphous intensity (I_a) from the powder pattern. Usually, the integrated line intensity (area under the curve), rather than the peak intensity (peak height), is measured. Second, there is a proportionality between the experimentally measured crystalline intensity and the crystalline fraction (x_c) in the sample. Finally, a proportionality exists between the experimentally measured amorphous intensity and the amorphous fraction (x_a) in the sample.

Therefore,

$$x_c = pI_c \qquad (2)$$
$$x_a = qI_a \qquad (3)$$

where p and q are the proportionality constants. The sum of the crystalline and amorphous fractions is

$$x = x_c + x_a \qquad (4)$$

Combining Eqs. (2)–(4),

$$qI_a = x - pI_c \qquad (5)$$

or

$$I_a = \frac{x}{q} - \frac{pI_c}{q} \qquad (6)$$

The values of I_a and I_c can be determined for samples of varying degrees of crystallinity. A plot of the measured values of I_a against those of I_c will result in a straight line whose slope is p/q. The intercepts on the y and x axes will provide the intensity values of the 100% amorphous and 100% crystalline materials, respectively.

The degree of crystallinity (or percent crystallinity), x_{cr}, is given by

$$x_{cr} = \frac{x_c \times 100}{x} = \frac{I_c \times 100}{I_c + qI_a/p} \tag{7}$$

If the value of q/p is known, the degree of crystallinity of an unknown sample can be calculated from the experimentally determined values of I_c and I_a. The above method was used by Nakai et al. [27] to estimate the degree of crystallinity of lactose that had been milled for various time periods. The crystallinity values obtained by this method were in reasonably good agreement with the values obtained by the more rigorous Ruland's method. The degree of crystallinity of microcrystalline cellulose milled for various time periods was also determined [31]. The powder x-ray diffraction patterns of these samples are given in Fig. 3a, while a plot of the integrated crystalline intensity as a function of the integrated amorphous intensity is given in Fig. 3b. As expected, a linear relationship was observed.

Often the degree of crystallinity is determined using the following approximate expression [1]:

$$x_{cr} = \frac{I_c \times 100}{I_c + I_a} \tag{8}$$

The degree of crystallinity of a number of compounds including digoxin has been determined using this equation [32].

There are two major problems associated with the x-ray method. The first problem is encountered during sample preparation. At this step, preferred orientation of the particles must be minimized [1]. Reduction of particle size is one of the most effective ways of minimizing preferred orientation, and this is usually achieved by grinding the sample. Grinding, however, can also disorder the crystal lattice. Moreover, decreased particle size can cause a broadening of x-ray lines, which in turn affects the values of I_c and I_a. The relationship between the crystallite size, t, and its x-ray line breadth, β, (assuming no lattice strain) is given by the Scherrer equation [2]:

$$t = \frac{0.9 \lambda}{\beta \cos \theta_B} \tag{9}$$

where λ is the wavelength of the x-rays used and θ_B is the angle of peak diffraction.

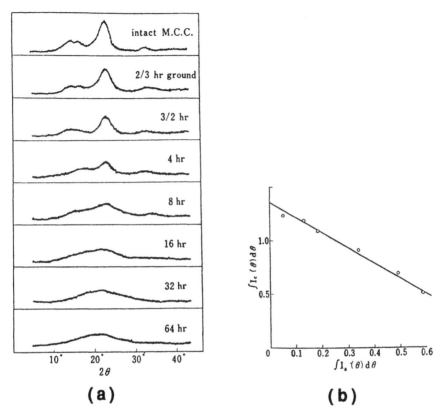

Fig. 3 (a) The powder x-ray diffraction patterns of microcrystalline cellulose milled for various time periods. (b) A plot of the integrated crystalline intensity as a function of the integrated amorphous intensity for each of the milled samples. (Reproduced with permission of the copyright owner, Pharmaceutical Society of Japan, from Ref. 31.)

From Eqs. (7) and (8) it is evident that this method requires separation of the crystalline (I_c) from the amorphous (I_a) intensities. It is a well-recognized fact that this separation is at best arbitrary [30]. Therefore, the crystallinity value obtained will be based on some subjective judgment.

B. Use of Peak Intensity (Peak Height) for Measuring the Degree of Crystallinity

If the shape of the diffraction peaks is not affected by variations in disorder and particle size, then the peak heights (rather than integrated intensities) may be used as a measure of diffraction line intensity. The use of an internal standard will permit measurement of relative intensities and will eliminate the need for the measurement of absolute intensities. Imaizumi et al. [33] developed a method to determine the degree of crystallinity of indomethacin using lithium fluoride as an internal standard (Fig. 4a). To indomethacin samples of various degrees of crystallinity, a constant proportion of lithium fluoride (20% w/w) was added. The peak intensities of one line unique to indomethacin (7.6 Å line; peak at 11.6° 2θ) and one line unique to lithium fluoride (2.4 Å line; peak at 37.8° 2θ) were measured. A linear relationship between the intensity ratio (intensity of indomethacin peak/intensity of lithium fluoride peak) and the degree of crystallinity of indomethacin was observed (Fig. 3b). This method was used to study the transition of amorphous indomethacin to crystalline indomethacin under different experimental conditions. A similar approach has been used to evaluate the degree of crystallinity of cephalexin and calcium gluceptate [34,35].

Ryan [36] developed a method to determine the relative degree of crystallinity (RDC) of imipenem in lypophilized combination products containing imipenem and amorphous cilastatin sodium. He used the relationship

$$RDC = \frac{I_{sa}}{I_{max}} \tag{10}$$

where I_{sa} = peak intensity at 27.5° 2θ of the sample under investigation, and I_{max} = peak intensity at the same angle for the sample with the highest intensity. The samples were compressed into pellets before the analyses.

Finally, the width of x-ray lines can be used as a measure of the degree

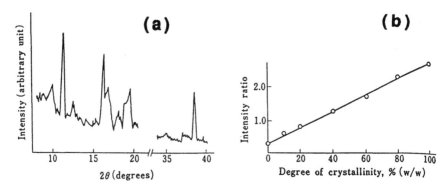

Fig. 4 (a) The powder x-ray diffraction pattern of 60% crystalline indomethacin. The powder also contained lithium fluoride as an internal standard (20% w/w). The 11.6° 2θ peak of indomethacin and the 37.8° 2θ peak of lithium fluoride were used for the analysis. (b) Relationship between the intensity ratio (intensity of indomethacin peak/intensity of lithium fluoride peak) and the degree of crystallinity of indomethacin. (Reproduced with permission of the copyright owner, Pharmaceutical Society of Japan, from Ref. 33.)

of crystallinity [37]. Although this method was not a quantitative indicator of the degree of crystallinity, it permitted the rank ordering of samples of varying degrees of crystallinity.

C. Comparison of Crystallinity Values Obtained by Different Methods

The degree of crystallinity of solids can be measured by a number of other techniques, including solution calorimetry, fusion calorimetry, infrared spectroscopy, and density. A comparison of the crystallinity values obtained by different methods reveals that the values are often in poor agreement (Table 3). All of these methods have implicitly assumed the two-state model of crystallinity. The lack of quantitative agreement between different methods can be predominantly attributed to the failure of this model [38]. Moreover, the reference standards selected (the 100% crystalline standard and the 0% crystalline standard) will also influence the numerical value of the percent crystallinity. Therefore, the degree of crystallinity values obtained using one set of standards and a particular experimental method may not agree with values obtained using other standards or another experimental method. As the data in the table reveal, even two x-ray diffraction methods resulted in different values for the crystallinity of cephalexin.

D. Relative Sensitivity of the X-Ray Method

The problems associated with the x-ray method were listed earlier. An added problem is that the x-ray methods can lack precision, and comparison with solution calorimetry suggests that the latter method is potentially more precise and less subject to artifacts [38]. Nail et al. [39] subjected freshly prepared aluminum hydroxide gel to x-ray and IR studies. The gel was x-ray amorphous, and the IR spectrum contained a broad absorption band in the 2900–3700 cm^{-1} region, indicating hydroxyl groups in many environments (thus confirming its highly disordered nature). As the gel aged and became partially crystalline, peaks appeared at 3520 cm^{-1} and 3740 cm^{-1}, with a shoulder at 3612 cm^{-1}. With aging, there was a gradual appearance of x-ray peaks. It was concluded, however, that IR spectroscopy was more sensitive to changes in crystallinity. With the IR method, crystallization was evident after about 42 days, whereas with x-ray diffractometry, crystallization was evident only after about 70 days. Recent studies suggest that water vapor sorption measurement is a very sensitive indicator of lattice disorder [40].

IV. QUANTITATIVE ANALYSIS

If there is a mixture of crystalline solids, each of these components will have a characteristic diffraction pattern, independent of the other components in the

Table 3 Comparison of the Degree of Crystallinity Values Obtained by Different Methods

| Compound | Method of sample preparation | | Percent crystallinity determined by | | | | | Reference |
| | | X-ray | | IR | Solution calorimetry | Stability at 50°C and 31% RH | | |
		Herman and Weidinger's method	Using internal standard					
Cellulose	Grinding 0 min	63		59				31
	40 min	49		57				
	8 h	14		38				
	32 h	0		10				
Cephalexin[a]	Grinding 15 min	60	28	12				34
	30 min	42	28	15				
	1 h	29	17	15				
	2 h	5	4	17				
Cephalothin	Commercial lots	72			93	100		38
	Freeze-dried	62			88	100		
	Freeze-dried (different batch)	47			54	85		
	Spray-dried	37			47	44		

[a]Interpolated from published figures.

mixture [1]. Provided appropriate absorption corrections can be made, the intensities of the peaks of each of the components will be proportional to the weight fraction of that component in the mixture. In the fields of geology and mineralogy, XPD has found extensive application for the quantitative analysis of inorganic solid mixtures. The quantification of crystalline sodium penicillin G was one of the early applications in the pharmaceutical field [41], and this was followed by several other pioneering studies [42,43]. Numerous other pharmaceutical applications have been reviewed by Zwell and Danko [44].

A. Quantitative Analysis Without an Internal Standard

1. Theory

The theoretical basis of the quantitative analysis of powder mixtures was developed by Alexander and Klug [1,45]. When an x-ray beam passes through any homogenous substance, the fractional decrease in the intensity, I, is proportional to the distance traveled, b. This relationship is given by

$$\frac{-dI}{I} = \mu \, db \tag{11}$$

The value of μ, the linear attenuation coefficient, depends on the substance, its density, and on the wavelength of the x-rays [2]. The mass attenuation coefficient, μ^*, is obtained by dividing μ by the density of the substance. Although a powder mixture may be composed of several components, it can be regarded as being composed of just two components: component J (which is the unknown), and the sum of the other components (which is designated as the matrix). The intensity of line i of component J in a powder mixture is given as

$$I_{iJ} = \frac{Kx_J}{\rho_J[x_J(\mu_J^* - \mu_M^*) + \mu_M^*]} \tag{12}$$

where K is a constant, x_J is the weight fraction of component J in the mixture, ρ_J is the density of component J, and μ_J^* and μ_M^* are the mass attenuation coefficients of component J and the matrix, respectively.

If we are dealing with a simple experimental system consisting of only two components, then one component is considered the unknown while the other is designated the matrix. The unknown component and the matrix are designated by the subscripts 1 and 2, respectively. Equation (12) can be rewritten as

$$I_{i1} = \frac{Kx_1}{\rho_1[x_1(\mu_1^* - \mu_2^*) + \mu_2^*]} \tag{13}$$

where x_1 and μ_1^* are respectively the weight fraction and mass attenuation coefficient of the unknown component, and μ_2^* is the mass attenuation coefficient of the matrix. The line i of component 1 should be chosen so that in the 2θ range where this peak occurs, the matrix does not exhibit any peaks. The intensity of peak i of a sample consisting of only 1 is given as

$$(I_{i1})_0 = \frac{K}{\rho_1 \mu_1^*} \tag{14}$$

Division of Eq. (13) by Eq. (14) yields

$$\frac{I_{i1}}{(I_{i1})_0} = \frac{x_1 \mu_1^*}{x_1(\mu_1^* - \mu_2^*) + \mu_2^*} \tag{15}$$

It is possible to calculate the mass attenuation coefficient value of a compound from its elemental composition. If $w_1, w_2, \ldots, w_n$ are the weight fractions of elements $1, 2, \ldots, n$ in the compound and $\mu_1^*, \mu_2^*, \ldots, \mu_n^*$ are their respective mass attenuation coefficients (for radiation of a particular wavelength), then the mass attenuation coefficient of the compound, μ^*, is [2]

$$\mu^* = \sum_{\kappa = 1}^{n} w_k \mu_k^* \tag{16}$$

The mass attenuation coefficient values of the elements are available in the literature [46]. Therefore, the mass attenuation coefficient of a compound can be calculated. Thus μ_1^* and μ_2^* (in Eq. 15) can be calculated provided the molecular formulas of components 1 and 2 are known. It is then possible to calculate the intensity ratio, $I_{i1}/(I_{i1})_0$, as a function of x_1. This ratio can also be experimentally obtained. The intensity of peak i of a sample consisting of only 1 is determined [$(I_{i1})_0$]. This is followed by the determination of the intensity of the same peak in mixtures containing different weight fractions of 1 and 2. This enables the experimental intensity ratio, $I_{i1}/(I_{i1})_0$, to be obtained as a function of x_1. The principles discussed above formed the basis for the successful analyses of quartz–beryllium oxide and quartz–potassium chloride binary mixtures [45].

2. Powder Mixtures

Equation (15) has been successfully used to determine the relative amounts of anhydrous carbamazepine ($C_{15}H_{12}N_2O$) and carbamazepine dihydrate ($C_{15}H_{12}N_2O \cdot 2H_2O$) when they occur as a mixture [47]. The powder x-ray patterns of anhydrous carbamazepine and carbamazepine dihydrate revealed that the lines with d-spacings of 6.78 Å (peak at 13.05° 2θ) and 9.93 Å (peak at 8.90° 2θ) were unique to anhydrous carbamazepine and carbamazepine dihydrate, respectively (Fig. 5). In mixtures containing these two phases, anhydrous carbamazepine was first considered the unknown component while carbamazepine dihydrate was designated the matrix. The intensity ratios, $I_{i1}/(I_{i1})_0$, were calculated (Table 4) for different values of x_1 (different weight fractions of

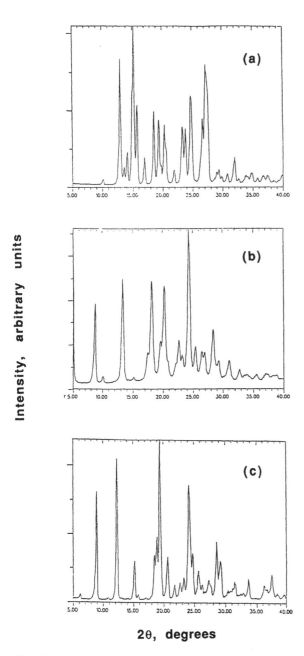

Fig. 5 The powder x-ray diffraction patterns of (a) anhydrous carbamazepine (β-form), (b) anhydrous carbamazepine (α-form), and (c) carbamazepine dihydrate. (Reproduced with permission of copyright owners, Plenum Press and ICDD, from Refs. 47 and 49.)

Table 4   The Intensity Ratios $[I_{i1}/(I_{i1})_0]$ Calculated Eq. (15)^a for Mixtures of Anhydrous Carbamazepine and Carbamazepine Dihydrate

Weight fraction		Calculated intensity ratio $[I_{i1}/(I_{i1})_0]$	
Anhydrous carbamazepine	Carbamazepine dihydrate	6.78 Å line^b	9.93 Å line^c
0.00	1.00	0.00	1.00
0.10	0.90	0.09	0.91
0.20	0.80	0.18	0.82
0.40	0.60	0.37	0.63
0.60	0.40	0.57	0.43
0.80	0.20	0.78	0.22
0.90	0.10	0.89	0.11
1.00	0.00	1.00	0.00

^{a} The 6.78 Å line is unique to anhydrous carbamazepine, and the 9.93 Å is unique to carbamazepine dihydrate.
^{b} In this series, μ_1^* and μ_2^* are respectively the mass attenuation coefficients of anhydrous carbamazepine and carbamazepine dihydrate. The calculated values were $\mu_1^* = 5.21$ cm^2 g^{-1} and $\mu_2^* = 5.87$ cm^2 g^{-1}. The theoretical line in Fig. 5 is based on these values.
^{c} In this series, μ_1^* and μ_2^* are respectively the mass attenuation coefficients of carbamazepine dihydrate and anhydrous carbamazepine. The calculated values were $\mu_1^* = 5.87$ cm^2 g^{-1} and $\mu_2^* = 5.21$ cm^2 g^{-1}. The theoretical line in Fig. 6 is based on these values.

anhydrous carbamazepine). Next, carbamazepine dihydrate was considered the unknown component, while anhydrous carbamazepine was designated the matrix, and similar calculations were performed. The integrated intensities of the 6.78 Å line (unique to anhydrous carbamazepine) and the 9.93 Å line (unique to carbamazepine dihydrate) were determined in mixtures containing different weight fractions of anhydrous carbamazepine and carbamazepine dihydrate as well as in pure samples of anhydrous carbamazepine and carbamazepine dihydrate $[(I_{i1})_0]$. The lines in Fig. 6 and Fig. 7 were based on the calculations given in Table 4, while the data points were experimental measurements. In each case, a good agreement was observed between the calculated and the experimental intensity measurements.

X-ray powder diffractometry offers a unique advantage in the quantitative analyses of mixtures consisting of the anhydrous and hydrated forms of a compound. In order to appreciate this, it is necessary to understand the interaction of water with solids. There are four possible states of water in solids [48]: (1) adsorbed on the solid surface, (2) absorbed or dissolved in the solid, (3) associated

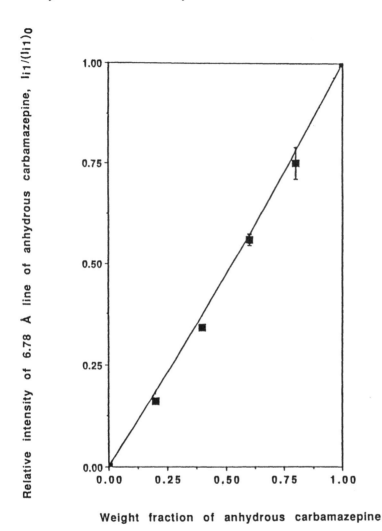

Fig. 6 The relative intensity of the 6.78 Å line of anhydrous carbamazepine (β-form) as a function of its weight fraction in binary mixtures of anhydrous carbamazepine (β-form) and carbamazepine dihydrate. The line is based on theoretical values (Table 5), while the data points are experimental measurements. (Reproduced with permission of the copyright owner, Plenum Press, from Ref. 47.)

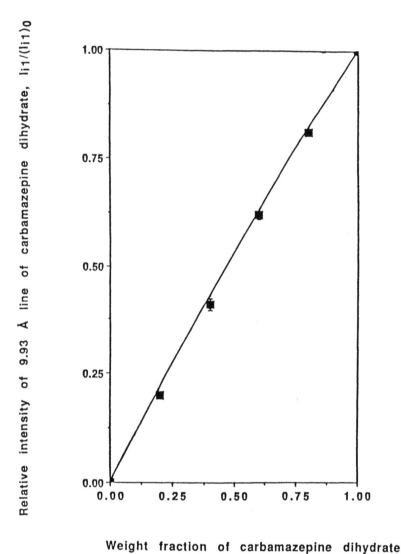

Weight fraction of carbamazepine dihydrate

Fig. 7   The relative intensity of the 9.93 Å line of carbamazepine dihydrate as a function of its weight fraction in binary mixtures of anhydrous carbamazepine (β-form) and carbamazepine dihydrate. The line is based on theoretical values (Table 5), while the data points are experimental measurements. (Reproduced with permission of the copyright owner, Plenum Press, from Ref. 47.)

as a hydrate, and (4) liquid water. The conventional methods of water content determination (Karl Fischer titrimetry and loss on drying) are often incapable of discriminating between these states of water in the solid. The x-ray method is based on the differences in the lattice structure between the anhydrous and hydrated forms of a compound. The method is therefore capable of quantifying the relative amounts of the two phases without any interference from the other types of water in the solid [47].

The equations developed by Alexander and Klug [1,45] have also been used in the quantitative analysis of polymorphic mixtures. If there is a mixture of two polymorphs, one is considered the unknown component while the other is designated the matrix. The intensity of line i of component 1 (the unknown component), I_{i1}, is determined in mixtures containing different weight fractions of components 1 and 2 (the matrix). When dealing with polymorphic mixtures, however, Eq. (15) is much simplified. Since $\mu_1^* = \mu_2^*$, Eq. (15) reduces to

$$\frac{I_{i1}}{(I_{i1})_0} = x_1 \tag{17}$$

Therefore, a plot of intensity ratio as a function of the weight fraction of the unknown component should result in a straight line with a slope of 1. This equation was used to determine the relative amounts of two polymorphs of anhydrous carbamazepine in a mixture [49].

The β-polymorphic form of anhydrous carbamazepine is official in the USP [3]. The USP stipulates that, "The X-ray diffraction pattern conforms to that of USP Carbamazepine Reference Standard, similarly determined." No limits have been set in the USP for the other polymorphs of anhydrous carbamazepine. Although several polymorphic forms of anhydrous carbamazepine have been reported, only the α- and β-forms have been extensively studied and characterized [49]. A comparison of the powder x-ray diffraction patterns of these two forms revealed that the 10.1 Å line (peak at 8.80° 2θ) was unique to α-carbamazepine, and so this line was used for the analysis (Fig. 5). It was possible to detect α-carbamazepine in a mixture where the weight fraction of α-carbamazepine was 0.02 at a signal-to-noise ratio of 2. Much greater sensitivity of this technique has been achieved in other systems. While studying the polymorphism of 1,2-dihydro-6-neopentyl-2-oxonicotinic acid, Chao and Vail [50] used x-ray diffractometry to quantify form I in mixtures of forms I and II. They estimated that form I levels as low as 0.5% w/w can be determined by this technique. Similarly the α-inosine content in a mixture consisting of α- and β-inosine was achieved with a detection limit of 0.4% w/w for α-inosine [51].

In x-ray diffractometry, quantitative analyses of mixtures of phases usually requires that at least one high-intensity peak unique to each of the phases be available for intensity measurement. Organic pharmaceutical compounds, however, often possess numerous x-ray peaks. Therefore, in a mixture of phases, even one peak unique to each of the phases may not be available for intensity measurement. The availability of profile-fitting computer software, however, permits quantitative analyses of areas under overlapping peaks. Figure 8a is the x-ray diffraction pattern of α- and δ-forms of prazosin hydrochloride [52]. Though there are pronounced differences in the powder patterns of the two forms, the authors were unable to find a high-intensity peak of α-prazosin that did not overlap with some peak of δ-prazosin. The peak at 27.5° 2θ was chosen for quantitative analysis, since it showed only a minor overlap with a peak of δ-prazosin (Fig. 8b). Using a mathematical profile-fitting program, the area under the curve of the α-prazosin was determined in mixtures containing different weight fractions of α- and δ-prazocin. In these mixtures, it was possible to detect α-prazocin down to a weight fraction of 0.005.

3. Drug Content in Tablets

The quantitative analysis of the active ingredient in multicomponent tablet formulations was accomplished using a modified form of Eq. (15) [53,54]. This analytical method had two advantages: (1) the intact tablet was analyzed directly, and (2) the technique was nondestructive. Carbamazepine was the model drug, and the other tablet ingredients were microcrystalline cellulose, starch, stearic acid, and silicon dioxide. The weight fraction of carbamazepine in these tablets ranged between 0.49 and 0.61. The intensities of 21 x-ray lines of carbamazepine were used in the analysis. There was a reasonably good agreement between the true weight fraction of carbamazepine in the tablets and the experimentally determined weight fraction. The relative error in the determination of drug content was quite high, however, making the method unsuitable for quality control purposes (Fig. 9). The method has potential application, however, for nondestructively distinguishing between tablets with and without (placebo) the active ingredient and to distinguish between formulations with different drug contents.

B. Quantitative Analysis Using an Internal Standard

1. Theory

Alexander and Klug [1,45] derived a general equation for a mixture of several components. This mixture can be regarded as being composed of just two components: component J (which is the unknown), and the sum of the other components (which is designated the matrix). Following the addition of an internal standard (component S) to the sample in known amount, the integrated intensities of line i of component J, I_{iJ}, and line k of component S, I_{kS}, are determined. The weight fraction of the unknown component in the original sample, x_J, is given as

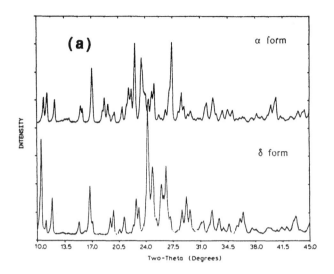

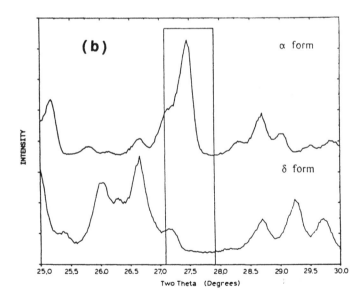

Fig. 8 (a) The powder x-ray diffraction patterns of the α- and δ-forms of prazosin hydrochloride. (b) Expanded view (25.0° to 30.0° 2θ) of the powder x-ray diffraction patterns of the α- and δ-forms of prazosin hydrochloride. Profile fitting was performed over the angular range of 27.1° to 27.8° 2θ. (Reproduced with permission of the copyright owner, Elsevier, from Ref. 52.)

$$x_J = \frac{kI_{iJ}}{I_{kS}} \tag{18}$$

where k is a constant. When the internal standard is added in a constant proportion, the concentration of component J is a linear function of the intensity ratio, I_{iJ}/I_{kS}. Alexander and Klug [45] prepared synthetic mixtures of quartz and calcium carbonate and determined the quartz content in these mixtures using fluorite (CaF_2) as an internal standard.

2. Mixture of Polymorphs

Using zinc oxide as the internal standard, the relative amounts of two polymorphic forms of fenretinide were quantified [55]. Mixtures containing 25, 50, and 75% w/w form I were prepared by mixing authentic standards of form I and form II fenretinide. After the addition of the internal standard, the maximum intensities (peak heights) of the 4.6 Å line (peak at 19.1° 2θ) of fenretinide form I ($I_{19.1}$) and the 2.8 Å line (peak at 31.8° 2θ) of zinc oxide ($I_{31.8}$) were determined. A plot of the intensity ratio ($I_{19.1}/I_{31.8}$) as a function of the weight percent form I was linear. The method was reported to be precise and accurate to within ±6%.

3. Active Ingredient in Tablets

The use of an internal standard has resulted in a precise and accurate method to quantify carbamazepine content in tablets [56]. Carbamazepine tablets were crushed into a powder and mixed with lithium fluoride (20% w/w), which was the internal standard. Five lines of carbamazepine with d-spacings of 3.38, 3.34, 3.28, 3.26, and 3.23 Å were chosen for the quantitative analysis, and the peak intensity was determined by integrating between 25.82° and 28.14° 2θ. The

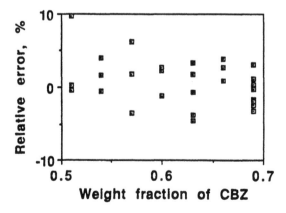

Fig. 9 The relative error in the determination of carbamazepine (CBZ) content in intact tablets. (Reproduced with permission of the copyright owner, Elsevier Science Publishers B. V., from Ref. 54.)

intensity of the 2.01 Å line of lithium fluoride was obtained by integrating between 44.05° and 45.69° 2θ (Fig. 10). After appropriate background subtraction, a plot of the intensity ratio (sum of the intensities of the five lines of carbamazepine/intensity of the lithium fluoride line) as a function of the weight fraction of carbamazepine in the tablets resulted in a straight line. Using this standard curve, the carbamazepine content in "unknown" tablets was determined and ranged between 98.6 and 101.6% of the actual drug content. The coefficient of variation in the determinations ranged between 0.28 and 2.1% (Table 5). This study demonstrated the utility of XPD for the precise and accurate determination of drug content in tablet formulations. Therefore the x-ray method is potentially useful for detecting and quantifying process-induced phase transitions occurring during the manufacture of pharmaceutical dosage forms.

4. Selection of Internal Standard

The selection of an appropriate reference standard is crucial for success in quantitative XPD. The following properties are desired in an internal standard [42]: (1) The compound must have high crystal symmetry so that strong, but few, diffraction peaks are produced. (2) The high-intensity lines in the active

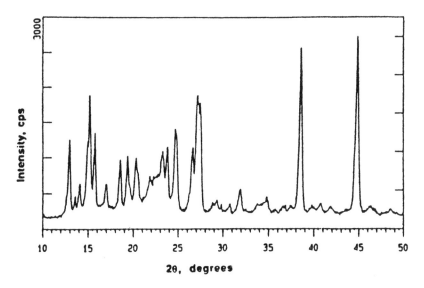

Fig. 10 The powder x-ray diffraction pattern of ground carbamazepine tablets containing lithium fluoride (20% w/w) as the internal standard. Each tablet weighed 333 mg and contained 200 mg of anhydrous carbamazepine (β-form). The peaks at 38.5° 2θ (2.34 Å line) and 45.0° 2θ (2.01 Å line) were due to lithium fluoride, while all other peaks were due to carbamazepine. (Reproduced with permission of the copyright owner, Plenum Press, from Ref. 56.)

Table 5 Accuracy in the Analysis of Carbamazepine Tablets ($n = 3$)

True wt. fraction of carbamazepine	Experimentally determined wt. fraction of carbamazepine, mean ± SD	Percent of true wt. fraction	Coefficient of variation, %
0.525	0.525 ± 0.008	100	1.48
0.555	0.564 ± 0.012	102	2.15
0.585	0.577 ± 0.011	98.7	1.82
0.600	0.601 ± 0.005	100	0.76
0.615	0.613 ± 0.003	99.6	0.40
0.645	0.646 ± 0.005	100	0.77
0.675	0.683 ± 0.007	101	1.00

Reproduced with permission of the copyright owner, Plenum Press, from Ref. 56.

ingredient and the internal standard that are to be used for quantitative purposes should not interfere with one another, and they should be close to one another. (3) The density of the internal standard should be close to that of the system ingredients so that homogenous mixtures can be prepared for analysis. (4) The internal standard must be chemically stable in the experimental system.

When dealing with organic pharmaceutical systems, an organic internal standard is preferred. The X-ray powder patterns of organic compounds, however, often contain numerous lines. As a result, it might not be possible to identify lines unique to the analyte and the internal standard that are completely separated from one another. Therefore, inorganic compounds are often used as internal standards.

C. Other Applications

The use of XPD has not been restricted to the quantitative analysis of the active ingredient in tablets. The technique has also been used to determine the salicylic acid content in plasters. This assay was successfully carried out by determining the intensity of the salicylic acid peak at 10.8° 2θ [57]. The results obtained from x-ray diffractometry were in excellent agreement with those obtained by chemical analysis. The solubility of salicylic acid in a complex, multicomponent hydrogel matrix was also estimated by x-ray diffractometry [58]. Formulations were prepared with high salicylic acid concentrations so that a fraction of the incorporated drug remained undissolved. The intensities of two peaks of salicylic were determined (Fig. 11). The first peak included lines with d-spacings of 3.58, 3.53, and 3.52 Å (integrated from 24.8° to 26.1° 2θ) and the second peak included lines and d-spacings of 2.95, 2.91, and 2.88 Å (integrated from 30.0° to 31.7°

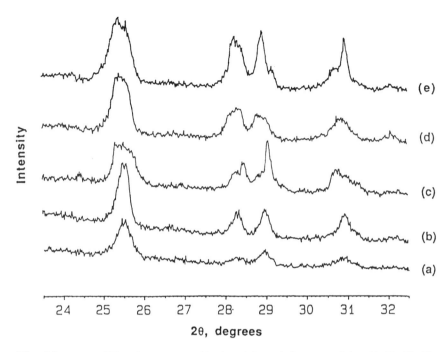

Fig. 11 X-ray diffraction patterns of hydrogel formulations containing (a) 26, (b) 32, (c) 37, (d) 41, and (e) 45% w/w salicylic acid.

2θ). The intensity of each peak was linearly related to the weight percent of salicylic acid in the formulation (Fig. 12). The intercept on the x axis was the solubility of salicylic acid in the matrix at room temperature and was found to be about 20% w/w. There was a large variance associated with the measured intensities of the x-ray lines. This was possibly due to the amorphous scattering of x-rays by the noncrystalline ingredients and the complex nature of the system. Based on scanning electron microscopic (SEM) studies of the formulations, the solubility of salicylic acid in the matrix at room temperature was estimated to be 19% w/w. Thus the results obtained by x-ray diffractometry and SEM were in good agreement.

D. Sources of Error

X-ray powder diffractometry is a relatively straightforward technique for phase identification. There are, however, numerous sources of error in quantitative XPD. The issues that are of greatest relevance to pharmaceutical systems are enumerated in the following.

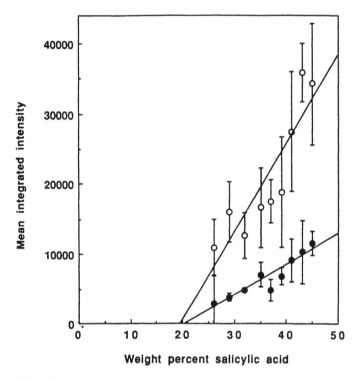

Fig. 12 Plot of the mean integrated intensity of the salicylic acid peaks as a function of the weight percent of salicylic acid in the hydrogel formulations. (○) Peak I contains lines with d-spacings of 3.58, 3.53, and 3.52 Å; (●) peak II contains lines with d-spacings of 2.95, 2.91, and 2.88 Å. (Reproduced with permission of the copyright owner, Elsevier, from Ref. 58.)

1. Preferred Orientation

Preferred orientation of the particles must be minimized. One of the most effective ways to achieve this is to reduce the particle size by grinding the sample [1]. As already discussed in Section III.A, however, grinding can disorder the crystal lattice. Grinding can also induce other undesirable transitions, such as polymorphic transformations [59]. In order to obtain reproducible intensities, there is an optimum crystallite size. The crystallites have to be sufficiently small so that the diffracted intensities are reproducible. Careful studies have been carried out to determine the desired crystallite size of quartz, but no such studies have been reported for pharmaceutical solids [60]. Care should be taken to ensure that the crystallites are not very small, since decreased particle size can cause a broadening of x-ray lines. This effect, discussed earlier (Eq. 9), usually becomes apparent when the particle size is below 0.1 μm.

The way the sample is filled in a holder affects the orientation of the crystallites. Numerous methods of sample preparation have been described in the literature [1]. Morris et al. [61] avoided preferred orientation by using a sample holder in which the top face had a rectangular cavity that extended to one end of the holder. Building on that concept, an x-ray holder has been fabricated wherein the powder is filled from the side [49]. By using this holder, reproducible and reliable intensity measurements have been obtained [49,56].

2. Measurement of Line Intensity

When measuring diffraction line intensity, the integrated line intensity (area under the curve), and not the maximum intensity (peak height), must be measured. Variations in lattice strain and particle size can have significant influences on the lineshape, but these will not affect the integrated intensity [2]. In some systems, the presence of overlapping peaks may make the determination of integrated intensity impossible. The maximum intensity may then be used for quantitative purposes provided the following conditions are met: (1) A constant proportionality between maximum and integrated intensity is evident; (2) The crystallite size is not so small as to cause particle-size-induced line broadening (Eq. 9); and (3) All the analyses are carried out under identical conditions [2]. The use of an internal standard is desirable under such circumstances.

3. Microabsorption

Let us consider diffraction in a powder mixture consisting of two crystalline phases, α and β [2]. The incident beam (on its way to a diffracting crystal) as well as the diffracted beam (on its way from the diffracting crystal) will pass through both α and β crystals. The intensity of both the beams will be decreased due to absorption (Eq. 11). When the beam passes through an α crystal, μ_α is the applicable mass attenuation coefficient. Similarly, μ_β is the mass attenuation coefficient when the beam passes through a β crystal. If $\mu_\alpha >> \mu_\beta$, or if the α particles are much larger than the β particles, the diffracted intensity of α crystals will be much less than the calculated intensity. This phenomenon is referred to as microabsorption. The accuracy of intensity measurements can be greatly affected by microabsorption.

4. Maximum Acceptable Particle Size

In samples consisting of fine particles, not only is preferred orientation minimized, but there is also satisfactory averaging of the absorption process. Assuming that absorption of x-rays within each particle is 1%, the maximum acceptable particle size, t_{max}, for quantitative studies can be calculated using the following equation [62]:

$$t_{max} = \frac{1}{100\bar{\mu}} \tag{19}$$

where $\bar{\mu}$ is the linear attenuation coefficient of the material composing the powder. Using this equation, the t_{max} values for anhydrous carbamazepine and carbamazepine dihydrate were calculated to be 15.4 and 14.1 μm, respectively [47].

5. Statistical Errors

The magnitude of the statistical errors depends on the total number of photons counted [1]. Therefore, the scanning must be carried out at an appropriately slow speed.

The other possible sources of error are primary and secondary extinction effects, inadequate sample thickness, and finally, instrument-related errors. These issues are addressed in the literature [1].

E. Use of Whole Profile for Quantitative Analysis

From the examples we have discussed, it is evident that quantitative x-ray diffractometric work has conventionally been carried out using a few high-intensity diffraction lines (one exception being the analyses of intact tablets, Section IV.A.3). With the advent of computer-controlled x-ray powder diffractometers, digital diffraction data are now routinely available [63]. This permits quantitative analysis using the *entire* powder pattern. In addition to the inherent advantage of using the entire powder pattern, this method minimizes the traditional problems of preferred orientation and overlapped x-ray lines. The whole profile fitting has been successfully used for quantitative analyses of multicomponent inorganic mixtures [63]. The different possible approaches, including the popular Rietveld method, are discussed in the x-ray literature [63–65]. These methods are potentially useful for studying complex pharmaceutical solid mixtures.

V. KINETICS OF SOLID STATE REACTIONS

X-ray powder diffractometry can be used to study solid state reactions, provided the powder pattern of the reactant is different from that of the reaction product. The anhydrous and hydrated states of many pharmaceutical compounds exhibit pronounced differences in their powder x-ray diffraction patterns. Such differences were demonstrated earlier in the case of fluprednisolone and carbamazepine. Based on such differences, the dehydration kinetics of theophylline monohydrate ($C_7H_8N_4O_2 \cdot H_2O$) and ampicillin trihydrate ($C_{16}H_{19}N_3O_4S \cdot 3H_2O$) were studied [66]. On heating, theophylline monohydrate dehydrated to a crystalline anhydrous phase, while the ampicillin trihydrate formed an amorphous anhydrate. In case of theophylline, simultaneous quantification of both the monohydrate and the anhydrate was possible. It was concluded that the initial rate of this reaction was zero order. By carrying out the reaction at several temperatures, the activation energy for the dehydration process was estimated from the Arrehenius plot to be 140 kJ mol^{-1}. In case of ampicillin trihydrate, the data was fit to a "diminishing sphere" model and yielded an activation energy value of approximately 95 kJ mol^{-1}.

The enantiomers of pseudoephedrine were observed to react in the solid state to form the racemic compound [67]. While the powder x-ray diffraction patterns of enantiomers were identical, the racemic compound exhibited a different powder pattern. The powder patterns revealed that the lines with d-spacings of 4.62 and 4.25 Å were unique to the enantiomers and to the racemic compound, respectively. Equimolar mixtures of the enantiomers were prepared, and the intensities of the 4.62 and 4.25 Å lines were obtained at regular intervals. The x-ray method permitted simultaneous quantification of the disappearance of the enantiomers and the subsequent appearance of the racemic compound. The rate of disappearance of the enantiomers followed a diffusion-controlled reaction model. Interestingly, it was observed during the kinetic experiment that the sum of the weight fractions of the enantiomers and the racemic compound progressively decreased from an initial value of unity. This suggested the existence of an intermediate noncrystalline phase, and this was confirmed by a steady increase in the background of the powder patterns.

VI. OTHER APPLICATIONS OF X-RAY DIFFRACTOMETRY IN PHARMACEUTICAL SYSTEMS

The mechanism of adsorption of drugs by montmorillonite has been studied using infrared (IR) spectroscopy and XPD [68,69]. X-ray diffractometry and a host of other techniques were used to study the interaction of bovine serum albumin with the surface of a microcrystalline aluminum oxide hydroxide compound [70].

VII. STANDARD REFERENCE MATERIALS

The National Institute of Standards and Technology provides Standard Reference Materials (SRMs) for calibrating powder x-ray diffractometers and also to measure instrument sensitivity [71]. In addition to being well characterized, these standards are stable under ambient conditions. Second, their microstructure is such that user-induced errors such as preferred orientation and statistical counting errors are minimal. Silicon powder and mica (fluorophlogopite) are available for calibration purposes, wherein the latter is suitable for large d-spacing measurements. By using these standards, a correction curve is generated, wherein the difference between the certified (calculated) and the experimentally measured peak position is plotted as a function of 2θ. Such a plot, obtained using silicon

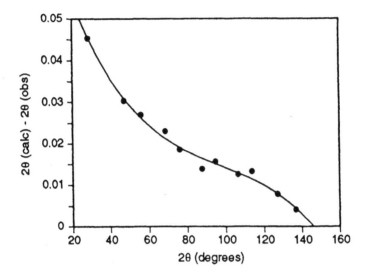

Fig. 13 Correction curve generated with silicon powder. The silicon powder is a Standard Reference Material available from the National Institute of Standards and Technology, Gaithersburg, Md. (Reproduced with permission of the National Institute of Standards and Technology, from Ref. 71.)

powder, is shown in Fig. 13. The instrument sensitivity can be measured using sintered alumina (α-Al_2O_3) and the instrument contribution to the observed line profiles can be characterized with lanthanum hexaboride (LaB_6). Several SRMs are also available for use in quantitative analysis.

REFERENCES

1. H. P. Klug and L. E. Alexander, *X-Ray Diffraction Procedures for Polycrystalline and Amorphous Materials*, Wiley, New York, 1974.

2. B. D. Cullity, *Elements of X-Ray Diffraction*, Addison-Wesley, Reading, Mass., 1978.

3. *The United States Pharmacopeia*, XXIII Revision, United States Pharmacopeial Convention, Rockville, Md., 1994, pp. 265–267, 1843–1844, 2216.

4. W. E. Garner, in *Chemistry of the Solid State*, W. E. Garner, ed., Academic Press, New York, 1955, pp. 213–221.

5. *Powder Diffraction File*, Inorganic Phases, Organic and Organometallic Phases Search Manual, W. F. McClune, editor-in-chief, International Centre for Diffraction Data, Newtown Square, Pa., 1993.

6. *Powder Diffraction File: Sets 31 to 32*, Organic Volume, W. F. McClune, editor-in-chief, pattern number 32-1723, International Centre for Diffraction Data, Swarthmore, Pa., 1988, p. 254.

7. *Powder Diffraction File: Set 43*, Inorganic and Organic Databook, W. F. McClune, editor-in-chief, International Centre for Diffraction Data, Swarthmore, Pa., 1993, pp. ix–xvi.
8. A. M. Bystrom-Asklund, *Am. Mineral.*, *51*, 1233 (1966).
9. J. K. Haleblian, R. T. Koda, and J. A. Biles, *J. Pharm. Sci.*, *60*, 1485 (1971).
10. R. R. Pfeiffer, K. S. Yang, and M. A. Tucker, *J. Pharm. Sci.*, *59*, 1809 (1970).
11. J. S. G. Cox, G. D. Woodard, and W. C. McCrone, *J. Pharm. Sci.*, *60*, 1458 (1971).
12. W. H. DeCamp, *J. Assoc. Off. Anal. Chem.*, *67*, 927 (1984).
13. J. T. R. Owen, J. E. Kountourellis, and F. A. Underwood, *J. Assoc. Off. Anal. Chem.*, *64*, 1164 (1981).
14. J. T. R. Owen, R. Sithiraks, and F. A. Underwood, *J. Assoc. Off. Anal. Chem.*, *55*, 1171 (1972).
15. A. De-Leenheer and A. Heyndrickx, *J. Assoc. Off. Anal. Chem.*, *54*, 625 (1971).
16. F. Sadik, J. H. Fincher, and C. W. Hartman, *J. Pharm. Sci.*, *60*, 916 (1971).
17. T. Layloff, *Pharm. Tech.*, *15*(Sept.), 146 (1991).
18. S. Boddapati, L. D. Butler, S. Im, and P. P. DeLuca, *J. Pharm. Sci.*, *69*, 608 (1980).
19. K. F. Landgraf, *Pharmazie*, *43*, 20, (1988).
20. H. G. Brittain, S. J. Bogdanowich, D. E. Bugay, J. DeVincentis, G. Lewen, and A. W. Newman, *Pharm. Res.*, *8*, 963, 1991.
21. T. G. Fawcett, E. J. Martin, C. E. Crowder, P. J. Kincaid, A. J. Strandjord, J. A. Blazy, D. N. Armentrout, and R. A. Newman, *Adv. X-Ray Anal.*, *29*, 323 (1986).
22. R. Jenkins, *Adv. X-Ray Anal.*, *35*, 653 (1992).
23. W. Parrish and K. Lowitzsch, *Amer. Mineralogist*, *44*, 765 (1959).
24. R. Jenkins and J. A. Nicolosi, "Long wavelength X-ray diffractometer," North American Phillips Corporation Patent Application, 1987.
25. R. L. Miller, in *Encyclopedia of Polymer Science and Technology*, Vol. 14, Wiley, New York, 1966, pp. 451–455.
26. W. Ruland, *Acta Cryst.*, *14*, 1180 (1961).
27. Y. Nakai, E. Fukuoka, S. Nakajima, and M. Morita, *Chem. Pharm. Bull.*, *30*, 1811 (1982).
28. M. Morita and S. Hirota, *Chem. Pharm. Bull.*, *30*, 3288 (1982).
29. P. H. Hermans and A. Weidinger, *J. Appl. Physics*, *19*, 491 (1948).
30. L. E. Alexander, *X-Ray Diffraction Methods in Polymer Science*, Wiley, New York, 1969, pp. 137–197.
31. Y. Nakai, E. Fukuoka, S. Nakajima, and J. Hasegawa, *Chem. Pharm. Bull.*, *25*, 96 (1977).
32. D. B. Black and E. G. Lovering, *J. Pharm. Pharmacol.*, *11*, 295, 1974.
33. H. Imaizumi, N. Nambu, and T. Nagai, *Chem. Pharm. Bull.*, *28*, 2565 (1977).
34. M. Otsuka and N. Kaneniwa, *Chem. Pharm. Bull.*, *31*, 4489, 1983.
35. R. Suryanarayanan and A. G. Mitchell, *Int. J. Pharm.*, *24*, 1 (1985).
36. J. A. Ryan, *J. Pharm. Sci.*, *75*, 805, 1986.
37. R. Suryanarayanan and S. Venkatesh, *Pharm. Res.*, *7*, S-104 (1990).

38. M. J. Pikal, A. L. Lukes, J. E. Lang, and K. Gaines, *J. Pharm. Sci.*, *67*, 767 (1978).
39. S. L. Nail, J. L. White, and S. L. Hem, *J. Pharm. Sci.*, *64*, 1166, 1975.
40. A. Saleki-Gerhardt, C. Ahlneck, and G. Zografi, *Int. J. Pharm.*, *101*, 237 (1994).
41. C. L. Christ, R. B. Barnes, and E. F. Williams, *Anal. Chem.*, *20*, 789 (1948).
42. J. W. Shell, *J. Pharm. Sci.*, *52*, 24 (1963).
43. G. J. Papariello, H. Letterman, and R. E. Huettemann, *J. Pharm. Sci.*, *53*, 663 (1964).
44. L. Zwell and A. W. Danko, *Appl. Spectroscopy Rev.*, *9*, 167 (1975).
45. L. Alexander and H. P. Klug, *Anal. Chem.*, *20*, 886 (1948).
46. *International Tables for X-ray Crystallography*, C. H. Macgillavry and G. D. Rieck, eds., Vol. III, 2nd ed., Mineralogical Society, London, 1961, p. 492.
47. R. Suryanarayanan, *Pharm. Res.*, *6*, 1017 (1989).
48. Report of the Advisory Panel on Moisture Specifications (Chair: G. Zografi), *Pharmacopeial Forum*, *17*, 1459 (1991).
49. R. Suryanarayanan, *Powder Diffraction*, *5*, 155 (1990).
50. R. S. Chao and K. C. Vail, *Pharm. Res.*, *4*, 429 (1987).
51. D. H. Doff, F. L. Brownen, and O. I. Corrigan, *Analyst*, *111*, 179 (1986).
52. V. P. Tanninen and J. Yliruusi, *Int. J. Pharm.*, *81*, 169 (1992).
53. R. Suryanarayanan and C. S. Herman, *Pharm. Res.*, *8*, 393 (1991).
54. R. Suryanarayanan and C. S. Herman, *Int. J. Pharm.*, *77*, 287 (1991).
55. F. A. Chrzanowski, B. J. Fegley, W. R. Sisco, and M. P. Newton, *J. Pharm. Sci.*, *73*, 1448 (1984).
56. R. Suryanarayanan, *Adv. X-Ray Anal.*, *34*, 417 (1991).
57. K. Kuroda, G. Hashizume, and F. Kume, *Chem. Pharm. Bull.*, *17*, 818 (1969).
58. R. Suryanarayanan, S. Venkatesh, L. Hodgin, and P. Hanson, *Int. J. Pharm.*, *78*, 77 (1992).
59. J. Haleblian and W. McCrone, *J. Pharm. Sci.*, *58*, 911 (1969).
60. L. E. Alexander, H. P. Klug, and E. Kummer, *J. Appl. Phys.*, *19*, 742 (1948).
61. M. C. Morris, H. F. McMurdie, E. H. Evans, B. Paretzkin, J. H. de Groot, C. R. Hubbard, and S. J. Carmel, *Standard X-Ray Diffraction Patterns*, *Monog. 25* (16), National Technical Information Service, Springfield, Va., 1975, pp. 1–5.
62. G. W. Brindley, in *The X-Ray Identification and Crystal Structure of Clay Minerals*, G. Brown, ed., 2nd ed., Mineralogical Society, London, 1961, p. 492.
63. R. L. Snyder and D. L. Bish, in *Modern Powder Diffraction*, D. L. Bish and J. E. Post, eds., Reviews in Mineralogy, Vol. 20, Mineralogical Society of America, Washington, D.C., 1989, pp. 101–144.
64. H. M. Rietveld, *J. Appl. Cryst.*, *2*, 65 (1969).
65. D. K. Smith, A. Scheible, A. M. Wims, J. L. Johnson, and G. Ullmann, *Powder Diffraction*, *2*, 73 (1987).
66. E. Shefter, H.-L. Fung, and O. Mok, *J. Pharm. Sci.*, *62*, 791 (1973).
67. S. P. Duddu, A. Khin-Khin, D. J. W. Grant, and R. Suryanarayanan, *Pharm. Res.*, *9*, S-149 (1992).
68. L. S. Porubcan, C. J. Cerna, J. L. White, and S. L. Hem, *J. Pharm. Sci.*, *67*, 1081 (1978).

69. L. S. Porubcan, G. S. Born, J. L. White, and S. L. Hem, *J. Pharm. Sci.*, *68*, 358 (1979).
70. M. Nishida, A. Ookubo, Y. Hashimura, A. Ikawa, Y. Yoshimura, K. Ooi, T. Suzuki, Y. Tomita, and J. Kawada, *J. Pharm. Sci.*, *81*, 828 (1992).
71. Brochure entitled "Powder Diffraction," Standard Reference Materials, National Institute of Standards and Technology, Gaithersburg, Md., 1993.

8

Thermal Methods of Analysis

James A. McCauley
Merck Sharp & Dohme Research Laboratories, Rahway, New Jersey

Harry G. Brittain
Ohmeda, Inc., Murray Hill, New Jersey

I. INTRODUCTION

Thermal analysis methods can be defined as those techniques in which a property of the analyte is determined as a function of an externally applied temperature. Dollimore [1] has listed three conditions that define the usual practice of thermal analysis:

1. The physical property and the sample temperature should be measured continuously.
2. Both the property and the temperature should be recorded automatically.
3. The temperature of the sample should be altered at a predetermined rate.

Measurements of thermal analysis are conducted for the purpose of evaluating the physical and chemical changes that may take place in a heated sample. This requires that the operator interpret the observed events in a thermogram in terms of plausible reaction processes. The reactions normally monitored can be endothermic (melting, boiling, sublimation, vaporization, desolvation, solid–solid phase transitions, chemical degradation, etc.) or exothermic (crystallization, oxidative decomposition, etc.) in nature.

Thermal methods can be extremely useful during the course of preformulation studies, since carefully planned work can be used to indicate the existence of possible drug–excipient interactions in a prototype formulation [2]. During the course of this aspect of drug development, thermal methods can be used to evaluate compound purity, polymorphism, solvation, degradation, drug–excipient compatibility, and a wide variety of other desirable characteristics. Several recent reviews have been written on such investigations [2–6].

II. DETERMINATION OF MELTING POINT

The melting point of a substance is defined as the temperature at which the solid phase exists in equilibrium with its liquid phase. This property is of great value as a characterization tool since its measurement requires relatively little material, only simple instrumentation is needed for its determination, and the information can be used for compound identification or in an estimation of purity. For instance, melting points can be used to distinguish among the geometrical isomers of a given compound, since these will not normally melt at equivalent temperatures. It is a general rule that pure substances will exhibit sharp melting points, while impure materials (or mixtures) will melt over a broad range of temperature.

When a substance undergoes a melting phase transition, the high degree of molecular arrangement existing in the solid becomes replaced by the disordered character of the liquid phase. In terms of the kinetic molecular approach, the melting point represents the temperature at which the attractive forces holding the solid together are overcome by the disruptive forces of thermal motion. The transition is accompanied by an abrupt increase in entropy, and often an increase in volume. The temperature of melting is not strongly affected by pressure, but the magnitude of effect is expressed by the Clausius–Clapeyron equation:

$$\frac{dT}{dp} = \frac{T(V_L - V_S)}{\Delta H} \tag{1}$$

where p is the environmental pressure, T is the temperature on the absolute scale, V_L and V_S are the molar volumes of liquid and solid, respectively, and ΔH is the molar heat of fusion. For most substances, the solid phase has a larger density than the liquid phase, making the term $V_L - V_S$ positive, and thus an increase in applied pressure usually raises the melting point. Water is one of the few substances that exhibit a negative value for $V_L - V_S$, which therefore yield a decrease in melting point upon an increase in pressure.

If a solid is heated at a constant rate, and its temperature monitored during the process, the melting curve as illustrated in Fig. 1 is obtained. Below the melting point, the added heat merely increases the temperature of the solid in accord with the heat capacity of the material. At the melting point, all heat introduced into the system is used to convert the solid phase into the liquid phase, and no increase in system temperature can take place as long as solid and liquid remain in equilibrium with each other. In the equilibrium condition, the system effectively exhibits an infinite heat capacity. Once all solid is converted to liquid, the temperature of the system again increases, but now in a manner determined by the heat capacity of the liquid phase. Measurements of melting curves can be used to obtain very accurate evaluations of the melting point of a compound when slow heating rates are used. The phase transition can also be monitored visually, with the operator marking the onset and completion of the melting process. This is most appropriately performed in conjunction with optical microscopy, thus yielding the combined method of thermomicroscopy [7].

A thorough discussion of apparatus suitable for the determination of melting points has been provided by Skau and Arthur [8]. One of the most common methods involves filling the analyte into a capillary tube, which is immersed in a batch whose temperature is progressively raised by an outside heating force. The Thiele arrangement, as illustrated in Fig. 2, is often used in this approach. The analyst observes the onset and completion of the melting process and then notes the temperatures of the ranges with the aid of the system thermometer. The thermometer should always be calibrated by observing the melting points of pure standard compounds, such as those listed in Table 1.

For pharmaceutical purposes, the melting range or temperature of a solid

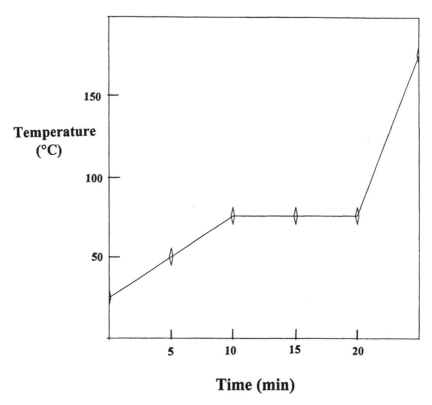

Fig. 1 Melting curve for a solid heated at a constant rate, with continual monitoring of the temperature being conducted during the process.

is defined as those points of temperature within which the solid coalesces or is completely melted. The general method for this methodology is given in the U.S. Pharmacopeia as a general test [9].

Stearate esters are unusual in that they exhibit substantial increases in volume upon solidification, which can be as large as 30% for very pure materials [10]. This property can have profound effects on formulations that are produced by the melting of stearate esters. During the development of an oil-in-water cream formulation, syneresis of the aqueous phase was observed upon using certain sources of glyceryl monostearate [11]. Primarily through the use of variable-temperature x-ray diffraction, it was learned that this material would undergo changes in phase composition upon melting and congealing. The ability of glyceryl monostearate to break the oil-in-water emulsion was directly related to the composition of the raw material, and in the degree of its expansion (or lack thereof) during the congealing process. Knowledge of the melting behavior of

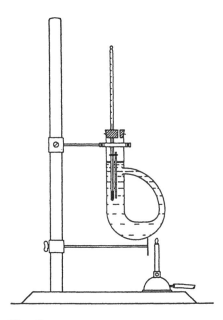

Fig. 2 Typical Thiele appratus for the determination of melting points.

this excipient, as influenced by its source and origin, proved essential to the transferability of the formulation in question.

III. DIFFERENTIAL THERMAL ANALYSIS

A. Background

Differential thermal analysis (DTA) consists of the monitoring of the differences in temperature existing between a solid sample and a reference as a function of temperature. Differences in temperature between the sample and reference are observed when a process takes place that requires a finite heat of reaction. Typical solid state changes of this type include phase transformations, structural conversions, decomposition reactions, and desolvation processes. These processes may require either the input or release of energy in the form of heat, which in turn translates into events that affect the temperature of the sample relative to a nonreactive reference.

Although it is possible to use DTA as a quantitative tool, such applications are not trivial. For this reason, DTA has historically been mostly used in a quantitative sense as a means to determine the temperatures at which thermal events takes place. Owing to the experimental conditions used for its measure-

Table 1 Corrected Melting Points of
Compounds Suitable as Reference Materials
in the Calibration of Thermometers

Melting point (°C)	Material
0	Ice
53	*p*-dichlorobenzene
90	*m*-dinitrobenzene
114	Acetanilide
122	Benzoic acid
132	Urea
157	Salicylic acid
187	Hippuric acid
200	Isatin
216	Anthracene
238	Carbanilide
257	Oxanilide
286	Anthraquinone
332	*N,N*-diacetylbenzidine

ment, the technique is most useful for the characterization of materials that evolve corrosive gases during the heating process. The technique is highly useful as a means for compound identification based on melting point considerations, and it has been successfully used in the study of mixtures.

B. Methodology

Methodology appropriate for the measuring of DTA profiles has been extensively reviewed [12,13]. A schematic diagram illustrating the essential aspects of the DTA technique is shown in Fig. 3. Both the sample and reference materials are contained within the same furnace, whose temperature program is externally controlled. The outputs of the sensing thermocouples are amplified, electronically subtracted, and finally shown on a suitable display device.

If the observed ΔH is positive (endothermic reaction), the temperature of the sample will lag behind that of the reference. If the ΔH is negative (exothermic reaction), the temperature of the sample will exceed that of the reference. Owing to a variety of factors, DTA analysis is not normally used for quantitative work; instead, it is used to deduce temperatures associated with thermal events. It can be a very useful adjunct to differential scanning calorimetry, since with most instrumentation DTA analysis can be performed in such a manner that corrosive

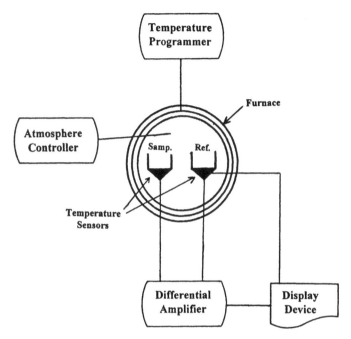

Fig. 3 Schematic diagram illustrating the essential aspects of the differential thermal analysis technique. The experimental observable is the differential temperature between sample and reference, which will be plotted as a function of the system temperature.

gases evolved from the sample do not damage expensive portions of the thermal cell assembly.

Wendlandt has provided an extensive compilation of conditions and requirements that influence the shape of DTA thermograms [14]. These can be divided into instrument factors (furnace atmosphere, furnace geometry, sample holder material and geometry, thermocouple details, heating rate, and thermocouple location in sample) and sample characteristics (particle size, thermal conductivity, heat capacity, packing density, swelling or shrinkage of sample, mass of sample taken, and degree of crystallinity). A sufficient number of these factors are under the control of the operator, thus permitting selectivity in the methods of data collection. The ability to correlate an experimental DTA thermogram with a theoretical interpretation is profoundly affected by the details of heat transfer between the sample and the calorimeter [15].

The calibration of DTA systems is dependent on the use of appropriate reference materials, rather than on the application of electrical heating methods. The temperature calibration is normally accomplished with the thermogram being obtained at the heating rate normally used for analysis [16], and the temperatures known for the thermal events used to set temperatures for the empirically observed features. Recommended reference materials that span melting ranges of pharmaceutical interest include benzoic acid (melting point 122.4°C), indium (156.4°C), and tin (231.9°C).

C. Applications

It was recognized quite some time ago that DTA analysis could be used to deduce the compatibility between a drug substance and its excipients in a formulation. The effect of lubricants on performance was as problematic then as it is now, and DTA proved to be a powerful method in the evaluation of possible incompatibilities. Jacobson and Reier used DTA to study the interaction between various penicillins and stearic acid [17]. For instance, the addition of 5% stearic acid to sodium oxacillin monohydrate completely obliterated the thermal events associated with the antibiotic. Since that time, many workers employed DTA analysis in the study of drug–excipient interactions, although the DTA method has been largely replaced by differential scanning calorimetry technology.

Proceeding along a parallel track, Guillory and coworkers used DTA analysis to study complexation phenomena [2]. Through the performance of carefully designed studies, they were able to prove the existence of association complexes and deduced the stoichiometries of these. In this particular work, phase diagrams were developed for 2:1 deoxycholic acid–menadione, 1:1 quinine–phenobarbital, 2:1 theophylline–phenobarbital, 1:1 caffeine–phenobarbital, and 1:1 atropine–phenobarbital. The method was also used to prove that no complexes were formed between phenobarbital and aspirin, phenacetin, diphenylhydantoin, and acetaminophen.

Differential thermal analysis proved to be a powerful tool in the study of compound polymorphism, and in the characterization of solvate species of drug compounds. In addition, it can be used to deduce the ability of polymorphs to thermally interconvert, thus establishing the system to be monotropic or enantiotropic in nature. For instance, form I of chloroquine diphosphate melts at 216°C, while form II melts at 196°C [18]. The DTA thermogram of form I consists of a simple endotherm, while the thermogram of form II is complicated (see Fig. 4). The first endotherm at 196°C is associated with the melting of form II, but this is immediately followed by an exotherm corresponding to the crystallization of form I. This species is then observed to melt at 216°C, establishing it as the thermodynamically more stable form at the elevated temperature.

Differential thermal analysis proved to be a powerful aid in a detailed study that fully explained the polymorphism and solvates associated with several sulfonamides [19]. For instance, three solvate species and four true polymorphs were identified in the specific instance of sulfabenzamide. Quantitative analysis

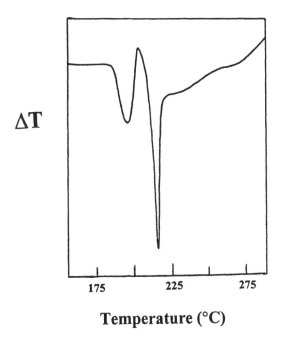

Temperature (°C)

Fig. 4 DTA thermogram of chloroquine diphosphate, form II, illustrating the melting endotherm of form II (196°C), recrystallization to form I, and final melting endotherm attributable to the converted form I (216°C). (Data adapted from Ref. 18.)

of the DTA thermograms was used to calculate the enthalpy of fusion for each form, with this information then being used to identify the order of relative stability. Some of these species were found to undergo phase conversions during the heating process, but others were noted to be completely stable with respect to all temperatures up to the melting point.

It is always possible that the mechanical effects associated with the processing of materials can result in a change in the physical state of the drug entity, and DTA has proven to be a valuable aid in this work. For instance, the temperature used in the spray-drying of phenylbutazone has been shown to determine the polymorphic form of the compound [20]. As is illustrated in Fig. 5, a lower melting form was obtained at reduced temperatures (30–40°C), while a higher melting material was obtained when the material was spray-dried at 100–120°C. This difference in crystal structure would be of great importance in the use of spray-dried phenylbutazone since the dried particles exhibited substantially different crystal morphologies.

The reduction of particle size by grinding can also result in significant alterations in structural properties, and DTA has been successfully used to follow

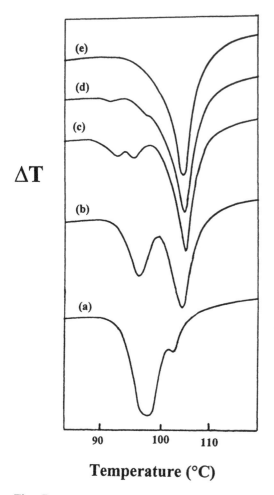

ΔT

Fig. 5 DTA thermograms of phenylbutazone, after being spray-dried at (a) 30°C, (b) 40°C, (c) 70°C, (d) 80°C, and (e) 100°C. (Data adapted from Ref. 20.)

these in appropriate instances. In one study, methisazone was found to convert from one polymorph to another upon micronization, and the phase transformation could be followed through a study of the thermal properties of materials ground for different lengths of time [21]. In another study, it was found that extensive grinding of cephalexin monohydrate would effectively dehydrate the material [22]. This physical change was tracked most easily through the DTA thermograms, since the dehydration endotherm characteristic of the monohydrate species became less prominent as a function of the grinding time. It was also concluded that grinding decreased the stability of cephalexin, since the temperature for the exothermic decomposition was observed to decrease with an increase in the grinding time.

Differential thermal analysis can also be conveniently used to determine the normal boiling point and vapor pressure of liquids. With a heating block that can accommodate glass capillaries whose size is about 4 mm in diameter and 30 mm in length, a liquid sample (5 to 10 μl in volume) is added to the sample capillary, which has been prefilled with glass beads. The reference capillary contains only glass beads. The external pressure can either be ambient or suitably reduced, but it should be measured. The DTA curve is be obtained at any reasonable heating rate (i.e., less than 20°C/min), with an endotherm being observed at the boiling point of the liquid in question. Superheating is sometimes seen, particularly when working at lower pressures and higher heating rates. The lowest temperature observed during the vaporization endotherm should be taken as the boiling point where superheating is encountered. Figure 6 contains boiling

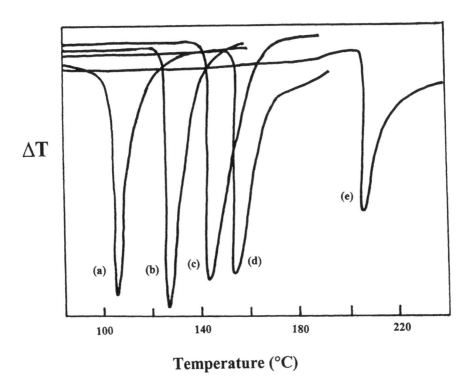

Fig. 6 Boiling point endotherms obtained for 2,4-difluoronitrobenzene, obtained at pressures of (a) 16 torr, (b) 44 torr, (c) 113 torr, (d) 162 torr, and (e) 759 torr.

point endotherms for 2,4-difluoronitrobenzene, obtained as a function of pressure. Figure 7 is a plot of vapor pressure (on a logarithmic scale) versus the reciprocal of the absolute temperature obtained for 2,4-difluoronitrobenzene and nitrobenzene on the basis of DTA data.

Differential thermal analysis can also be used to construct binary phase diagrams on the basis of observed melting points. This information is of importance since the nature of the phase diagram as would exist for an enantiomeric pair can be instrumental in choosing a resolution strategy [23,24]. When a drug candidate contains one or more chiral centers, it is frequently

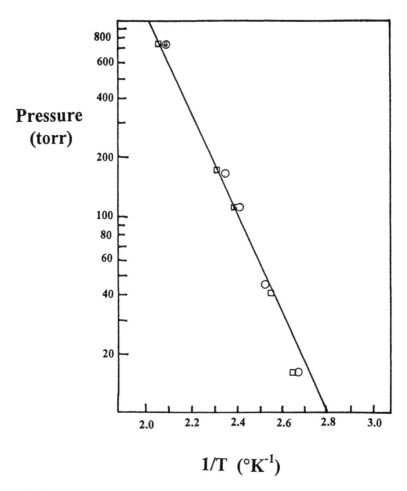

Fig. 7 Relation between the vapor pressure and DTA vaporization endotherm of nitrobenzene (□) and 2,4-difluoronitrobenzene (○).

desirable to develop a single enantiomer rather than the racemic mixture. This can be accomplished either through stereospecific synthesis or through the resolution of the racemate. Enantiomer resolution can be accomplished with a resolving agent, or upon spontaneous crystallization of the desired enantiomer, but the latter is possible only if the compound can be shown to crystallize as a conglomerate. The nature of the compound (i.e., whether it crystallizes as a conglomerate or as a racemate) can be ascertained by using DTA as the means to determine the enantiomer binary phase diagram. Mixtures of the two enantiomers, or the racemate plus one of the enantiomers, can be used. The initial extrapolated onset temperature and the final peak temperature are plotted against the initial enantiomer composition to construct the phase diagram. If a simple eutectic diagram having its eutectic composition at 50% is obtained, then the racemate is a conglomerate and can be resolved through spontaneous crystallization. If the behavior is more complex, such as has been illustrated in Fig. 8 for the *N*-acetyl analogue of methyldopa, then the material in question is identified as a racemate. Such compounds are ordinarily resolved through the formation and separation of dissociable diastereomers, but chiral chromatography methods can also be used.

IV. DIFFERENTIAL SCANNING CALORIMETRY

A. Background

In many respects, differential scanning calorimetry (DSC) is similar to the DTA method, and analogous information about the same range of thermal events can be obtained. However, DSC is far easier to use routinely on a quantitative basis, and for this reason it has become the most widely used method of thermal analysis. The relevance of the DSC technique as a tool for pharmaceutical scientists has been amply documented in numerous reviews [3–6,25–26], and a general chapter on DSC is documented in the U.S. Pharmacopeia [27].

In the DSC method, the sample and reference are maintained at the same temperature and the heat flow required to keep the equality in temperature is measured. Hence DSC plots are obtained as the differential rate of heating (in units of watts/second, calories/second, or Joules/second) against temperature. The area under a DSC peak is directly proportional to the heat absorbed or evolved by the thermal event, and integration of these peak areas yields the heat of reaction (in units of calories/second · gram or Joules/second · gram).

When a compound is observed to melt without decomposition, DSC analysis can be used to determine the absolute purity [28]. If the impurities are soluble in the melt of the major component, the van't Hoff equation applies:

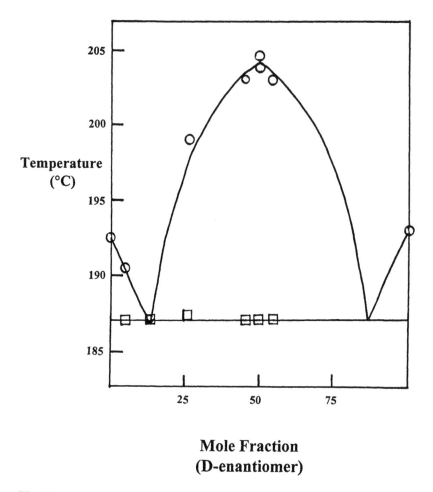

Fig. 8 Melting point phase diagram obtained for the N-acetyl analogue of methyldopa, which exhibits typical racemate behavior.

$$T_s = T_0 - \frac{R(T_0)^2 X_i}{F \, \Delta H_f} \tag{2}$$

In Eq. (2), T_s is the sample temperature, T_0 is the melting point of the pure major component, X_i is the mole fraction of the impurity, F is the fraction of solid melted, and ΔH_f is the enthalpy of fusion of the pure component. A plot of T_s against $1/F$ should yield a straight line whose slope is proportional to X_i. This method can therefore be used to evaluate the absolute purity of a given compound without reference to a standard, with purities being obtained in terms of mole percent. The method is limited to reasonably pure compounds that melt without decomposition. The assumptions justifying Eq. (2) fail when the compound purity is below approximately 97 mole percent, and the method cannot be used in such instances. The DSC purity method has been critically reviewed, with the advantages and limitations of the technique being carefully explored [29–31].

B. Methodology

Two types of DSC measurement are possible, which are usually identified as power-compensation DSC and heat-flux DSC, and the details of each configuration have been fully described [1,14]. In power-compensated DSC, the sample and reference materials are kept at the same temperature by the use of individualized heating elements, and the observable parameter recorded is the difference in power inputs to the two heaters. In heat-flux DSC, one simply monitors the heat differential between the sample and reference materials, with the methodology not being terribly different from that used for DTA. Schematic diagrams of the two modes of DSC measurement are illustrated in Fig. 9.

When displaying DTA data, an exothermic reaction is invariably plotted

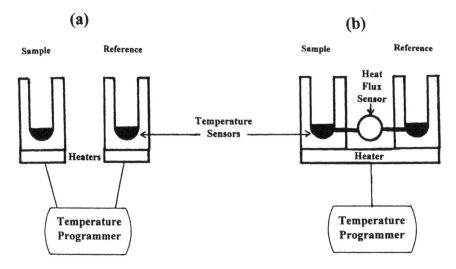

Fig. 9 Schematic diagrams illustrating the sample cell configurations for (a) power-compensation and (b) heat-flux modes of DSC detection. Each cell system is contained in the furnace assembly, and the differential heat flow between sample and reference is monitored as the experimental observable and ultimately is plotted as a function of the system temperature.

as a positive thermal event, while an endothermic reaction is usually displayed as a negative event. Unfortunately, this simple situation has not carried over to DSC methodology, since different methods of data acquisition promote alternative fashions for display of the data. The use of power-compensation DSC results in endothermic reactions being displayed as positive events, a situation that is counter to the latest IUPAC recommendations [32]. When the heat-flux method is used to detect the thermal phenomena, the signs of the DSC events concur with those obtained using DTA, and they also agree with the IUPAC recommendations.

The calibration of DSC instruments is normally accomplished through the use of compounds having accurately known transition temperatures and heats of fusion. A list of the standards currently supplied by the National Technical Information Service (NTIS) [33] is provided in Table 2. Once the DSC system is properly calibrated, it is trivial to obtain the melting point and enthalpy of fusion data for any compound upon integration of its empirically determined endotherm and application of the calibration parameters. The current state of methodology is such, however, that unless a determination is repeated a large number of times, the deduced enthalpies must be regarded as being accurate only to within approximately 5%.

C. Applications

Owing to the vast number of applications that have been published over the years, it is not possible to provide a comprehensive review of the pharmaceutical applications of DSC analysis in the present chapter. Interested readers should consult the leading reviews and books that are devoted solely to this topic [2–6,23–24]. Instead, certain well-established uses for the technique will be discussed to illustrate some of the applications that have been made.

Table 2 Melting Temperatures and Enthalpies of Fusion for Compounds Suitable as Reference Materials in Differential Scanning Calorimetry

Material	Melting point (°C)	Enthalpy of fusion (kJ/mol)
Naphthalene	80.2	19.05
Benzil	94.8	23.35
Acetamide	114.3	21.65
Benzoic acid	122.3	18.09
Diphenylacetic acid	148.0	31.27
Indium	156.6	3.252

In a manner similar to that just described for differential thermal analysis, DSC can be used to obtain useful and characteristic thermal and melting point data for crystal polymorphs or solvate species. This information is of great importance to the pharmaceutical industry since many compounds can crystallize in more than one structural modification, and the FDA is vitally concerned with this possibility. Although the primary means of polymorph or solvate characterization is centered around x-ray diffraction methodology, in suitable situations thermal analysis can be used to advantage.

In Fig. 10, illustrative DSC curves are shown, which had been obtained during the development of diflunisal, a nonsteroidal anti-inflammatory agent. Both samples were chemically identical, but they exhibited different x-ray powder diffraction (XRD) patterns and different DSC thermograms. The DSC thermogram of form II (the lower melting polymorph) contained the melting transition of this form, but this was immediately followed by conversion to and melting of the higher melting polymorph (designated as form I). As shown in the figure, pure form I exhibits only a single melting endotherm, indicating it to be the thermodynamically more stable form around the melting points of the two solids. The XRD powder pattern of the form II sample was not judged to contain any evidence for the existence of form I in the solid, and so the DSC features obtained for form II were assigned as being a kinetic interconversion of the metastable form II into the more stable form I during the assay.

Another example of the use of DSC in the characterization of a polymorphic system is given in Fig. 11, which contains the DSC thermograms of two different samples of Proscar (a benign prostatic hypertrophy medicinal agent). Both samples were chemically identical, but they exhibited different XRD powder patterns. In the DSC of one modification, a minor endotherm was observed that was interpreted as the solid–solid transition of form I to form II. The major feature in the thermogram of this material then consisted of the melting of form II. In the DSC of the other modification, only a single endotherm was observed, and this was assigned to the melting of form II. Both samples were recrystallized from their respective melts, and it was then found that the DSC curves of these processed materals were identical to each other and to the lower curve in Fig. 11. This finding indicated that the solid–solid phase conversion was irreversible. Additional experiments were conducted in which it was confirmed that after the transformation corresponding to the minor endotherm took place, no chemical change in the material composition occurred and the XRD pattern converted to form II.

Differential scanning calorimetry can also supply valuable information regarding solvate species, and it is particularly useful with respect to temperature and energetics of the desolvation process Two samples of the developmental compound L-706000-001T were shown to be chemically identical, and each contained two moles of water. The XRD powder patterns for the two samples were found to be quite different, demonstrating the existence of polymorphism

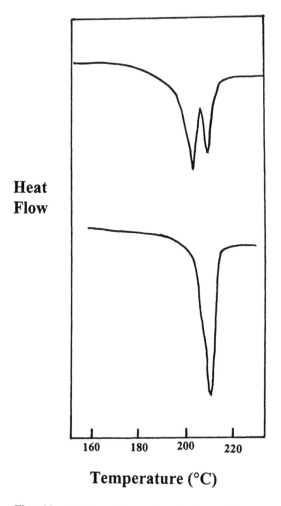

Fig. 10 Differential scanning calorimetry thermograms obtained for diflunisal, form I (lower trace) and form II (upper trace).

within the dihydrate system. The DSC curve for one dihydrate phase showed only a single desolvation endotherm, while the corresponding thermogram for the other dihydrate phase showed two water-loss endotherms. The endothermic heat for the loss of two moles of water from one dihydrate corresponded to 2.48 KJ/mole, while the analogous heat losses of the water from the other were 2.54 and 2.44 KJ/mole for each bound water. Since the heat associated with the initial water loss from the second dihydrate was somewhat larger than that of the first, it appeared that the second dihydrate polymorph was the more thermodynamically

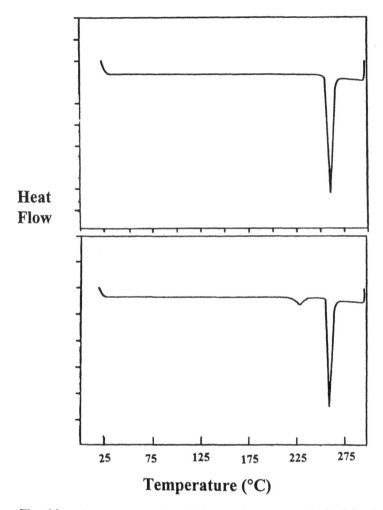

Fig. 11 Differential scanning calorimetry thermograms obtained for the two poly-morphs of Proscar.

stable polymorph. This conclusion was confirmed by water solubility studies that showed the second dihydrate to be less soluble in water at 25°C than was the first (6.1 vs. 7.1 mg/g).

Some compounds are capable of forming solvates with a variety of solvents, and DSC can be effectively used to screen the various species that may be isolated from the different systems. Piretanide [34] provides a typical example of this behavior, for which representative DSC thermograms have been illustrated in Fig. 12. After recrystallization of this compound from 27 different solvents, six

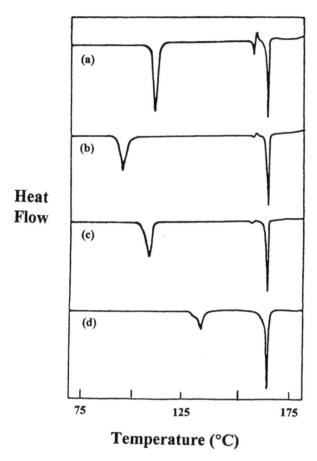

Fig. 12 Differential scanning calorimetry thermograms obtained for piretanide, as recrystallized from (a) *t*-butanol, (b) *n*-propanol, (c) *i*-propanol, and (d) *N,N*-dimethyl-formamide. (Data adapted from Ref. 34.)

solvate and two anhydrous forms were obtained. These were all distinct crystal forms, as evidenced by the nonequivalent XRD powder patterns obtained for each form. Upon desolvation, the solvate species were observed to convert into one of the two anhydrous structures, and one of these anhydrous forms was shown by DSC to be irreversibly convertible into the other.

Since DSC is a quantitative technique, the enthalpies determined for reactions can be used for analytical purposes. For instance, a method has been described whereby the water content in hydrate species can be determined using DSC techniques [35]. In this method, it was assumed that the enthalpy of binding *n* moles of water molecules in a hydrate is the same as that of *n* moles of water molecules in liquid water. From the literature value known for the enthalpy change associated with liquid water, it was possible to use the enthalpies of dehydration obtained on hydrate species to deduce the number of bound water molecules. Fairly good agreement between the number of calculated water molecules and the authentic degree of hydration was obtained for nine compound systems.

One of the most classic applications of DSC analysis is in the determination of possible interactions between a drug entity and the excipients in its formulation [2,17,23]. It is vitally important to establish the existence of any incompatibilities during the preformulation stage so that there would be no disasters waiting to be discovered during subsequent stability studies. In one variation of this approach, binary mixtures of each ingredient are mixed, and changes in melting endotherms are monitored, which would indicate either a physical or chemical interaction between the two. A much better approach to compatibility studies consists of stressing mixtures of the appropriate materials for short time periods (such as 55°C for three weeks), and then using DSC to monitor any changes that might have taken place [36]. The stressing of the formulation may be conducted either by isothermal or nonisothermal means, and the advantages of each approach have been discussed [37].

Since the number of compatibility studies that have used DSC methodology are countless, one example will be cited to illustrate the approach. Based on DSC work, it was concluded that famotidine was compatible with talc, magnesium stearate, and avicel PH 101, and that definite interactions took place with kollidon, primojel, emcompress, crospovidone, and lactose [38]. Representative DSC thermograms are shown in Fig. 13, where it may be seen that no change in thermal features was detected in binary mixtures of famotidine and talc, but that a definite change in famotidine melting point was observed in a binary mixture with emcompress. Other incompatible excipients led to the complete suppression of all thermal features, undoubtedly through the formation of an amorphous solid solution.

V. THERMOGRAVIMETRY

A. Background

Thermogravimetry (TG) is a measure of the thermally induced weight loss of a material as a function of the applied temperature [39]. Thermogravimetric analysis is restricted to studies involving either a mass gain or loss, and it is most commonly used to study desolvation processes and compound decomposition. Thermogravimetric analysis is a very useful method for the quantitative determination of the total volatile content of a solid, and it can be used as an adjunct to Karl Fischer titrations for the determination of moisture.

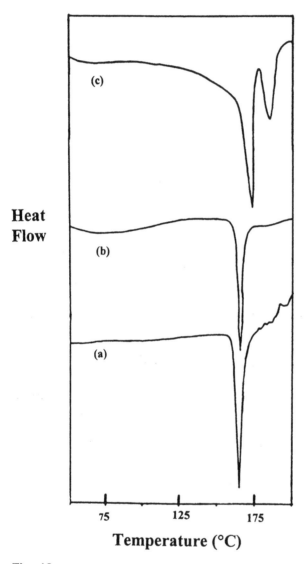

Fig. 13 Differential scanning calorimetry thermograms obtained for (a) famotidine, (b) 1:1 famotidine:talc, and (c) 1:1 famotidine:emcompress. (Data adapted from Ref. 38.)

Thermogravimetric analysis also represents a powerful adjunct to DTA or DSC analysis, since a combination of either method with a TG determination can be used in the assignment of observed thermal events. Desolvation processes or decomposition reactions must be accompanied by weight changes, and they can be thusly identified by a TG weight loss over the same temperature range. On the other hand, solid–liquid or solid–solid phase transformations are not accompanied by any loss of sample mass and would not register in a TG thermogram.

When a solid is capable of decomposing by means of several discrete, sequential reactions, the magnitude of each step can be separately evaluated. Thermogravimetric analysis of compound decomposition can also be used to compare the stability of similar compounds. The higher the decomposition temperature of a given compound, the more positive would be the DG value and the greater would be its stability.

B. Methodology

Thermogravimetry consists of the continual recording of the mass of the sample as it is heated in a furnace, and a schematic diagram of a TG apparatus is given in Fig. 14. The weighing device used in most systems is a microbalance, which permits the accurate determination of milligram changes in the sample mass. The

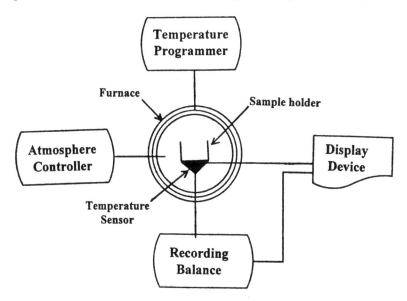

Fig. 14 Schematic diagram of apparatus suitable for thermogravimetric analysis. The experimental observable is the percent weight loss of the sample, which will be plotted as a function of the system temperature.

balance chamber itself is constructed so that the atmosphere may be controlled if desired, and this is normally accomplished by means of a flowing gas stream. The furnace must be constructed so that its internal temperature can be totally programmed in a reproducible fashion, and its inside surfaces must be resistant to any and all gases evolved during the TG experiment.

It is most essential in TG design that the temperature readout pertain to that of the sample, and not to that of the furnace. To achieve this end, the thermocouple or resistance thermometer is normally mounted as close to the sample pan as possible, in direct contact if this can be achieved.

Although the TG methodology is conceptually simple, the accuracy and precision associated with the results are dependent on both instrument and sample factors [40]. The furnace heating rate used for the determination will greatly affect the transition temperatures, while the atmosphere within the furnace can influence the nature of the thermal reactions. The sample itself can play a role in governing the quality of data obtained, with factors such as sample size, nature of evolved gases, particle size, heats of reaction, sample packing, and thermal conductivity all influencing the observed thermogram.

C. Applications

Thermogravimetry is particularly suited to the study of desolvation. As an illustration of this application, Fig. 15 contains the weight loss–temperature profiles for the two dihydrate polymorphs of the experimental compound L-706000-001T. The TG curve for type A shows a single weight loss step of 6.36%, which corresponds to the loss of two moles of water (the theoretical weight loss for a dihydrate of L-706000-001T is 6.7%). The TG thermogram for type D consisted of a two-step weight loss, each of which was equivalent to the loss of one mole of water. The TG data for type D therefore imply that a relatively stable monohydrate may exist, and this has been confirmed through exposure of material at relative humidities less than 50%.

The combination of TG studies with DSC work can lead to unambiguous assignment of the observed thermal events. For instance, a simultaneous TG and DSC study of the low-temperature endothermic features observed for certain crystal modifications of mefloquinine hydrochloride permitted the assignment of these crystal forms as being solvate species of the drug entity [41]. The use of DSC/TG methodology has proven to be of particular importance in the characterization of the pseudopolymorphs of magnesium stearate, where the melting and dehydration endotherms of the hydrate species are sufficiently close in temperature to require the use of both techniques for absolute assignments [42].

Thermogravimetry can be used in a more traditional analytical way for the determination and quantification of volatile materials present in final products and process intermediates. Not only can water and other solvents be determined by TG,

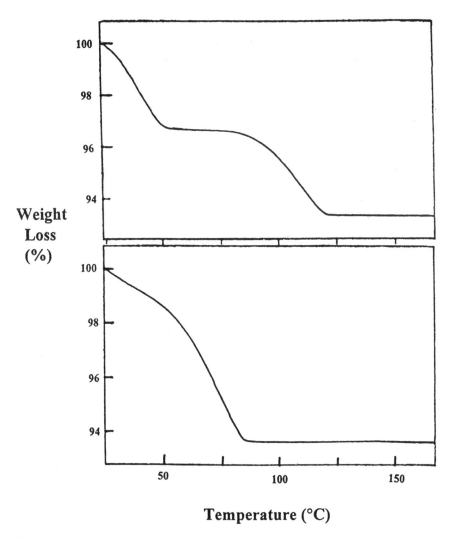

Fig. 15 Thermogravimetric weight loss–temperature profiles for the two dihydrate polymorphs of the experimental compound L-706000-001T. Upper plot: type D; lower plot: type A.

but any volatile component that is separable from other volatiles can be determined. A particularly appropriate and concrete example occurred during the process development of diflunisal. A key process intermediate, sodium phenolate, was known to exist as a hydrate in the solid state, and for processing reasons it was necessary to determine both the water content and the amount of free phenol. As is evident in Fig. 16, a single TG run could be used to quantitatively estimate both water and free phenol. Complicating the situation is the fact that sodium phenolate is thermally unstable at elevated temperatures, and its decomposition could compromise the deduced free phenol content values. For this reason, a more precise TG method for free phenol was developed. The free phenol was extraced from the phenolate with toluene, and then an aliquot of the toluene solution was evaporated at 80°C under nitrogen in the TG apparatus. The extracted free phenol was then measured as the weight loss detected between 100 and 200°C. If necessary, the sublimed material could be collected and its identity verified by thin-layer chromatography and mass spectral analysis. The method of standard additions to the phenolate was also used to quantitate the free phenol with the extractive procedure, and this has been illustrated in Fig. 17.

VI. ALTERNATIVE METHODS OF THERMAL ANALYSIS

Most workers in the pharmaceutical field identify thermal analysis with the melting point, differential thermal analysis, differential scanning calorimetry,

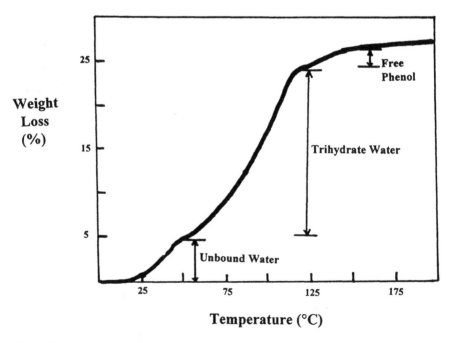

Fig. 16 Typical thermogravimetric analysis of sodium phenolate, illustrating the temperature resolution associated with the separate liberation of water and free phenol.

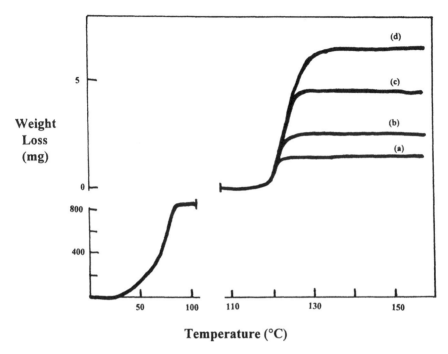

Fig. 17 Application of the method of standard additions to the quantitation of free phenol in sodium phenolate. Thermograms are shown for (a) the initial sample, (b) the sample with 1 mg free phenol added, (c) the sample with 3 mg free phenol added, and (d) the sample with 5 mg free phenol added.

and thermogravimetry methods just described. There are several other, less well-known, techniques available for the characterization of solid materials, each of which can be particularly useful to deduce certain types of information. Although it is beyond the scope of this chapter to delve into each type of methodology in great detail, short summaries will be provided. As in all thermal analysis techniques, the observed parameter of interest is obtained as a function of temperature, while the sample is heated at an accurately controlled rate.

A. Evolved Gas Analysis

In this technique, both the amount and composition of the volatile component are measured as a function of temperature. The composition of the evolved gases can be determined using gas chromatography, mass spectrometry, or infrared spectroscopy.

B. Thermomechanical Analysis

The deformation of the analyte, under the influence of an externally applied mechanical stress, is followed as a function of temperature. When the deformation of the sample is followed in the absence of an external load, the technique is identified as *thermodilatometry*.

C. Thermoptometry

This category refers to a variety of techniques in which some optical property of the sample is followed during the heating procedure. Observable quantities could be the absorption of light at some wavelength (*thermospectrophotometry*), the emission of radiant energy (*thermoluminescence*), changes in the solid refractive index (*thermorefractometry*), or changes in the microscopic particle characteristics (*thermomicroscopy*). The latter state is often referred to as *hot-stage microscopy*.

D. Dielectric Analysis

As applied to thermal analysis, dielectric analysis consists of the measurement of the capacitance (the ability to store electric charge) and conductance (the ability to transmit electrical charge) as functions of applied temperature. The measurements are ordinarily conducted over a range of frequencies to obtain full characterization of the system. The information deduced from such work pertains to mobility within the sample, and it has been extremely useful in the study of polymers.

REFERENCES

1. D. Dollimore, "Thermoanalytical Instrumentation," in *Analytical Instrumentation Handbook*, G. W. Ewing, ed., Marcel Dekker, New York, 1990, pp. 905–960.
2. J. K. Guillory, S. C. Hwang, and J. L. Lach, *J. Pharm. Sci.*, *58*, 301 (1969).
3. D. Ghiron, *J. Pharm. Biomed. Anal.*, *4*, 755 (1986).
4. D. Ghiron-Forest, C. Goldbronn, and P. Piechon, *J. Pharm. Biomed. Anal.*, *7*, 1421 (1989).
5. I. Townsend, *J. Therm. Anal.*, *37*, 2031 (1991).
6. A. F. Barnes, M. J. Hardy, and T. J. Lever, *J. Therm. Anal.*, *40*, 499 (1993).
7. M. Kuhnert-Bradstätter, *Thermomicroscopy in the Analysis of Pharmaceuticals*, Pergamon Press, Oxford, 1971.
8. E. L. Skau and J. C. Arthur, "Determination of Melting and Freezing Temperatures," in *Physical Methods of Chemistry*, Vol. I, part V, A. Weissberger and B. W. Rossiter, eds., Wiley-Interscience, New York, 1971, pp. 137–171.
9. *United States Pharmacopeia XXIII*, general test <741>, 1995, pp. 1805–1806.
10. A. P. Simonelli and T. Higuchi, *J. Pharm. Sci.*, *51*, 584 (1962).

11. R. O'Laughlin, C. Sachs, H. G. Brittain, E. Cohen, P. Timmins, and S. Varia, *J. Soc. Cosmet. Chem.*, *40*, 215 (1989).
12. W. Smykatz-Kloss, *Differential Thermal Analysis*, Springer-Verlag, Berlin, 1974.
13. M. I. Pope and M. D. Judd, *Differential Thermal Analysis*, Heyden, London, 1977.
14. W. W. Wendlandt, *Thermal Analysis*, 3rd ed., Wiley-Interscience, New York, 1986.
15. G. M. Lukaszewski, *Lab. Pract.*, *15*, 664 (1966).
16. M. J. Richardson and P. Burrington, *J. Therm. Anal.*, *6*, 345 (1974).
17. H. Jacobson and G. Reier, *J. Pharm. Sci.*, *58*, 631 (1969).
18. Ph. Van Aerde, J. P. Remon, D. DeRudder, R. Van Severen, and P. Braeckman, *J. Pharm. Pharmacol.*, *36*, 190 (1984).
19. S. S. Yang and J. K. Guillory, *J. Pharm. Sci.*, *61*, 26 (1972).
20. Y. Matsuda, S. Kawaguchi, H. Kobayshi, and J. Nishijo, *J. Pharm. Sci.*, *73*, 173 (1984).
21. K. C. Lee and J. A. Hersey, *J. Pharm. Pharmacol.*, *29*, 249 (1977).
22. M. Otsuka and N. Kaneniwa, *Chem. Pharm. Bull.*, *32*, 1071 (1984).
23. J. Jacques, A. Collet, and S. H. Wilen, *Enantiomers, Racemates, and Resolutions*, John Wiley & Sons, New York, 1981.
24. H. G. Brittain, *Pharm. Res.*, *7*, 683 (1990).
25. J. L. Ford and P. Timmins, *Pharmaceutical Thermal Analysis*, Ellis Horwood, Ltd., Chichester, U. K., 1989.
26. D. Giron, *Acta Pharm. Jugosl.*, *40*, 95 (1990).
27. *United States Pharmacopeia XXIII*, general test <891>, 1995, pp. 1837–1838.
28. R. L. Blaine and C. K. Schoff, *Purity Determinations by Thermal Methods*, ASTM Press, Philadelphia, 1984.
29. F. F. Joy, J. D. Bonn, and A. J. Barnard, *Thermochim. Acta*, *2*, 57 (1971).
30. E. G. Palermo and J. Chiu, *Thermochim. Acta*, *14*, 1 (1976).
31. A. A. van Dooren and B. W. Muller, *Int. J. Pharm.*, *20*, 217 (1984).
32. R. C. Mackenzie, *Pure Appl. Chem.*, *57*, 1737 (1985).
33. E. L. Charsley, J. A. Rumsey, and S. B. Warrington, *Anal. Proc.*, *5* (1984).
34. Y. Chikaraishi, A. Sano, T. Tsujiyama, M. Otsuka, and Y. Matsuda, *Chem. Pharm. Bull.*, *42*, 1123 (1994).
35. R. K. Khankari, D. Law, and D. J. W. Grant, *Int. J. Pharm.*, *82*, 117 (1992).
36. A. A. van Dooren, *Drug Dev. Ind. Pharm.*, *9*, 43 (1983).
37. C. Ahlneck and P. Lundgren, *Acta Pharm. Suecica*, *22*, 305 (1985).
38. G. Indrayanto, A. Mugihardjo, and R. Handayani, *Drug Dev. Ind. Pharm.*, *20*, 911 (1994).
39. C. J. Keattch and D. Dollimore, *Introduction to Thermogravimetry*, 2nd ed., Heyden, London, 1975.
40. C. Duval, *Inorganic Thermogravimetric Analysis*, 2nd ed., Elsevier, Amsterdam, 1963.
41. S. Kitamura, L.-C. Chang, and J. K. Guillory, *Int. J. Pharm.*, *101*, 127 (1994).
42. K. D. Ertel and J. T. Carstensen, *Int. J. Pharm.*, *42*, 171 (1988).

9

Micromeritics

Ann W. Newman
*Bristol-Meyers Squibb Pharmaceutical Research Institute, New Brunswick,
New Jersey*

I. INTRODUCTION

The term *micromeritics* was proposed by J. M. Dallavalle [1] to refer to the science of small particles. It is derived from the Greek words for small and part. Dallavalle found that the behavior and characteristics of small particles brought together widely scattered information on particle measurements, size distributions, packing arrangements, and the general theory of the physical properties of finely divided substances. He also discovered that the uses of fine particles ranged from manufacturing processes and commercial products to atmospheric studies and soil science [2].

Micromeritics has become an important area of study in the pharmaceutical industry because it influences a large number of parameters in research, development, and manufacturing. Specific micromeritic topics important to pharmaceuticals were summarized by Parrott [3] and are listed in Table 1.

The study of fine particles in pharmaceutical applications involves a number of different techniques. Micromeritic investigations involve surface areas, particle sizes and their distributions, the nature of solid surfaces, and particle shapes [4]. Scientists working in this field realize that a number of techniques are necessary to fully investigate a system and that an interdisciplinary approach is essential. This ability to correlate data from different techniques allows a more thorough understanding of the system, process, or problem being investigated.

This chapter on micromeritics will deal specifically with surface area, porosimetry, and density measurements. It is designed to introduce the importance of the specific technique in pharmaceutics and briefly describe the theory, instrumentation, and data collection involved. Examples are presented to

Table 1 Micromeritic Topics Important to Pharmaceuticals

Particle shape
Diameters and size distributions
Methods of size measurement
Characteristics of particulate solids (packing, void, density)
Surface, surface energy, adsorption, and solubility
Chemical properties and stability
Sedimentation and aggregation
Sieve and sizing
Dynamics (pneumatic transport, flow rate)
Separation of particles from air (pollution, toxicity, explosion)
Fine grinding and blending
Sampling
Release of medicinal compound from dosage form and subsequent therapy

Source: Ref. 3.

illustrate the investigation of various systems using data from the technique in combination with other data. Due to the extensive literature available on the other micromeritic techniques of particle size and powder characteristics measurements, reviews on these subjects have been included as separate topics in Chapters 6 and 10, respectively.

II. SURFACE AREA

In the pharmaceutical industry, surface area is becoming more important in the characterization of materials during development, formulation, and manufacturing. The surface area of a solid material provides information about the void spaces on the surfaces of individual particles or aggregates of particles [5]. This becomes important because factors such as chemical activity, adsorption, dissolution, and bioavailability of the drug may depend on the surface on the solid [3,5]. Handling properties of materials, such as flowability of a powder, can also be related to particle size and surface area [4].

The adsorption of inert gases onto solid materials represents the most widely used method for the determination of surface area, although other methods are available [6,7]. The BET method, developed by Brunauer, Emmett, and Teller [8], is generally used for gas adsorption surface area measurements.

A. BET Surface Area Measurements

The BET method is based on the monolayer adsorption of an inert gas on the solid surface at reduced temperatures. Any condensible gas can be used for BET measurements, although the preferred gases are nitrogen and krypton. Nitrogen is used as the adsorbate gas for most samples exhibiting surface areas of approximately 1.0 m^2/g or greater, while materials with smaller surface areas should be measured using krypton. The lower vapor pressure of krypton causes a larger amount of gas to be adsorbed on the solid, resulting in more accurate values at low surface area values.

The way in which a material adsorbs a gas is referred to as an adsorption isotherm. All adsorption isotherms can be described by five representative curves, given in Fig. 1. The isotherm shapes reflect specific conditions for adsorption, such as pore size and heats of adsorption [6]. The most common type of isotherm and the most useful for BET measurements is the Type II isotherm. The inflection point of this isotherm usually indicates monolayer coverage of the adsorbate [9].

B. BET Theory and Equations

The BET surface area equation is based on Langmuir's kinetic theory of monolayer gas adsorption on surfaces [6]. Langmuir theorized that the collision

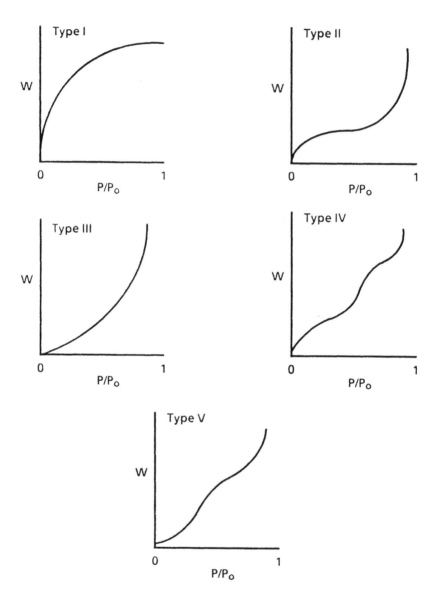

Fig. 1 The five types of adsorption isotherms. W, weight adsorbed; P/P_0, relative pressure. Condensation occurs at $P/P_o \geq 1$.

of a gas molecule with a solid is inelastic and the gas molecules remain in contact with the solid before returning to the gas phase, resulting in adsorption. The following equation relates the rate at which molecules strike a surface and the rate at which they leave the surface or evaporate [2]:

$$V = \frac{V_m bP}{1 + bP} \tag{1}$$

where

V = volume of gas adsorbed at pressure P
V_m = the volume adsorbed when the entire surface is covered by a monomolecular layer
b = a constant

This equation can be rewritten in a linear form:

$$\frac{P}{V} = \frac{1}{V_m b} + \frac{P}{V_m} \tag{2}$$

When P/V is plotted versus P, a straight line should result if the equation applies. The agreement of experimental data and theory is poor at low pressures and pressures approaching saturation [6]. This has been attributed to the formation of multimolecular layers of adsorbate on the surface caused by the condensation of gas molecules onto the adsorbed monolayer.

Brunauer, Emmett, and Teller extended the Langmuir theory to multi-molecular layer adsorption [8]. They related the condensation rate of gas molecules onto an adsorbed layer and the evaporation rate from that layer for an infinite number of layers. The linear form of the relationship is called the BET equation:

$$\frac{P}{V(P_0 - P)} = \frac{1}{V_m C} + \frac{C - 1}{V_m C}\left(\frac{P}{P_0}\right) \tag{3}$$

where

V = volume of gas adsorbed at pressure P
P = partial pressure of adsorbate
V_m = volume of gas adsorbed in monolayer
P_0 = saturation pressure of adsorbate at experimental temperature
C = BET constant exponentially relating the heats of adsorption and condensation of the adsorbate

For multipoint BET measurements, a plot of $P/V(P_0 - P)$ versus P/P_0 is obtained using various concentrations of adsorbate and measuring the adsorbed volume. Since the BET equation is in the form of a straight line:

$$\text{slope} = \frac{C - 1}{V_m C} \tag{4}$$

$$\text{intercept} = \frac{1}{V_m C} \tag{5}$$

The monolayer of adsorbed gas (V_m) and the BET constant (C) are calculated from the slope and intercept:

$$V_m = \frac{1}{\text{slope} + \text{intercept}} \tag{6}$$

$$C = \frac{\text{slope}}{\text{intercept}} + 1 \tag{7}$$

The total surface area (S_t) of the sample is calculated from

$$S_t = \frac{V_m N_0 A_{cs}}{M} \tag{8}$$

where

S_t = total surface area
N_0 = Avogadro's number
A_{cs} = cross-sectional area of the adsorbent
M = molecular weight of the adsorbent

The specific surface area (S) of the solid is obtained from

$$S = \frac{S_t}{m} \tag{9}$$

where

S = specific surface area
m = mass of powdered sample measured

The theory for single-point BET measurements is similar; however, an assumption is made that the intercept of the $P/V(P_0-P)$ versus P/P_0 plot equals zero. This assumption is not always valid and can result in surface area values that differ from multipoint values. When the assumption is valid, the single-point method is simpler and faster, and it results in accurate surface area values.

From Eqs. (4) and (5) it can be seen that

$$\frac{\text{slope}}{\text{intercept}} = C - 1 \tag{10}$$

From this equation, it is evident that with high values of C the intercept will be small compared with the slope, and in many cases the intercept may be assumed as zero. Using this assumption, Eq. (3) becomes

$$\frac{P}{V(P_0 - P)} = \frac{C - 1}{V_m C}\left(\frac{P}{P_0}\right)$$ (11)

By assuming the intercept, $1/V_m C$, is zero, the BET equation simplifies to

$$V_m = V\left(1 - \frac{P}{P_0}\right)$$ (12)

Substituting the value of V_m into Eq. (8), the total surface area for a single-point BET measurement can be calculated from

$$S_t = V\left(1 - \frac{P}{P_0}\right)\frac{N_0 A_{cs}}{M}$$ (13)

The specific surface for the single-pont BET measurement is then calculated from Eq. (9).

A number of the assumptions used in the BET theory have been questioned for real samples [6]. One assumption states that all adsorption sites are energetically equivalent, which is not the case for normal samples. The BET model ignores lateral adsorbate interactions on the surface, and it also assumes that the heat of adsorption for the second layer and above is equal to the heat of liquefaction. This assumption is not valid at high pressures and is the reason for using adsorbate pressures less than 0.35. In spite of these concerns, the BET method has proven to be an accurate representation of surface area for the majority of samples [9,10].

C. Instrumentation and Data Collection

A number of commercial surface area instruments are available, and varying levels of sophistication exist. In this section, a general overview of the instrumentation will be presented, and options will be mentioned when applicable.

The instrument setup consists of

A pure adsorbate (nitrogen, krypton)
A carrier gas (helium)
A sample holder
A liquid nitrogen dewar
A detector

The adsorbate gas must be mixed with the carrier gas in the required concentrations for analysis. This can be done prior to analysis and a number of tanks for various concentrations can be kept, or the mixing can be done during the analysis with a gas mixer. The sample holder can allow the gas to flow through the sample (such as a modified U-tube), or a vacuum can be pulled on the sample, which requires a sample holder consisting of a single stem with a bulb at the bottom to hold sample. The most common type of detector is the thermal conductivity detector, which detects the amount of gas desorbed from the sample. The ambient pressure will need to be measured either manually or by the instrument.

For BET measurements, preadsorbed gases and vapors must first be removed from the surface of the solid using gas flow or a vacuum [6], and this procedure is called outgassing the sample. Elevated temperatures can also be used to decrease outgassing time, but care should be taken when performing this procedure [11]. Increased temperatures, especially under vacuum, can change the properties of organic pharamceuticals (i.e., dehydration, conversion to other forms, decomposition). Once the sample is outgassed, the solid sample is cooled to the boiling point of the adsorbate gas by immersing the sample holder in a dewar filled with liquid nitrogen. The sample is exposed to partial pressures of the adsorbate gas until the surface is covered by the gas. The liquid nitrogen dewar is then removed from the sample cell, and the amount of previously adsorbed gas is measured by the detector as the sample warms to room temperature. Surface areas can obtained from single-point or multipoint measurements. Single-point measurements require only one partial pressure of the adsorbent gas, which is usually 30% for nitrogen. Multipoint measurements require more than one partial pressure of nitrogen, which usually range from 5 to 30% for nitrogen. The volume of gas adsorbed is converted to the surface area (m^2/g) of the sample using the BET equation.

When analyzing data, a number of parameters should be checked to ensure the validity of the data. The correlation coefficient should be as close to 1 as possible, with a range of 0.999 to 1 being acceptable in most cases. If a linear relationship is not achieved for a sample, more material may be needed for the analysis. When using nitrogen as an absorbate, very low surface areas cannot be measured, and switching to another adsorbate, such as krypton, may be required. The C constant is a measure of the affinity of the adsorbate to the material, and for nitrogen the C value can range from 3 to 1000 [6], depending on the adsorption isotherm of the material. For low C constants, higher adsorbate pressures may be required to obtain a good correlation coefficient [6]. If possible, collecting and identifying the type of adsorption isotherm could also give information on the nonlinearity of a sample.

Figure 2 shows a typical multipoint BET plot for bendroflumethiazide, and Table 2 summarizes the parameters and calculations used to obtain the surface area for this compound. The single-point surface area calculated from a P/P_o of 0.3108 is also presented in Table 2. Good agreement is evident between the two methods, considering the C constant given in Table 2 for this sample.

D. Applications of Surface Area Measurements

Surface area measurements have been used in a wide variety of pharmaceutical applications, ranging from dissolution to manufacturing [12–25], Table 3 is a

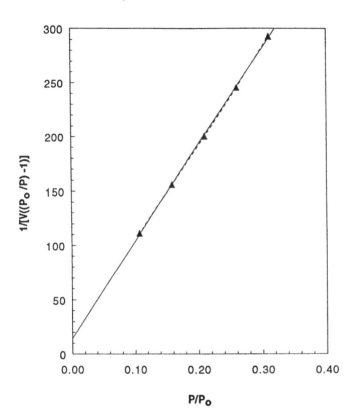

Fig. 2 Multipoint BET plot for bendroflumethiazide. Solid line represents theoretical curve; dashed line represents experimental curve.

short listing of compounds and excipients related to pharmaceuticals. Most of the organic materials listed exhibit a relatively low surface area (below 2 m²/g), which in certain cases increases substantially upon freeze-drying. The inorganic compounds listed tend to show higher surface areas than the organic compounds. In the case of caffeine, it is evident that the preparation of the material will also alter the surface area of a material.

Many surface area studies have investigated various properties of excipients. The relationship between excipient grades, flow properties, particle sizes, and surface areas have been studied. Fast-flo lactose is manufactured to contain larger particles (approximately 50 μm) than hydrous lactose (approximately 20 μm) [23]. This has been done to improve the flow properties of the fast-flo

Table 2 Multipoint BET Data for Bendroflumethiazide

$\dfrac{P}{P_0}$	Volume (cm^3/g)	$\dfrac{1}{V[(P_0/P) - 1]}$
0.1072	0.865	1.111×10^2
0.1591	0.971	1.559×10^2
0.2101	1.065	1.999×10^2
0.2605	1.149	2.452×10^2
0.3108	1.232	2.928×10^2

Multipoint surface area = 3.85 m^2/g
Slope = 8.903×10^2
Y intercept = 1.442
Correlation coefficient = 0.9998
C = 6.27

Single-point surface area = 3.70 m^2/g at P/P_0 = 0.3108

material for its use in direct compression granulations. From Table 3, it is evident that the fast-flo grade exhibits a lower surface area than the hydrous lactose. A similar situation exists for PH101 and PH102 grades of microcrystalline cellulose [24]. The PH102 grade is composed of larger particles of microcrystalline cellulose to again give it better flow properties. Although the surface area differences are not as prominent as for lactose, the surface area of the PH102 grade microcrystalline cellulose is slightly lower than for the PH101 material.

Other studies have been performed to investigate the effect of surface area and tablet lubricant efficacy. In a comparison study between sodium stearyl fumarate and magnesium stearate, it was found that sodium stearyl fumarate was effective as a lubricant to about the same degree as magnesium stearate [15]. It was also reported that the lubricating properties correlated better to the surface area of the lubricant than to the amount of lubricant used. A micronized lubricant was more efficient than a coarse fraction, and it was suggested that the surface area be standardized to obtain reproducible effects.

Surface area and moisture uptake have been related to the disintegration properties of excipients such as crosspovidone, starch, and alginic acid [17]. The surface areas of the three materials were measured, and a linear correlation was found between the maximum moisture sorption and specific surface area for the three disintegrants. The greater the surface area of the material, the more numerous were the sites for capillary attraction of water to its surface. It was postulated that the capillary action appears to be responsible for the disintegration properties of the materials.

Table 3 Surface Areas Reported for Selected Pharmaceutical Materials

Compound	Form/preparation	Surface area (m^2/g)	Ref.
Alginic acid	—	0.75	[17]
Bentonite	—	3.25	[24]
Caffeine	Anhydrous—prep 1	0.94	[21]
	Anhydrous—prep 2	1.98	[21]
	Anhydrous—prep 3	0.74	[21]
	Freeze-dried	7.39	[21]
	Hydrous—prep 1	0.89	[21]
	Hydrous—prep 2	0.98	[21]
	Hydrous—prep 3	0.48	[21]
Chloramphenicol palmitate	Polymorph A	1.4	[16]
	Freeze-dried	15.6	[16]
Dicalcium phosphate	Dihydrate	1.6	[18]
Erythromycin	Anhydrate	1.0	[22]
	Monohydrate	1.3	[22]
	Dihydrate	6.6	[22]
Lactose	Anhydrous form	0.38	[23]
	Hydrous form	0.53	[23]
	Fast-flo	0.34	[23]
Magnesium oxide	—	13.1	[24]
Microcrystalline cellulose	PH 101 grade	1.00	[25]
	PH 102 grade	0.97	[25]
Polyvinyl-pyrrolidone (crosslinked)	—	1.03	[17]
Starch	—	0.59	[17]
Titanium dioxide	—	13.7	[5]

When evaluating the effect of binder concentration on a number of tablet properties, surface area measurements were used to investigate the bond strength of the binder with the other particles [18]. A steady reduction in the surface area of the granules with increasing binder concentration indicated that the binder had covered or penetrated the particles, with the formation of particle–binder bonds. This was related to friability, and the increased bond strength was related to the decreased surface areas.

Methods for synthesizing highly porous microspheres were investigated, and surface area measurements were used to confirm the porous nature of the samples [19]. A high surface area was measured and was compared with the calculated surface area value. The measured value was 35 times that of a nonporous particle, indicating the extensive porosity of the spheres. The surface area was also used to explain the drug release mechanisms in the pores of these systems.

Polyacrylic nanoparticles have been prepared to investigate their use as colloidal drug carriers, and surface area measurements were used for characterization [14]. When the experimental surface area values were compared with calculated values, it was found that the measured surface area value was 10 times smaller than the calculated value. The discrepancy was explained by the surfactant used in the nanoparticle preparation. The surfactant appears to coat the particles when it is not completely removed, resulting in the low surface areas observed for the particles. Particles prepared without surfactant showed good agreement with the calculated values.

In summary, surface area measurements are relatively easy to obtain, and the technique can be applied to a wide variety of pharmaceutical applications. When used in conjunction with other techniques, it is a powerful problem solving technique.

III. POROSIMETRY

The pore structure of a solid can contribute to the disintegration, dissolution, adsorption, and diffusion of a drug material [26,27]. Because of this, porosity and pore size distribution measurements have been used extensively to study tablets [28–30], granules [31,32], and excipients [33]. The following classification system of pore sizes has been developed based on the average pore radii [6]:

Pore designation	Pore size (Å)
Macropores	>1000
Transitional (Mesopores)	15–1000
Micropores	<15

The porosity of a sample is a measure of the void spaces in a material, and it can be calculated using data from a number of techniques, including density, gas adsorption, water displacement, and porosimetry [34]. Pore size measurements, on the other hand, provide information about the actual pore structure of a material, including the pore radii and volume, and these measurements are usually obtained using gas adsorption and mercury porosimetry. Studies comparing and contrasting the two techniques have reported that gas adsorption is limited to pore radii smaller than 1000 Å, whereas mercury porosimetry is capable of measuring much larger pores, such as interparticle spaces [35,36]. The broad range of pore sizes available from mercury porosimetry makes it more applicable to many pharmaceutical applications.

A. Mercury Porosimetry

Mercury porosimetry is based on the fact that mercury behaves as a nonwetting liquid toward most substances and will not penetrate the solid unless pressure is applied. To measure the porosity, the sample is sealed in a sample holder that is tapered to a calibrated stem. The sample holder and stem are then filled with mercury and subjected to increasing pressures to force the mercury into the pores of the material. The amount of mercury in the calibrated stem decreases during this step, and the change in volume is recorded. A curve of volume versus pressure represents the volume penetrated into the sample at a given pressure. The intrusion pressure is then related to the pore size using the Washburn equation [37], and the amount of mercury intruded is indicative of the number of pores in the system.

B. Porosimetry Theory and Equations

The affinity of a liquid for a solid can be described as wetting. A liquid that spreads spontaneously along a solid surface is described as a wetting liquid. If the liquid remains stationary and appears spherical, it is nonwetting. A measure of the degree of wetting is the contact angle (θ), where $\theta < 90°$ is considered a wetting liquid, and $\theta > 90°$ is considered a nonwetting liquid, as shown in Fig. 3. The use of mercury in porosimetry is based on its high contact angle of approximately 140° and its nonwetting characteristics. However, the applicability of mercury porosimetry to materials wetted by mercury has also been investigated [38].

The measured contact angle of mercury on various samples can range from 112 to 170° [39], but for most applications the average value of 140° is used. It should be noted, however, that the accuracy of the pore radii measurement is limited by the accuracy of the contact-angle measurement [40]. Contact angles can readily be measured on flat surfaces or compacts of powders [6], and the measurement of contact angles with powder systems has also been reported [41].

(a)

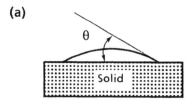

(b)

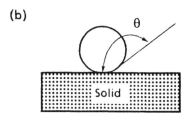

Fig. 3   Diagram of the contact angle used in porosimetry measurements: (a) Wetting liquid, $\theta < 90°$; (b) nonwetting liquid, $\theta > 90°$.

For a circular pore opening, the Washburn equation [37] relates the applied pressure and the radius of the pores intruded with a nonwetting liquid:

$$r = \frac{-2\gamma \cos \theta}{P} \qquad (14)$$

where

r = pore radius (μm)
γ = surface tension of the liquid
θ = contact angle between the liquid and the sample
P = applied pressure (psi)

The inverse relationship between pore radius and pressure indicates that low pressures are utilized when measuring large pore sizes and high pressures are necessary when measuring small pore sizes.

Using a value of 480 dynes/cm for the surface tension of mercury (γ) and an angle of 140° for the contact angle (θ), Eq. (14) reduces to

$$r = \frac{106.7}{P} \tag{15}$$

This abbreviated equation can be used for pore size distribution comparisons of similar samples when absolute pore size is not necessary. If, however, comparison of absolute pore sizes is required, the contact angles should be measured for all samples [42].

C. Instrumentation and Data Interpretation

A mercury porosimeter consists of three parts:

Sample container, known as a dilatometer
Source of pressure
Monitoring equipment

The dilatometer is generally made of glass and is the vessel where the mercury is intruded into the sample pores. The design is dependent on the pressure source and monitoring system of the instrument. The dilatometer consists of a sample holder and a calibrated stem, which is used to measure the amount of mercury intruded into the sample. The sample in the dilatometer must be cleaned from adsorbed species by degassing the material in a vacuum [42]. Most commercial instruments degas the sample in the instrument before mercury intrusion. Once the sample is degassed, the dilatometer (sample holder and stem) are filled with mercury.

Porosimeters fall into two groups depending on the pressure used for the measurements. Low-pressure, or subambient, units operate from 0.5 psi to ambient pressure to measure large pores. High-pressure porosimeters operate from ambient pressure to as high as 60,000 psi [43] to measure much smaller pores. These elevated pressures are achieved in a number of ways, such as pressurizing in a hydraulic pump oil medium.

Various systems have been used to measure the mercury level change in the stem during intrusion. Indirect methods include resistance or capacitance measurements along the stem of the dilatometer [39]. These readings are taken in conjunction with pressure readings, to correlate the number of pores at a specific pore size.

A plot of the volume of mercury versus pressure is a common way to display the raw data, as shown in Fig. 4. When increasing the pressure, mercury is forced into the pores and an intrusion curve is produced from the increased mercury volume in the sample. When decreasing the pressure, mercury will leave the pores, causing a decrease in volume, and an extrusion curve is observed. The instrusion and extrusion curves will not be the same due to hysteresis, which is caused by mercury being permanently trapped in the pores of the sample

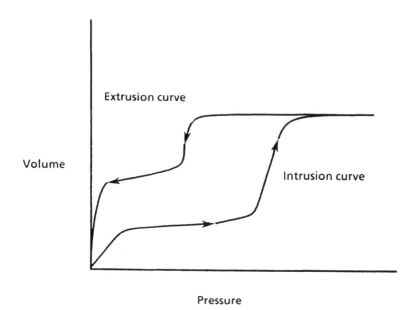

Fig. 4 Schematic of a porosimetry curve showing intrusion, extrusion, and hysteresis.

[44–47]. The volume measured at a specific pressure is directly related to the number of pores with that radius.

The steps displayed in the porosimetry curves are attributed to the pores of the sample. Very low pressures will fill the interparticle spaces when the sample is a powder. Increasing pressures will cause the mercury to penetrate into the pores, with higher pressures corresponding to smaller pores. For each step in the intrusion curve, there will be a corresponding step in the extrusion curve at a lower pressure.

The shape of the porosimetry curve provides information about the pores. The diagram in Fig. 5a represents a sample that contains essentially one pore size, as indicated by only one increase in volume. As diagrammed in Fig. 5b, two volume increases in the intrusion curve are observed, which is indicative of a bimodal pore distribution. Figure 5c is an example of a curve demonstrating a continuous range of pore sizes.

Porosimetry data can be graphed in a variety of ways and can be tailored to the purpose of the study. Plotting volume versus pore size will easily display the pore sizes observed in the sample. Pore size distributions can be calculated from the raw data and plotted to give the pore volume per unit radius interval. Other parameters can be calculated from porosimetry data, including average pore radius [40,48], surface area [7,39,40], pore surface area [6], particle size [40], and density [6,49].

(a)

(b)

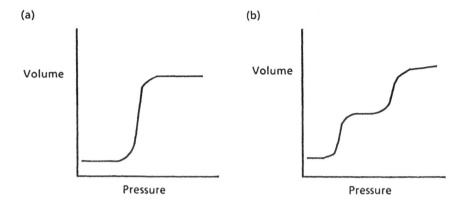

(c)

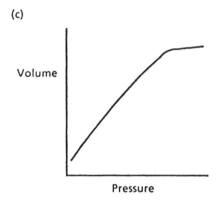

Fig. 5 Examples of porosimetry curves: (a) Sample with one pore size; (b) sample with bimodal pore size; (c) sample containing continuous pore sizes.

A number of factors can contribute to errors in mercury porosimetry [34]. The sample should be sufficiently outgassed for the type of data required. Some samples may require heating, whereas others may need to be outgassed at ambient temperatures due to possible degradation. The Washburn equation assumes a cylindrical pore shape, but this may not accurately represent the pores of most samples. Using a constant value for the surface tension and contact angle can also introduce error depending on the sample and the required accuracy and precision of the measurements. Compressibility of the solid at high intrusion pressures can occur, especially with highly porous samples. However, the compressibility of the sample can be measured with mercury porosimetry if needed [6]. High intrusion pressures can also result in the breakdown of larger pores, especially in sponge or foam type materials. With careful experimental design, porosimetry will provide quality data for most samples.

D. Applications

Many of the pharmaceutical applications of mercury porosimetry involve tablets [28–30,50–52]. A study of lactose granulations [30] reported that granules with varying porosities and compressive strengths will result in different tablet pore size distributions when using the same compression conditions. It was found that some granulations were fractured during compaction, resulting in a unimodal pore size distribution. Other granulations exhibited a bimodal curve, attributed to the presence of both intra- and intergranular pores in the compaction. The effect of varying compression parameters on the tablet porosity has also been studied. It has been reported that increasing the compression pressures can cause a decrease in the porosity of a tablet and a shift to smaller pore sizes [50]. Relationships between the pore surface area and tablet strength have also been observed [51].

Studies have shown that the pore size distribution can influence liquid penetration into tablets [28,29,52]. The accessibility of water vapor to tablet components has been related to chemical stability, especially in the case of hydrolyzable drugs [28]. Solvent penetration, which can affect tablet disintegration and dissolution, has been related to pore size [52].

The granulation procedure can also determine the porosity of tablets and granules. In a comparison study between wet and dry granulations [29], the wet granulation parameters, such as the water concentration, were found to influence the pore structure of the compression. The dry granulations, however, resulted in intense granule fragmentation under the same compression conditions. In the case of granules, changing wet granulation parameters resulted in a bimodal granule pore size distribution due to the presence of micropores and macropores [31]. The overall study supported the theory that microporous granules formed first during the process, followed by agglomeration, which formed the macropores. Another study on wet granulation showed that increasing the massing time produced more compact granules [53], as evidenced by the decreased intragranular porosity and increased granular strength.

Inert polymer matrices, studied for use in possible controlled release applications, have used porosimetry to investigate a number of properties [54–56]. The kinetics of liquid capillary penetration into these matrices was explored using a modified Washburn equation [54]. It was shown that water penetration depends on the matrix wettability and mean pore radius, as well as the pore size distribution. Modification of the polymer matrices has been attempted by sintering the material. One study [55] showed an initial expansion of the pore structure followed by a shrinking process for one polymer mix and a progressive reduction of the pore volume for a second polymer matrix. These changes in pore structure were found to affect the drug release rate. Another study [56] reported an increase in tablet tensile strength upon sintering the polymer as well as an increase in porosity and mean pore radius. This change in pore structure led to an increase in drug release rate.

Mercury porosimetry data has been used to calculate parameters other than pore size. Mean particle sizes have been obtained from mercury intrusion curves using the Mayer–Stowe equation and have been compared with sieving and electrical sensing zone data [57]. When an accurate measurement of the contact angle was obtained and possible aggregation was eliminated, good correlation was found between the techniques. Another study showed good correlation between particle sizes obtained from mercury porosimetry, image analysis, and sedimentation [58]. Surface area distributions have been obtained from porosimetry data using the Rootare–Prenzlow equation [57]. Analysis of pharmaceutical compounds, such as indoprofen and crosslinked povidone, resulted in comparable surface area data obtained from porosimetry and gas adsorption methods. Porosimetry has also been used in the characterization of magnesium stearate to calculate the surface area and particle size of materials from different vendors [59]. By using the one technique, it was found that one lot contained substantially smaller particles, which resulted in a higher surface area when compared with the other lot.

Overall, porosimetry measurements have been applied to a variety of pharmaceutical systems and have been used to investigate a number of material properties.

IV. DENSITY

Density is defined as the ratio of the mass of an object to its volume. It is dependent on the type of atoms in the molecule, as well as the arrangement of the atoms in the molecule and the arrangement of molecules in the sample. In a solid, the arrangement of the molecules, and therefore the density, is related to the crystalline nature of the compound. The density of a powder or granulation can affect a number of pharmaceutical processes, including flow, mixing, and tableting.

A number of different types of densities are reported in the literature [5,48,60,61], and some confusion is evident when comparing data in the literature. A summary of the most common terminology is given here:

Bulk density: The mass of particles composing the bed divided by the volume of the bed [49]. Also referred to as the apparent density [5,49].

Particle density: The mass of particles divided by the volume as determined by the displacement of mercury (pores greater than 10 μm). For nonporous solids, similar to true density [62]. Also known as granule density [49].

Tap density: The mass of particles divided by the volume after tapping the powder in a container. Also referred to as the drop density [61]. The volume measurement is similar to that of bulk density.

True density: The mass of the particles divided by the solid volume.

The distinction between the different types of densities is attributed to the volume, as shown schematically in Fig. 6. For the bulk density, the pore volume of the

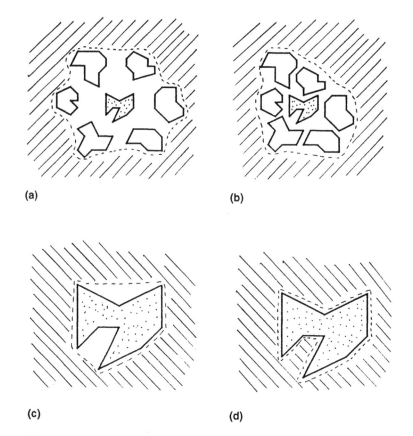

Fig. 6   The different types of densities. (a) Bulk density. (b) Tap density. (c) Particle density. (d) True density (Adapted from Ref. 49.)

particles and the interparticle spaces are included in the volume measurement. As illustrated for the tap density, when the sample is tapped, the interparticle space decreases, but the measurement is similar to the bulk density. For particle density, the interparticle spaces are not measured as part of the volume, but the pore volume is included. For true density, only the particle volume is measured.

The discussion in this section will focus on true, bulk, and tap densities. The procedure and instrumentation used for each type of density measurement will be outlined, and pharmaceutical applications involving density will be presented.

A. True Density

The three most common ways of obtaining true density measurements are gas pycnometry (gas displacement), liquid displacement, and flotation in a liquid. These three techniques have been compared based on accuracy, ease of use, and instrumentation [63], and the results are summarized in Table 4. Gas pycnometry will be discussed in this section because of its wide use and ease of operation.

The gas pycnometer is based on the ideal gas law. A known quantity of gas (characterized by a determined temperature and pressure) is allowed to flow from a calibrated reference volume into a calibrated sample cell containing the solid. A second pressure reading is obtained, and the sample volume is calculated.

The sample cell at ambient pressure can be defined as

Table 4 Comparison of True Density Techniques

Technique	Advantages	Disadvantages
Gas pycnometry (gas displacement)	Rapid Nondestructive Easy to use	Expensive instrumentation Large sample size Accuracy limited by precision of volume measurements
Liquid displacement	Glass pycnometers available	Tedious Tends to underestimate density Large sample size
Flotation in liquid	Simple operation Inexpensive Demonstrates density distribution Accurate to 4 significant figures	More time consuming than gas pycnometry Density vs. temperature curve suggested for each liquid Sample should be insoluble in liquid

$$P_A(V_C - V_P) = n_A R T_A \tag{16}$$

where

P_A = ambient pressure
V_C = cell volume
V_P = powder volume
n_A = number of moles occupying cell volume
R = gas constant
T_A = ambient temperature

When the reference volume is pressurized above the ambient pressure, the state of the reference volume (V_R) can be expressed as

$$P_1 V_R = n_1 R T_A \tag{17}$$

where

P_1 = pressure above ambient
V_R = reference volume
n_1 = number of moles of gas in V_R

When the sample cell is connected to the reference volume, the pressure will fall to a lower pressure (P_2), given by

$$P_2(V_C - V_P + V_R) = n_A R T_A + n_1 R T_A \tag{18}$$

Substituting $P_A(V_C - V_P)$ for $n_A R T_A$ and $P_1 V_R$ for $n_1 R T_A$ in Eq. (18) yields

$$P_2(V_C - V_P + V_R) = P_A(V_C - V_P) + P_1 V_R \tag{19}$$

or

$$(P_2 - P_A)(V_C - V_P) = (P_1 - P_2)V_R \tag{20}$$

Then,

$$V_C - V_P = \frac{P_1 - P_2}{P_2 - P_A} V_R \tag{21}$$

Allowing P_A to be zero, because all measurements are made above ambient and are relative to P_A, the equation becomes

$$V_C - V_P = \frac{P_1 - P_2}{P_2} V_R \tag{22}$$

which rearranges to

$$V_P = V_C - V_R \left(\frac{P_1}{P_2} - 1 \right) \tag{23}$$

The true density is then calculated from

$$\rho_{true} = \frac{W}{V_P} \tag{24}$$

where

ρ_{true} = true density
W = sample mass

When measuring the true density of a powder, it is important to use a gas that is not adsorbed by the material being measured. Both helium and nitrogen obey the ideal gas law at ambient pressures and temperatures; however, helium is preferred because of its smaller size [6]. It is also important to outgas the sample before the measurement to obtain consistent and reliable results.

B. Bulk and Tap Density

The bulk density (ρ_{bulk}) of a material depends on a number of factors, such as the size, shape, and cohesion of the particles [60,62]. Large particles, which may leave spaces, will result in a lower bulk density than smaller particles, which pack closer together and leave smaller spaces. The way in which the measurement is done can also affect the bulk density.

The most common method for measuring bulk density is to pour the powder into a tared graduated cylinder and measure the bulk volume (V_b) and mass of the material. Other methods, however, have been employed to ensure reproducibility, and a standard procedure has been reported [62,64]. For this testing, a sample of 50 g is passed through a U.S. Standard No. 20 sieve and is poured into a 100-ml graduated cylinder. The cylinder is then dropped from a height of 1 inch onto a hard surface three times at 2-second intervals. The volume of the powder is then read and used to calculate the bulk density. This three-tap method was found to give consistent results between laboratories.

Tap density (ρ_{tap}) is an extension of bulk density measurements, and the procedure used to measure the tap density again varies from lab to lab and can affect the final results. To measure tap density, a graduated cylinder is filled with powder and the weight and volume are recorded. The difference observed between procedures is the number of taps used for the measurement. In some cases, a particular number of taps is used, such as 200 [65], 500 [66], or even 1000 taps [67]. Other procedures involve tapping the cylinder for a number of times, recording the volume, and repeating the procedure until the volume remains constant [68]. This method ensures more consistent results.

C. Other Parameters

A number of other parameters can be calculated from true, bulk, and tap density measurements. These include void volume, porosity, bulkiness, and compressibility.

The volume of the spaces between particles is called the void volume (v) and can be calculated from the bulk and true volumes [62]:

$$v = V_b - V_p \tag{25}$$

Large or irregular particles will result in a large void volume, whereas small particles will pack closer together and exhibit a decreased void volume.

The total porosity (ϵ) is a combination of the void volume and the pores within the particles [5,62]. It can be calculated using volume measurements:

$$\epsilon = 1 - \frac{V_p}{V_b} \tag{26}$$

or density values:

$$\epsilon = 1 - \frac{\rho_{bulk}}{\rho_{true}} \tag{27}$$

The porosity value can be multiplied by 100 to obtain the percent porosity of the material.

Bulkiness, or specific bulk volume, is defined as the reciprocal of the bulk density [5,62]. It is often measured for the packaging of powders and, as expected, is affected by particle size. Bulkiness will increase with a decrease in overall particle size. A sample exhibiting a wide particle size distribution, however, will result in a lower bulkiness because the small particles will occupy the void spaces formed by the larger particles.

The compressibility of a material can be estimated from the tap and bulk density measurements [69]:

$$\text{compressibility} = \frac{\rho_{tap} - \rho_{bulk}}{\rho_{tap}} \times 100\% \tag{28}$$

Compressibility is indirectly related to the relative flow rate, cohesiveness, and particle size of a powder. A compressible material will be less flowable, and powders with compressibility values greater than 20–21% have been found to exhibit poor flow properties [69].

D. Applications

Density measurements have been correlated to a number of physical properties. The crystallinity and purity of powders have been investigated using a variety of density measurements [63]. Particle shape has been shown to influence bulk and tap densities [66], and the calculation of the shape coefficient was found to correlate linearly with bulk density [20].

A number of controlled release applications have used density to estimate the porosity of materials such as microcapsules and microsponges [70–73]. The preparation of microcapsules of theophylline [70] and ibuprofen [71] relied on porosity values to control the process by investigating the effects of solvent concentration, copolymer concentration, and pH. The density was also used to calculate the wall thickness of the microcapsules, and it was found that the decreased porosity could be correlated to a greater wall thickness. Investigations of hollow microspheres [72] for use as a floating controlled drug delivery system in the stomach also extensively employed density measurements for porosity and wall thickness calculations. The study showed that the floatability of the hollow microspheres was dependent on the bulk density, with the larger-diameter spheres resulting in a lower density and better floating behavior. The tap density of the material was used to investigate the packing properties of the microspheres in gelatin capsules. In the case of prolonged release microsponges [73], the internal porosity was calculated and was found to be easily controlled by changing the concentration of the drug and the polymer in the emulsion droplet. The porous internal structure resulted in stronger tablets and was correlated to the drug release rate for the materials.

Porosity calculations from density measurements have also been applied to granulations prepared using different processes. The method of granulation, such as the type of adjuvant used [64] or the amount of granulation liquid [74], was found to change the bulk density and porosity of the material. Consequently, the compression and flow properties of the materials were also different.

The flow properties of powders have been correlated to bulk density measurements. It was reported that an increase in bulk density will increase flow properties, as predicted by both the flow factor and shear index of the material [75]. During the granulation of micronized active drugs, the granules with the highest bulk density resulted in the best flow rate [64]. In a study on the lubrication properties of magnesium stearate, it was concluded that low bulk densities of the granules resulted in poor flow properties, which impeded the formation of a lubricant film during mixing [76].

The various types of density measurements have been used in a variety of applications, ranging from synthesis to granulation and manufacturing. Although the techniques are relatively simple, valuable information is obtained about the system being studied.

V. CONCLUSIONS

Surface area, porosimetry, and density measurements have become an integral part of the characterization of pharmaceutical systems. These techniques can be applied to all stages of research, development, and manufacturing to better understand the material, system, or process being investigated. The combination of a number of characterization techniques will result in a strong interdisciplinary approach to effectively solve problems and help prevent difficulties in the future.

REFERENCES

1. J. M. Dallavalle, *Micromeritics*, 2nd ed., Pitman Publishing Company, New York, 1948.
2. C. Orr and J. M. Dallavalle, *Fine Particle Measurements*, The Macmillan Company, New York, 1959.
3. E. L. Parrott, *Pharm. Manuf.*, *2*, 30–37 (1985).
4. J. T. Carstensen, *Pharmaceutical Principles of Solid Dosage Forms*, Technomic Publishing Co., Lancaster, Pa., 1993.
5. E. L. Parrott, in *Pharmaceutical Technology: Fundamental Pharmaceutics*, Burgess, Minneapolis, Minn., 1970, pp. 1–36.
6. S. L. Lowell and J. E. Shields, *Powder Surface Area and Porosity*, Chapman and Hall, New York, 1984.
7. H. M. Rootare and C. F. Prenzlow, *J. Phys. Chem.*, *71*(8), 2733–2736 (1967).
8. S. Brunauer, P. H. Emmett, and E. Teller, *J. Am. Chem. Soc.*, *60*, 309 (1938).
9. S. J. Gregg and K. S. W. Sing, *Adsorption, Surface Area, and Porosity*, Academic Press, New York, 1967.
10. K. S. W. Sing, in *Particle Size Analysis*, N. G. Stanley-Wood and R. W. Lines, eds., Royal Society of Chemistry, Cambridge, U.K., 1992, pp. 13–32.
11. R. Antila and J. Yliruusi, *Acta Pharm. Nord.*, *3*(1), 15–18 (1991).
12. E. Suzuki, K.-I. Shirotani, Y. Tsuda, K. Sekiguchi, *Chem. Pharm. Bull.*, *27*(5), 1214–1222 (1979).
13. A. Sano T. Kuriki, Y. Kawashima, H. Takeuchi, T. Hino, and T. Niwa, *Chem. Pharm. Bull.*, *38*(3), 733–739 (1990).
14. J. Kreuter, *Int. J. Pharm.*, *14*, 43–58 (1983).
15. A. Holzer and J. Sjogren, *Int. J. Pharm.*, *2*, 145–153 (1979).
16. E. Suzuki, K. Chihara, and K. Sekiguchi, *Chem. Pharm. Bull.*, *31*(2), 632–638 (1983).
17. S. S. Kormblum and S. B. Stoopak, *J. Pharm. Sci.*, *62*(1), 43–49 (1973).
18. N. G. Stanley-Wood and M. S. Shubair, *J. Pharm. Pharmacol.*, *31*, 429–433 (1979).
19. T. Sato, M. Kanke, H. G. Schroeder, and P. P. Deluca, *Pharm. Res.*, *5*(1), 21–30 (1988).
20. K. Ridgway and R. Rupp, *J. Pharm. Pharmac.*, *21*(Suppl.), 30S–39S (1969).
21. E. Suzuki, K.-I. Shirotani, Y. Tsuda, and K. Sekiguchi, *Chem. Pharm. Bull.*, *33*(11), 5028–5035 (1985).
22. P. V. Allen, D. Rahn, A. C. Sarapu, and A. J. Vanerwielen, *J. Pharm. Sci.*, *67*(8), 1087–1093 (1978).
23. H. Brittain, S. J. Bogdanowich, D. E. Bugay, J. DeVincentis, G. Lewen, and A. W. Newman, *Pharm. Res.*, *8*(8), 963–973 (1991).
24. N. G. Stanley-Wood and M. E. Johansson, *Zbl. Pharm.*, *117*, 568–582 (1978).

25. M. Whiteman and R. J. Yarwood, *Powder Tech.*, *54*, 71–74 (1988).
26. P. J. Dees, *Powder Tech.*, *29*, 187–197 (1981).
27. W. Lowenthal, *J. Pharm. Sci.*, *61*(11), 1695–1711 (1972).
28. H. Gucluyildiz, G. S. Banker, and G. E. Peck, *J. Pharm. Sci.*, *66*(3), 407–414 (1977).
29. A. B. Selkirk and D. Ganderton, *J. Pharm. Pharmacol.*, *22*, 86S–94S (1970).
30. M. Wikberg and G. Alderborn, *Int. J. Pharm.*, *84*, 191–195 (1992).
31. M. A. Zoglio and J. T. Carstensen, *Drug Dev. Ind. Pharm.*, *9*(8), 1417–1434 (1983).
32. W. O. Opakunle and M. S. Spring, *J. Pharm. Pharmacol.*, *28*, 806–809 (1976).
33. I. Colombo and F. Carli, *Farm. Ed. Prat.*, *39*(10), 329–341 (1984).
34. H. M. Rootare, in *Advanced Experimental Techniques in Powder Metallurgy*, Plenum, New York, 1970, pp. 225–252.
35. N. G. Stanley-Wood, N. Osborne, and M. Till, in *Particle Size Analysis*, N. G. Stanley-Wood and R. W. Lines, eds., Royal Society of Chemistry, Cambridge, U.K. 1992, pp. 48–57.
36. S. Lowell and J. E. Shields, *Powder Tech.*, *29*, 225–231 (1981).
37. E. W. Washburn, *Proc. Natl. Acad. Sci. USA*, *7*, 115–116 (1921).
38. M. Svata, *Powder Tech.*, *29*, 145–149 (1981).
39. J. Van Brakel, S. Modry, and M. Svata, *Powder Tech.*, *29*, 1–12 (1981).
40. C. Orr, *Powder Tech.*, *3*, 117–123 (1969/70).
41. N. W. F. Kossen and P. M. Heertjes, *Chem. Eng. Sci.*, *20*, 593–599 (1965).
42. L. Moscou and S. Lub, *Powder Tech.*, *29*, 45–52 (1981).
43. Autoscan-60 Instrument Manual, Quantochrome Corporation, Syosset, N.Y.
44. S. Lowell and J. E. Shields, *Powder Tech.*, *38*, 121–124 (1984).
45. M. L. Shively, *J. Pharm. Sci.*, *80*(4), 376–379 (1991).
46. S. Lowell and J. E. Shields, *J. Colloid Interface Sci.*, *83*(1), 273–278 (1981).
47. S. Lowell and J. E. Shields, *J. Colloid Interface Sci.*, *80*(1), 192–196 (1981).
48. H. L. Ritter and L. C. Drake, *Ind. Eng. Chem. Anal. Ed.*, *17*, 782–786 (1945).
49. K.-I. Mukaida, *Powder Tech.*, *29*, 99–107, (1981).
50. F. Carli, I. Colombo, L. Simioni, and R. Bianchini, *J. Pharm. Pharmacol.*, *33*, 129–135 (1981).
51. A. H. De Boer, H. Vromans, C. F. Lerk, G. K. Bolhuis, K. D. Kussendrager, and H. Bosch, *Pharm. Weekbl. Sci. Ed.*, *8*, 145–150 (1986).
52. D. Sixsmith, *J. Pharm. Pharmacol.*, *29*, 82–85 (1977).
53. W. O. Opakunle and M. S. Spring, *J. Pharm. Pharmacol.*, *28*, 508–511 (1976).
54. F. Carli and L. Simioni, *Pharm. Acta Helv.*, *53*(11), 320–326 (1978).
55. F. Carli and L. Simioni, *Int. J. Pharm. Tech. Prod. Mfr.*, *2*(1), 23–28 (1981).
56. R. C. Rowe, P. H. Elworthy, and D. Ganderton, *J. Pharm. Pharmacol.*, *25*, 12P–16P (1973).
57. F. Carli and A. Motta, *J. Pharm. Sci.*, *73*(2), 197–202 (1984).
58. R. Pospech and P. Schneider, *Powder Tech.*, *59*, 163–171 (1989).
59. I. Colombo and F. Carli, *Il Faarm. Ed. Pr.*, *39*(10), 329–341 (1985).
60. J. E. Rees, *Boll. Chim. Farm.*, *116*, 125–141 (1977).
61. E. L. Parrott, *J. Pharm. Sci.*, *70*(3), 288–291 (1981).
62. A. Martin, J. Swarbrick, and A. Cammarata, *Physical Pharmacy*, 3rd ed., Lea and Febiger, Philadelphia, 1983.

63.   W. C. Duncan-Hewitt and D. J. Grant, *Int. J. Pharm.*, *28*, 75–84 (1986).
64.   A. Soinen, *Acta Pharm. Fenn.*, *90*, 117–127 (1981).
65.   L. C. Li, and G. E. Peck, *Drug Dev. Ind. Pharm.*, *16*(9), 1491–1503 (1990).
66.   E. Shotton and B. A. Obiorah, *J. Pharm. Pharmacol.*, *25*(Suppl.), 37P–43P (1973).
67.   M. K. Doelling and R. A. Nash, *Pharm. Res.*, *9*(11), 1493–1501 (1992).
68.   K. A. Khan and C. T. Rhodes, *Can. J. Pharm. Sci.*, *11*, 109–112 (1976).
69.   R. L. Carr, *Chem. Eng.*, *72*(2), 163–168 (1965).
70.   S. Y. Lin, J. C. Yang, and S. S. Jiang, *J. Taiwan Pharm. Assoc.*, *37*(1), 1–10 (1985).
71.   H. Takenaka, Y. Kawashima, and S. Y. Lin, *J. Pharm. Sci.*, *69*(5), 513–516 (1980).
72.   Y. Kawashima, T. Niwa, H. Takeuchi, T. Hino, and Y. Itoh, *J. Pharm. Sci.*, *81*(2), 135–140 (1992).
73.   Y. Kawashima, T. Niwa, H. Takeuchi, T. Hino, and Y. Itoh, *Chem. Pharm. Bull.*, *40*(1), 196–201 (1992).
74.   M. Wikberg and G. Alderborn, *Int. J. Pharm.*, *62*, 229–241 (1990).
75.   C. F. Harwood, *J. Pharm. Sci.*, *60*(1), 161–163 (1971).
76.   C. E. Bos, H. Vromans, and C. F. Lerk, *Int. J. Pharm.*, *67*, 39–49 (1991).

10

Physical and Mechanical Property Characterization of Powders

Gregory E. Amidon
The Upjohn Company, Kalamazoo, Michigan

I. INTRODUCTION

A great many investigations have dealt with the effects of the physical and chemical properties of materials on powder processing. While physical properties clearly influence powder flow and compaction, systematic research on the effects of the mechanical properties of materials is limited. This chapter describes the importance of the physical and the mechanical properties of materials as well as some basic principles and methodologies that can be used to investigate the influence of these properties on powder flow and compaction. For the purposes of this discussion, physical properties are substantially independent of energy and fundamental force considerations and are therefore limited to those properties that are "perceptible especially through the senses" (i.e., properties such as particle size and shape). In contrast, mechanical properties are those which deal with the energy and forces that influence the properties of materials. Elasticity, plasticity, viscoelasticity, hardness, brittleness, etc., all involve energy considerations and the fundamental forces of nature.

II. FACTORS INFLUENCING POWDER PROPERTIES

Table 1 lists a number of factors that may influence powder flow and compaction. The list is long and includes physical and mechanical properties as well as

Table 1 Factors That Influence Powder Flow and Compaction

1. Purity	14. True density
2. Crystallinity	15. Bulk density
3. Surface energy	16. Moisture content
4. Electrostatic charge	17. Humidity
5. Elastic deformation properties	18. Adsorbed air, water, impurities
6. Plastic deformation properties	19. Packed density
7. Brittleness	20. Consolidation load
8. Viscoelastic properties	21. Consolidation time
9. Particle density	22. Direction of shear
10. Particle size	23. Rate of shear
11. Particle size distribution	24. Storage container dimensions
12. Surface area	25. Particle/wall interactions
13. Particle shape	

environmental effects. Surface energy changes and elastic deformation properties, for example, influence individual particle true areas of contact. Plastic deformation likely occurs to some extent in powder beds depending on the applied load, and almost certainly it occurs during the compaction of powders into tablets. Certainly at asperities, local regions of high pressure can lead to localized plastic yielding. Electrostatic forces can also play a role in powder flow depending on the insulating characteristics of the material and environmental conditions. Particle size, shape, and size distribution have all been shown to influence flow and compaction as well. A number of environmental factors such as humidity, adsorbed impurities (air, water, etc.), consolidation load and time, direction and rate of shear, and storage container properties are also important. With so many variables, it is not surprising that a wide variety of methods have been developed to characterize materials. A number of these methods of physical and mechanical property characterization are discussed in this chapter.

III. IMPORTANT PHYSICOCHEMICAL PROPERTIES

A. Particle Size

As we just suggested, particle size and shape are important physical properties influencing powder flow and compaction. Particle size is a simple concept and yet a difficult one to quantitate. Feret's diameter, Martin's diameter, projected area diameter, specific surface diameter, Stokes diameter, and volume diameter are but several of the measurements that have been used to quantify particle size using a variety of methods.

The importance of particle size in determining powder flow has long been recognized. Nelson [1] first utilized angle of repose measurements to demonstrate that decreasing particle size resulted in an increased angle of repose, suggesting that the coefficient of interparticle friction increased with decreasing particle size. Still others [2–4] have demonstrated particle-size effects with flow-through-an-orifice experiments. In general the flow rate was observed to increase with decreasing particle size (in contrast to angle of repose observations) until a maximum is reached; followed by a relatively rapid decrease in flow at smaller particle size. The decreased flow for small particles has been ascribed to the increasing importance of van der Waals, electrostatic, and surface tension forces [2,5], while the decreased flow at larger particle sizes is attributable, at least in part, to limitations due to the size of the orifice relative to that of the particles.

More fundamental papers have assessed the effects of particle size on the tensile strength and flow of powders. Rumpf [6], for example, developed a simple model of particle–particle interactions and tensile strength. Taking account of geometrical considerations, the tensile strength was predicted to be inversely proportional to the square of the diameter and directly proportional to the bonding force per point of contact. Cheng and coworkers [7–9] also derived an expression for the tensile strength of a powder as a function of particle size, surface area and volume, and the interparticle forces. The tensile strength depends on the number of particle–particle contacts, which is a function of the particle size, and the interparticle forces. Once again, the tensile strength was predicted to increase with decreasing particle size.

B. Particle Shape

A variety of measures of particle shape have also been devised and have been reviewed in the literature [10–12]. Ridgway and Rupp [13] studied the effects of size and shape (as characterized by Heywood shape parameters) on the flow of sand and showed that the angle of repose increased while the flow rate through an orifice and bulk density decreased with increasing angularity. The same authors [14] showed that angular materials reduced the amount of mixing of material moving down an inclined chute. Others [15] have done similar work using a variety of methods for quantifying particle shape. Harwood and Pilpel [16] determined that the friction coefficient (determined in a shear cell similar to that of Jenike) not only varied with particle size, but also with particle shape: angular griseofulvin had a larger coefficient of friction than rounded griseofulvin.

As with particle size, particle shape can influence the compaction properties of solids. While work in this area is limited, some work has been reported. Ridgway and Scotton [17], for example, investigated the effect of particle shape on die-fill weight and found that more angular materials had a greater weight variation. Also, Rupp [18] investigated the effects of particle shape on tablet strength and observed that tablets made with more angular material had higher strength; the author ascribed this to increased particle interlocking.

C. Moisture

Other factors that can influence particle cohesion include moisture adsorption and capillary condensation. While the significance of these factors is difficult to assess, they have been the subject of some detailed theoretical and experimental analysis. Moisture adsorption and capillary condensation at points of particle contact have been examined by several authors [6,19–21]. In general, water can play a role in one of two ways. At low levels of adsorption (i.e., 1–2 molecular layers), adsorbed water is generally considered to be immobile and therefore will not act to form a liquid bridge. Rumpf [6] concluded that, to a first approximation, the adsorbed water layer can be added to the solid, in effect bringing the particles closer together. The water can fill in the valleys, which also increases the contact area. This of course means greater van der Waals interactions. When adsorbed water begins to act as bulk water, liquid bridging can occur, and added strength is then associated with capillary condensation.

Coelho and Harnby [20] have shown, theoretically, that the relative humidity at which liquid bridges form is a function of the affinity of the material for water adsorption. The same authors [21] also concluded that as surface roughness increases, the significance of the adsorbed moisture layer decreases until, with very rough surfaces, it plays no role until capillary condensation occurs. While the thickness of adsorbed moisture layers is frequently unknown, Kontny et al. [22] have shown that the water vapor adsorption onto water-soluble, crystalline, materials (NaCl and sodium salicylate) consisted of at most a single layer of water. For cellulosic materials up to 70% relative humidity, the sorbed moisture falls in to the "nonfree" category [23]. In general, then, this would suggest that at low to moderate levels of humidity, capillary condensation will not influence particle adhesion significantly for many materials.

D. Electrostatic Effects

Electrostatic effects on particle–particle interactions, too, are difficult to assess accurately. Electrostatic charging of powders during processing steps such as milling, micronizing, flow, and tableting or capsule filling do occur. For pharmaceutical materials, which are insulators, this can be due to particle–wall collisions, particle–particle collisions, or particle fragmentation. Harper [24] concludes that it is generally the latter two that are important unless an effort is made to maximize particle–wall interactions. Rumpf [6] concludes, however, that electrostatic forces are dependent on the charge density of the material but that they should have a negligible influence on the strength of particle–particle bonding. Tablets are not held together by electrostatic charges. Staniforth [25] showed that the flow rate through an orifice of material charged in an air cyclone and tested within 15 minutes of charging did show poorer flow than "uncharged" material. In practice, though, it is possible to minimize the electrostatic effects by permitting the charge to decay. The presence of moisture in the air may also provide a conducting medium through which the electrostatic charge can easily dissipate.

E. Fundamental Forces

What, then, actually holds particles together? A detailed discussion is beyond the scope of this chapter, and the reader is referred to the literature for more detailed discussions [6]. It is important, however, to realize that the forces that hold particles together in a tablet or powder bed are the very same forces discussed in detail in introductory physical chemistry courses. There is nothing magical about these interactions. Among the very important forces involved are London dispersion forces, dipole interactions, surface energy considerations, and perhaps hydrogen bonding. Ionic and covalent bonding are generally not thought to be important. Electrostatic interactions may be important under some circumstances such as during powder flow and handling. As we will see, the consolidation of powders brings particles into close proximity, where these fundamental forces can begin to act effectively. Elastic and plastic deformation of material establishes greater areas of true contact between particles; the end result is cohesion.

III. IMPORTANT MECHANICAL PROPERTIES

Materials we use in the pharmaceutical industry can be elastic, plastic, visco-elastic, hard, tough, or brittle in the same sense that construction materials are; the same concepts that mechanical engineers use to explain or characterize tensile, compressive, or shear strength are relevant to pharmaceutical materials. These mechanical properties of materials, though not often studied in detail, can have a profound effect on solids processing. Clearly, tableting properties are influenced by the elastic and plastic deformation properties as well as the viscoelastic properties of a material. Also affected, though, can be the powder flow properties as well as the tendency of materials to "set up" on storage.

The mechanical properties of a material play an important role in powder flow and compaction by influencing particle–particle interaction and cohesion, that is to say, by influencing the true area of contact between particles. For example, Hertz [26] demonstrated that both the size and shape of the zone of contact followed simply from the elastic properties of a material. Clearly then, the true area of contact is affected by elastic properties. From the laws of elasticity, one can predict the area of contact between two elastic bodies. More recent work has demonstrated, however, that additional factors must be taken into consideration. Johnson and coworkers [27] discussed the influence of surface energy in addition to elastic properties on the area of contact between elastic spheres. The authors' model predicts a finite area of contact at zero applied load in contrast to the work of Hertz, which predicts zero area of contact at zero applied load. The theory was shown to apply to the contact of gelatin spheres on flat acrylic plastic under small applied loads.

Because of the importance of mechanical properties, it is important to be able to quantitatively characterize materials. Reliable mechanical property information can be useful in (1) helping to choose a processing method such as granulation or direct compression, (2) selecting excipients with properties that will mask the poor properties of the drug, and (3) helping to document what went wrong, for example, when a tableting process is being scaled up or when a new bulk drug process is being tested. Since all of these things can influence the quality of the final product, it is to the formulator's advantage to understand the importance of these issues and to be able to quantitate them.

A. Stress, Strain, and Deformation

To characterize the mechanical properties of a material, one first needs a basic understanding of the concepts of stress, strain, and deformation, as they provide the tools necessary. The deformation properties of a material can be determined by applying a stress, either in compression or tension, and determining dimensional changes in the specimen. The applied stress will result in an elongation of the specimen, $\epsilon = \Delta l / l_0$. The elongation is called the strain, while the stress is defined as the applied load divided by the area over which it is applied.

B. Elastic Deformation

In general, during the initial stages of deformation, a material is deformed elastically. That is to say, any change in shape caused by the applied stress is completely reversible, and the specimen will return to its original shape upon release of the applied stress. During elastic deformation, the stress–strain relationship for a specimen is described by Hooke's law:

$$\sigma = E\epsilon \tag{1}$$

where σ is the applied stress, E is referred to as Young's modulus of elasticity, and ϵ is the strain.

For materials with a large modulus of elasticity, relatively high stresses yield only small changes in length (strain); materials with low elastic moduli are normally thought of as elastic materials.

The elastic properties of materials can be understood, at least qualitatively, by considering the attractive and repulsive forces between atoms and molecules.

Elastic strain results from a change in the intermolecular spacing and, at least for small deformations, is reversible.

C. Plastic Deformation

Plastic deformation is the permanent change in shape of a specimen due to applied stress. The onset of plastic deformation is seen as curvature in the stress–strain curve. Plastic deformation is important because it allows pharmaceutical excipients and drugs to establish large true areas of contact during compaction that can remain on decompression. In this way, good, intact, tablets can be prepared.

Plastic deformation, unlike elastic deformation, is not accurately predicted from atomic or molecular properties. Rather, plastic deformation is determined by the presence of crystal defects such as dislocations and grain boundaries. While it is not the purpose of this chapter to discuss this in detail, it is important to realize that dislocations and grain boundaries are influenced by things such as the rate of crystallization, particle size, the presence of impurities, and the type of recrystallization solvent used. Processes that influence these can be expected to influence the plastic deformation properties of materials, and hence the processing properties.

The plastic properties of a material are often determined by an indentation test [28]. Both static and dynamic test methods are available, but all generally determine the pressure necessary to cause permanent and nonrecoverable deformation.

D. Viscoelastic Properties

In addition to elastic and plastic deformation, the viscoelastic properties may be important. Viscoelasticity reflects the time-dependent nature of strain. Once again, a basic understanding of viscoelasticity can be gained by considering processes that occur at a molecular level when a material is under stress. An applied stress, even when in the elastic region, effectively moves atoms or molecules from their lowest energy state. With time, then, the rearrangement of atoms or molecules can occur. The stress–strain relationship can therefore depend on the time frame over which the test is conducted. In compacting tablets, for example, it is frequently noted that higher compaction forces are required to make a tablet with a given hardness when the compaction speed is high compared with when the speed is low. This is consistent with the view that pharmaceutical materials are viscoelastic. Viscoelastic properties clearly show up when comparing pseudostatic and dynamic deformation testing.

E. Brittle and Ductile Fracture

In addition to plastic deformation, materials may fail by either brittle fracture or ductile fracture; fracture being the separation of a body into two or more parts.

Brittle fracture occurs by the rapid propagation of a crack throughout the specimen. The propagation is generally very rapid. Conversely, ductile fracture is characterized by fracture following extensive plastic deformation. Ductile fracture is not usually seen with compacts of organic materials. Brittle fracture is often seen, however. The characteristic snap of a tablet during hardness testing is brittle fracture.

IV. METHODS OF CHARACTERIZING THE MECHANICAL PROPERTIES OF MATERIALS

Methods for characterizing the elastic, plastic, and brittle properties of compacts of organic materials have been developed by Hiestand and coworkers [29–33]. These indices of tableting performance measure the mechanical properties of compacted materials.

A. Pendulum Impact Device

A simple schematic of a pendulum impact device (PID) is given in Fig. 1. This equipment permits the permanent deformation pressure (H) of a compact of material to be determined [30,31]. Flat-faced tablets of the test substance are compressed at different compression forces and then subjected to impact with a

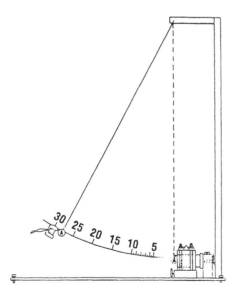

Fig. 1 Schematic diagram of the pendulum impact device (PID) used to determine the indentation hardness of materials. (Adapted from Ref. 30.)

stainless ball in the PID. The rebound height of the ball and the chordal radius of the dent are carefully measured and used to calculate the permanent deformation pressure. In a simple sense, one is measuring the energy necessary to make the permanent deformation (the difference between the initial height of the ball and the rebound height). By measuring the volume of the dent, one can calculate the deformation pressure, which is the energy divided by the volume. The permanent deformation pressure, then, is the pressure (i.e., stress) necessary to cause plastic deformation.

B. Tensile Strength Determination

The tensile strength of compacts [30] also provides useful information. Excellent specimens of square compacts are necessary to conduct the tensile testing. For this reason, a split die [31] (Fig. 2) is used to make compacts that are not flawed. The split die permits triaxial decompression, which relieves the stresses in the compact more uniformly in three dimensions and minimizes cracking. These specimens are then compressed with platens 0.4 times the width of the square compacts in the tensile testing apparatus. (Fig. 3). Occasionally nylon platens and side supports are used to reduce the tendency to fail in shear rather than tension. The force necessary to cause tensile failure (tensile forces are a maximum

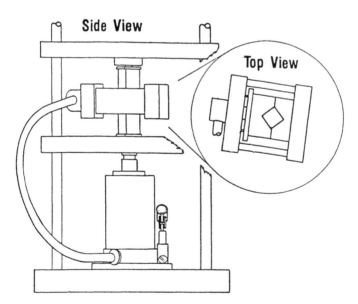

Fig. 2 Schematic diagram of the split die apparatus, which allows for the compaction of materials for characterization. (Adapted from Ref. 31 with permission of the publisher.)

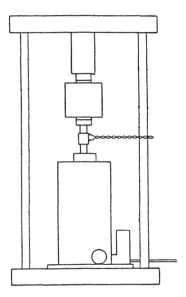

Fig. 3 Schematic diagram of the tensile testing apparatus, which allows for the determination of the tensile strength of compacts of materials. (Adapted from Ref. 31 with permission of the publisher.)

at the center of the tablet) is monitored by a load cell, and the magnitude of the force at fracture is determined. Testing of square compacts has advantages over the testing of circular compacts; however, circular compacts can be used; conventional hardness testing of tablets can result in a measurement of tensile strength.

V. TABLETING INDICES

Using the methodology we have described, several indices of tableting performance have been developed by Hiestand and coworkers [29,31–33]. These indices provide relative measures of properties that are considered important and that reflect the performance of materials during processing.

A. Bonding Index

The purpose of the Bonding Index is to estimate the survival of strength during decompression [31]; it is defined as

$$\mathrm{BI} = \frac{\sigma_T}{H} \tag{2}$$

where

σ_T = the tensile strength of the compact at a given solid fraction, and
H = the permanent deformation pressure (i.e., hardness) of a compact at a given solid fraction

At maximum compression, the bonded areas are at a maximum because the true areas of contact are maximized. During decompression, some of that area is "lost" due to elastic recovery. A high Bonding Index (BI) indicates that, relatively speaking, a larger portion of the strength remained intact during decompression. A low Bonding Index indicates that less of the strength remains. The term *Bonding Index*, then, is a good description since it, in effect, characterizes the tendency of the material to remain intact after it has been compressed. Tablets made of materials with poor bonding characteristics may be quite friable. Compacts made of materials with good bonding indices may, conversely, make strong tablets.

Hiestand and coworkers have further refined the concept of Bonding Index to include both a worst case and a best case Bonding Index [31]. The Bonding Index is determined under different experimental conditions (i.e., the rate at which the permanent dent is made in a compact is varied) such that the viscoelastic properties of the material are assessed. If a material is very viscoelastic, there is substantial stress relaxation with time. It is reasonable to expect, then, that tablets that are slowly deformed during the determination of the hardness, H, may retain more of the bonded area than tablets that are rapidly deformed (i.e., as in the pendulum impact device) since some of the stresses developed during compaction will have a chance to be relieved. The Worst Case Bonding Index is measured using a very high speed, dynamic method, while the Best Case Bonding Index is measured using a "quasi-static" method. The Bonding Index is calculated as previously described for both.

B. Brittle Fracture Index

Just as metals can be ductile or brittle, so can organic materials. The Brittle Fracture Index is a measure of the brittleness of a material. It is a measure of the ability of a compact of material to relieve stress by plastic deformation. The Brittle Fracture Index (BFI) is determined [29,31] by comparing the tensile strength of a compact, σ_T, with that of a compact with a small hole (stress concentrator) in it, σ_{T0}, using the tensile test we have described. A hole in the center of the compact generally weakens a tablet. If a material is very brittle, theoretical considerations show that the tensile strength of a tablet with a hole in it will be about one-third that of a "solid" tablet. If, however, the material can relieve stress by plastic deformation, then the strength of the compact with a hole in it will approach that of a compact with no hole. The Brittle Fracture

Index is defined such that very brittle compacts have a BFI of 1 and very nonbrittle materials have a BFI close to 0; it is calculated as follows:

$$BFI = \frac{1}{2} \left(\frac{\sigma_T}{\sigma_{T_0}} - 1 \right) \tag{3}$$

C. Strain Index

The Strain Index is an indicator of the relative strain (change in size) during decompression (assuming that plastic deformation has occurred). Some materials require large stresses to cause a given dimensional change; other materials require relatively less stress to give the same change in dimension. The latter have a low modulus of elasticity and therefore, for a given stress, are going to show a larger change in dimension. The Strain Index is defined as

$$SI = \frac{H}{E'} \tag{4}$$

where E' = the reduced elastic modulus [29].

The Strain Index is determined from data obtained using the pendulum impact device.

VI. METHODS OF DETERMINING POWDER FLOW

Powder flow is most frequently thought of as relevant to formulation development, and there are numerous references attempting to correlate any one of a number of measures of powder flow to the manufacturing properties of a formulation [34–40]. In particular, the importance of physical properties in affecting powder flow has been well documented. Research into the effect of the mechanical properties on powder flow has, however, been very limited. It is, of course, important to be able to determine and quantitate the powder flow properties of formulations. It is of equal importance, however, to determine the powder flow characteristics of bulk drug early in the development process (preformulation phase). Often, the preformulation or formulation scientist is constrained by time, materials, and manpower. Yet certainly the preformulation studies carried out should be meaningful. Well-defined experimental methods and procedures should be used; the information generated should be reproducible and permit useful predictions to be made.

A. Angle of Repose

The angle of repose has been used in several branches of science to characterize the flow properties of solids. Nelson [1] was one of the first to use angle of repose measurements to determine the flow properties of pharmaceutical materials by studying the influence of particle size, lubricant levels, and percentage fines on the flow of sulfathiazole granulations. Train [41] evaluated several methods of determining angle of repose and concluded that it was a measure of interparticulate friction, or resistance to movement between particles. He also concluded that results were very dependent upon the method used. Simple geometric considerations led Train to conclude that, for noncohesive materials, "theoretical" upper and lower limits for the angle of repose were 45°, and 19.3°, respectively.

Carr [42–44] proposed that the angle of repose should be just one of several methods used in characterizing powder flow. On the other hand, Jenike states that angle of repose "is not a measure of the flowability of solids" [45] and is popular primarily because of its ease of determination. He points out that experimental difficulties arise due to segregation of material and consolidation or aeration of the powder as the cone is formed. Poor correlations between angle of repose and other measures of flow such as flow through an orifice [4,46,47] have been observed. In general, the reproducibility of the measured angle is on the order of 1°.

Amidon and Houghton [48] completed a comparative study of several common methods of characterizing powder flow. Table 2 contains experimental results for a number of commonly used pharmaceutical excipients. Compressibility index, angle of repose, flow rate through an orifice, and shear cell data are presented.

Angle of repose measurements have generally not been considered reliable predictors of powder flow. In part, this is due to a lack of sensitivity of the method to distinguish between two "slightly" different flowing materials. This is illustrated in Table 2 by comparing sodium chloride, spray-dried lactose, and fast-flo lactose. Similar angles of repose were observed, yet quite different flow rates through an orifice (methodology to be described) were observed. This is consistent with the observations of Gold and coworkers [46]. Unavoidable experimental difficulties such as powder-bed consolidation, particle segregation, and aeration in forming the powder cone cannot easily be avoided [45]. This, in addition to measurements being accurate only to within approximately 1°, makes predictions based on angle of repose unreliable. Even for poor flowing materials, the correlation between other methods of determining powder flow such as compressibility index [49,50] and angle of repose appears poor, although the anticipated trend is apparent. The correlation between the angle of repose and compressibility index was quite good, however, for mixtures of bolted and spray-dried lactose (Table 3).

B. Compressibility Index

In recent years the compressibility index C_i, has become a simple, fast, and popular method of predicting powder flow characteristics [40,42–44,51,52]. Carr [42–44] proposed its use as an indirect measure of bulk density, size and shape,

Table 2 A Comparison of Several Methods of Measuring Powder Flow

Material	C_i %	Angle of repose	Flow rate through a 6-mm orifice	n (SE)[l]	Simplified shear cell S (SE[l]), kdyne	δ'	f_c, kdyne
NaCl[a]	15.2	33.7 (0.5)	3.75	0.99 (0.02)	13 (13)	32.9	29.6
Spray-dried lactose	16.0	33.5 (0.5)	2.05	1.02 (0.006)	2.9 (4.2)	33.1	78
Fast-flo lactose[c]	17.6	33.5 (0.6)	1.53	1.03 (0.007)	13.7 (4.3)	32.9	40.6
Microcrystalline cellulose—coarse[d]	31.2	38.2 (0.9)	Plugged	1.10 (0.01)	73 (10)	37.3	347
Microcrystalline cellulose—medium[e]	34.4	39.2 (0.6)	Plugged	1.17 (0.02)	58 (14)	40.5	480
Cornstarch[f]	38.4	57.1 (1.5)	Plugged	1.27 (0.03)	100 (21)	34.0	645
Lactose[g]	44.4	50.5 (0.7)	Plugged	1.38 (0.01)	6.5 (4.7)	37.5	289
Talc[h]	57.2	45.6 (0.8)	Plugged	1.42 (0.05)	0.0 (17)	42.5	—
Dicalcium phosphate[i]	49.6	51.1 (0.9)	Plugged	1.59 (0.04)	0.0 (9.4)	39.5	—
Sucrose[j]	49.2	56.3 (1.9)	Plugged	1.62 (0.02)	33.2 (7.8)	42.0	1640
Magnesium stearate[k]	57.6	48.8 (2.3)	Plugged	—	—	—	—

[a]Sodium chloride USP (granular), Morton Salt.
[b]Lactose USP—hydrous spray process standard, McKesson Chemical.
[c]Fast-flo lactose USP, McKesson Chemical.
[d]Microcrystalline cellulose—coarse NF powder (Avicel PH102), FMC.
[e]Microcrystalline cellulose—medium powder (Avicel PH101), FMC.
[f]Cornstarch NF powder, bolted, National Starch & Chemical.
[g]Bolted lactose USP hydrous, processed by The Upjohn Company.
[h]Bolted talc, processed by The Upjohn Company.
[i]Dicalcium phosphate dihydrate USP hydrous, Monsanto.
[j]Sucrose NF powder, processed by the Upjohn Company.
[k]Magnesium stearate NF powder flood grade, Mallinckrodt.
[l]SE = standard error.
Adapted from Ref. 48 with permission of the publisher.

Table 3 Comparison of Several Measures of the Powder Flow of Mixtures of Bolted Lactose and Spray-Dried Lactose

Percent bolted lactose	C_i %	Angle of repose	n (SE^a)	Simplified shear cell		
				S (SE^a), kdynes	δ'	f_c, kydnes
0	16.0	33.5 (0.5)	1.02 (0.006)	2.9 (4.2)	33.1	78
10	19.2	38.0 (0.9)	1.06 (0.005)	24.9 (2.9)	33.9	91.7
20	24.0	41.2 (0.3)	1.10 (0.014)	51.2 (9.1)	34.4	219
30	28.4	45.0 (0.4)	1.16 (0.017)	55.0 (10)	35.3	320
50	36.8	49.8 (0.9)	1.31 (0.022)	6.9 (8.9)	36.6	191
75	42.0	52.0 (0)	1.38 (0.018)	0.0 (6.2)	37.4	—
100	46.4	55.2 (0.6)	1.34 (0.035)	0.0 (12)	37.6	—

^{a}SE = standard error.
Adapted from Ref. 48 with permission of the publisher.

surface area, moisture content, and cohesiveness of materials since all of these can influence the observed compressibility index. The compressibility index, C_i, is defined as

$$C_i = \frac{\text{Initial Volume} - \text{Final Volume}}{\text{Initial Volume}} \times 100 \tag{5}$$

where the volume of powder is measured in a graduated cylinder before and after tapping.

Carr also proposed it as a direct measure of the "potential strength that a material could build up in its arch in a hopper and also the ease with which such an arch could be broken [44]. The author was clear in pointing out that compressibility should be used with other measures of flow since no single type of measurement adequately assesses all the factors influencing flow.

The compressibility index values reported in Table 2 were obtained by tapping powder until a plateau value was obtained. As can be seen from Fig. 4, 2000 to 3000 taps were sufficient to obtain plateau conditions in most cases when the material was relatively free flowing. For poor flowing materials, however, 7000 taps were required.

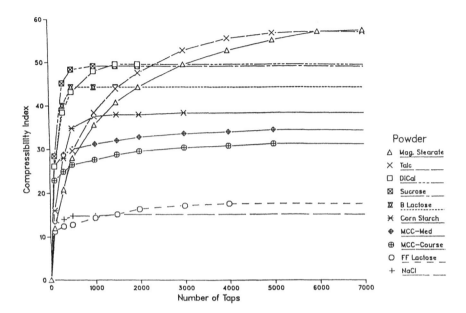

Fig. 4 Influence of number of taps on the compressibility index. (Adapted from Ref. 48 with permission of the publisher.)

Because the compressibility index can be thought of as a measure of the likelihood of arch formation and the ease with which arches will fail [44], and because it can be rapidly completed, it appears to be a useful measure of flow. Obviously, exceptions will occur, and caution should be exercised in its interpretation. As we will dicuss, the compressibility index values reported in Tables 2 and 3 and Fig. 4 are quite consistent with shear cell characterization.

C. Flow Through an Orifice

Gold and coworkers [46,53–55] proposed monitoring the rate of flow of material through an orifice as a better measure of powder flowability. Using this approach the authors were able to show the effects of glidants, granule size, and type of granulating agent on flow. Of particular importance is the utility of monitoring flow continuously, since pulsating flow patterns were observed even for free flowing materials. Changes in flow rate as the hopper empties can also be observed [56]. Empirical equations relating flow rate to opening diameter, particle size, and particle density have been determined [2–4]. Dahlinder and coworkers concluded that flow through an orifice is useful only with free flowing materials [57]. Gioia [38] demonstrated a practical use by showing a good correlation between the coefficient of variation of tablet weight or capsule fill weight and a "viscosity value" determined by measuring flow through an orifice; powder flow testing is now done on all powders used at the author's manufacturing facility.

Amidon and Houghton [48] evaluated several materials using flow-rate-through-an-orifice methodology. Flow profiles of these materials (sodium chloride, spray-dried lactose, and fast-flo lactose) are shown in Fig. 5 using a 16-mm diameter orifice, and the flow rate through the 6-mm orifice is reported in Table 2. In general, flow profiles show an initial linear region and then, at least for the two lactoses, a notable increase in flow rate as the cylinder emptied. Note the uneven flow for the lactoses through the 16-mm diameter orifice, indicating a pulsating flow pattern even for these free flowing materials. Most materials studied could not be tested using this method since the orifice plugged; hence, this method can be used only for free flowing materials. This methodology may be useful for formulation development but is not practical for assessing

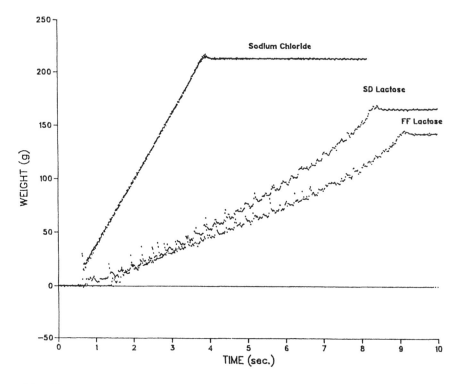

Fig. 5 Flow through a 16-mm orifice. (Adapted from Ref. 48 with permission of the publisher.)

materials with a wide range of flow. As such, it is of use in formulation development but is generally not applicable to preformulation characterization of drugs.

D. Shear Cell Methods

In an effort to put powder flow studies and hopper design on a more fundamental basis, Jenike [45] developed a powder shear tester and methodology that permits an assessment of powder flow properties as a function of consolidation load and time as well as powder–hopper material interactions. The methodology has been used extensively in the study of pharmaceutical materials [39,58–61]. From the yield loci obtained using this method, several parameters can be determined that influence powder flow, and discussions of these points are well documented in the literature [49,62,63].

Annular shear cell designs [47,49,62,64,65] offer some advantages over the Jenike shear cell design, including the need for less material. A disadvantage, however, is that because of its design, the powder bed is not sheared uniformly; material on the outside of the annulus is sheared more than material in the inner region. Triaxial shear cell testing is commonly used in soil mechanics. It has the advantage of better defining the stresses within the powder bed and permits hydrostatic pressure to be applied. While this is most useful in soil mechanics, it is also of some advantage in testing pharmaceutical powders. It has not found wide acceptance in pharmaceutics, however, and it, too, has the disadvantage of requiring large amounts of material for testing.

Roscoe [66] developed a simple shear cell that is unique in design. It forces uniform shear on "all planes" of the powder bed so that all powder is effectively in the same state of shear. It has the advantage of permitting precise and meaningful measurement of powder bed density during the shearing process, and the shear plane is known based on the geometry of the shear cell. It, too, requires large amounts of material for testing and has not been extensively used in the pharmaceutical field.

Yet another type of shear cell (translational or plate-type) consists of a thin sandwich of powder between a lower stationary rough surface and an upper rough surface (Fig. 6) that is movable [67–70]. The powder bed can then be sheared, and the force necessary to produce shear can be determined as a function of applied load. Simplified shear cell measurements utilizing equipment based on the design of Nash [67] and modified by Hiestand et al. [68–70] are reported in Table 2. Experimental procedures have been described extensively in the literature and consist of the following: A thin layer of powder is formed between a lower stationary rough surface and an upper movable rough surface. The layer of powder is then brought to a uniform state of

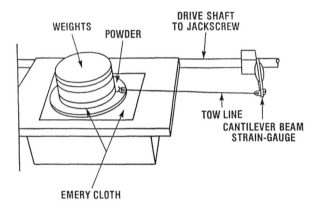

Fig. 6 Schematic diagram of the simplified shear cell of Hiestand. (Adapted from Ref. 48 with permission of the publisher.)

consolidation by (1) applying a consolidation load to the powder bed by adding weights to the upper movable surface, (2) applying a shear force to the powder bed by pulling the upper movable surface, (3) monitoring the force applied to the powder bed until shear is initiated, and (4) rapidly returning the shear force to zero after shear in the powder bed is induced. Steps 2 through 4 of this procedure are repeated until a series of nearly identical shear force measurements are obtained. Once the uniform state of consolidation is obtained, some weight is removed and a single determination of the shear strength at a reduced load can be measured. The consolidation process just outlined is to be repeated for each reduced load.

In general, a yield locus is obtained that relates the shear strength of the powder bed to the consolidation load and reduced load. The yield locus has been found to take the following form [71]:

$$\frac{\tau}{\tau_p} = \left(\frac{\sigma + S}{\sigma_p + S}\right)^{1/n} \tag{6}$$

where τ_p is the shear strength of the powder bed when the consolidation load, σ_p, is applied; τ is the shear strength of the powder bed at the reduced load, σ. The value of S corresponds to the tensile strength of the powder (the negative load at which the shear strength is zero), and n is commonly termed the shear index (related to the curvature of the yield locus).

In a slightly different form, Eq. (6) is commonly referred to as the Warren spring equation. Representative yield loci determined utilizing the simplified shear cell are shown in Fig. 7 for spray-dried lactose, bolted lactose, and sucrose. The yield locus for each material relates the shear strength to the applied load.

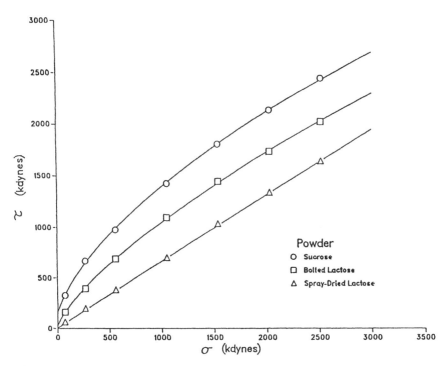

Fig. 7 Yield loci of spray-dried lactose, bolted lactose, and sucrose as determined using the simplified shear cell apparatus. (Adapted from Ref. 48 with permission of the publisher.)

The precise relationship is, however, very dependent on the conditions under which the powder bed was formed. In all cases the powder bed was brought to a uniform state of consolidation before determining the shear strength at a reduced load.

Shear cell measurements are more time consuming, in general, but offer much more experimental control than the other methods. In particular, the degree of consolidation is well controlled, the location of the shear plane is more precisely defined, and the rate and forces causing shear are controlled. Control of these "environmental" factors permits more direct assessment of these individual parameters and also should lead to more reproducible and consistent results. Materials with a wide range of flowability can also be done under identical experimental conditions and meaningful results obtained. This is particularly important when materials are being compared that were tested several months or years apart or whose flow properties may vary significantly due to changes in processing or supplier.

VII. APPLICATIONS

A. Powder Flow Characterization

The range of application of shear cell testing methodology is seen in Tables 2–6. Table 3 relates the flow properties of mixtures of spray-dried lactose and bolted lactose. These mixtures, in combination with the excipients tested, cover a broad range of flow. Tables 4 and 5, for example, show lot to lot variations in the flow properties of several materials, and Table 6 shows the variation in flow properties of bolted starch, sucrose, and phenacetin at different relative humidities (RH). Figure 8 presents the yield loci of sucrose at four different consolidation loads. Also shown in the figure are the shear indices determined at each consolidation load.

Shear cell measurements offer several pieces of information that permit a better understanding of the material flow characteristics. Two parameters, the shear index, n, and the tensile strength, S, determined by fitting simplified shear cell data to Eq. (6), are reported in Table 2. Because of the experimental method, only a poor estimate of the tensile strength is obtained in many cases. The shear index estimate, however, is quite reliable based on the standard error of the estimate shown in parenthesis in Table 2. The shear index is a simple measure of the flowability of a material and is used here for comparison purposes because it is reasonably reliable [50] and easy to determine. The effective angle of internal

Table 4 Powder Flow Properties of Several Excipients

| Material | Lot | Simplified shear cell | |
		n (SE[a])	S (SE[a])
Microcrystalline	A	1.10 (0.01)	73.0 (10)
cellulose—coarse	B	1.05 (0.02)	90.0 (18)
Microcrystalline	A	1.17 (0.02)	58.0 (14)
cellulose—medium	B	1.12 (0.01)	100 (11)
Lactose	A	1.46 (0.01)	9.0 (3.6)
	B	1.38 (0.01)	6.5 (4.7)
	C	1.35 (0.02)	66.5 (9.4)
	D	1.34 (0.03)	0.0 (12.3)
Fast-flo	A	1.03 (0.007)	13.7 (4.3)
lactose	B	1.02 (0.009)	5.6 (5.1)
Spray-dried	A	1.05 (0.02)	14.0 (12)
lactose	B	1.04 (0.016)	4.0 (9.4)
	C	1.02 (0.006)	29 (4.2)

[a]SE = standard error.
Data from Ref. 48.

Table 5 Powder Flow Properties of Several Drugs Showing Lot to Lot Variations

Material	Lot	n (SE[a])	S (SE[a])
		Simplified shear cell	
Ibuprofen	A	1.75 (0.13)	0.9 (34)
	B	1.63 (0.10)	0.0 (26)
	C	1.32 (0.03)	14.6 (9.9)
Sucrose	A	1.62 (0.02)	33.2 (7.8)
	B	1.56 (0.006)	50.4 (2.3)
	C	1.45 (0.04)	4.0 (13)
Spectinomycin	A	1.46 (0.04)	32.0 (18)
	B	1.27 (0.01)	56.6 (7.8)

[a]SE = standard error.
Data from Ref. 48.

friction, δ', and the unconfined yield stress, f_c, are two other commonly used parameters [49,63] and are shown schematically in Fig. 9. Others include the specific tensile strength and specific cohesion [49].

The unconfined yield stress is a measure of the stress necessary to cause a material unsupported in two directions to fail in shear. This is what must happen when an arch fails within the powder or at the hopper opening. The effective

Table 6 Powder Flow Properties of Several Materials as a Function of Relative Humidity

Material	% RH	n (SE[a])	S (SE[a])
		Simplified shear cell	
Starch	20	1.12 (0.02)	42.0 (14)
	40	1.12 (0.02)	69.0 (17)
	60	1.21 (0.03)	62.0 (20)
Sucrose	20	1.50 (0.03)	44.0 (12)
	40	1.56 (0.006)	50.4 (2.3)
	60	1.77 (0.03)	47.0 (10)
Phenacetin	20	1.39 (0.02)	32.8 (7.9)
	40	1.41 (0.02)	53.7 (9.7)
	60	1.50 (0.04)	14.0 (29.)

[a]SE = standard error.
Data from Ref. 48.

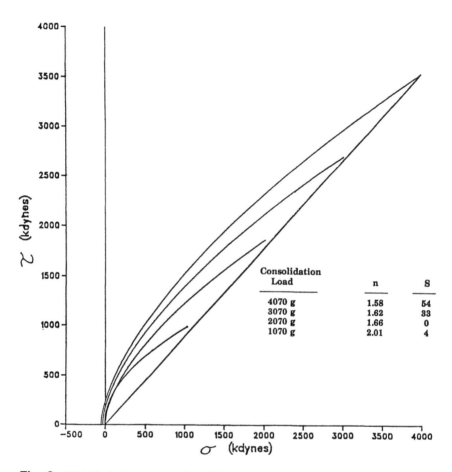

Fig. 8 Yield loci of sucrose at four different states of consolidation using simplified shear cell methodology. (Adapted from Ref. 48 with permission of the publisher.)

angle of internal friction is a measure of the shear strength of a material per unit consolidation force and is therefore useful for comparison purposes. Obviously, tensile strength and cohesive strength are also useful parameters to characterize materials. These points have been discussed in detail in the references already cited [72]. A wealth of information dealing with several aspects of powder flow and interparticulate movement can be obtained, however, using shear cell data, and a wide range of flow properties can effectively be determined. This advantage is not shared by non–shear cell methods.

An analysis of the experimental errors associated with shear cell measurements indicates that, in general, neither the standard deviation nor the percent

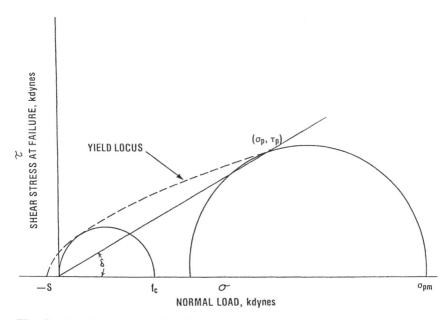

Fig. 9 Graphical representation of several parameters that can be obtained using shear cell characterization. (Adapted from Ref. 48 with permission of the publisher.)

relative standard deviation are a constant. Because the variance appears to be proportional to the observed shear strength, all shear strength values were weighted in inverse proportion to the observed shear strength [73] and the data fit to Eq. (6) using a nonlinear least squares fitting routine. Estimates of the tensile strength, S, and the shear index, n, were obtained in this manner. Also, the effective angle of internal friction, δ', and the unconfined yield stress, f_c, for each material are defined in Fig. 9 and reported in Table 2 for many of the materials tested. The unconfined yield stress was calculated by the method of Hiestand and Wells [74].

It is important to keep in mind that powder flow is a complicated matter, and no single parameter is likely to meet every need. Some authors have combined several parameters that are indicative of powder flow (tensile strength, shear strength, likelihood of flooding, etc.) into one "index of flowability." These include the Jenike flow factor [45], the general index of flowability by Stainforth and coworkers [75], and the flowability index by Hiestand and coworkers [74]. These indices can certainly be useful as a guide to flowability, though they, too, do not assess all the important parameters influencing powder flow (Table 1).

Lot to lot variations in excipients and bulk drugs are seen in Tables 4 and 5. The ibuprofen lots in Table 5 represent three lots specially recrystallized to produce different particle size material. Not surprisingly, very significant differences were observed in the flow properties. Quantitating these changes provides useful information for troubleshooting formulation problems. Smaller variations in the flow properties were observed for the excipients studied, but lot to lot variability can occur.

Table 6 shows the usefulness of shear cell data in formulation development. Variations in relative humidity can profoundly influence flow; this is a valuable piece of information for formulation development. Shear cell methodology thus provides useful data for optimizing the flow of formulations as well.

The influence of consolidation load on the flowability of sucrose is shown in Fig. 8. For this material, the effective angle of internal friction is nearly constant yet the shear index is seen to change with state of consolidation. Apparently, for sucrose, increased consolidation results in a somewhat more free flowing although still cohesive material. As such, sucrose can be considered a complex powder [49] with perhaps somewhat better flow characteristics when consolidated (as might occur in a hopper).

A direct comparison of the results obtained by the simplified shear cell methodology and the Flowfactor Tester of Jenike and associates is difficult since the experimental procedures differ. Figures 10 and 11 show the yield loci obtained using both methods. For free flowing spray-dried lactose the yield loci are very similar (Fig. 10) and nearly linear, as is generally observed [45]. Bigger

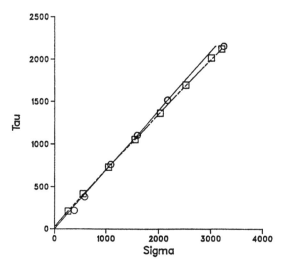

Fig. 10 Yield loci of spray-dried lactose determined using the Jenike shear cell and the simplified shear cell. Open circles: Jenike shear cell; open squares: simplified shear cell. (Adapted from Ref. 48 with permission of the publisher.)

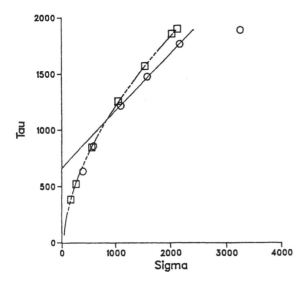

Fig. 11 Yield loci of sucrose determined using the Jenike shear cell and the simplified shear cell. Open circles: Jenike shear cell; open squares: simplified shear cell. (Adapted from Ref. 48 with permission of the publisher.)

deviations are observed when comparing poor flowing materials such as sucrose (Fig. 11). In general, however, while results from each method differ quantitatively, a similar rank ordering of materials has been observed. Hiestand and Wells [74] observed a similar rank order when comparing two lots of spectinomycin. They determined the minimum hopper opening and hopper angle using the Flowfactor Tester. The more freely flowing lot (shear index = 1.29) required a minimum hopper opening of 23.8 cm, while the more cohesive material (shear index = 1.43) required a hopper opening of 77.4 cm.

B. Use of Tableting Indices in Powder Flow Characterization

The importance of the physical properties of materials such as particle size and shape on powder flow is well documented in the literature. The mechanical properties, however, have been largely ignored. Recently, Amidon [76,77] has investigated the influence of mechanical properties on powder flow. The physical and mechanical properties of several excipients and seven lots of ibuprofen bulk drug were analyzed. For the purposes of their analysis, the shear cell index value was used as a simple measure of powder flow [48]. It was obtained by fitting the shear cell data to the Warren spring equation using nonlinear least squares methods. The shear index value is a measure of the degree of curvature of the yield locus and is one of many parameters obtained from the yield locus. For free flowing materials, the shear index is close to 1.0, while for poor flowing materials the shear index increases and can approach a value of 2. More details of the methods have been reported by Amidon and Houghton [48].

Shown in Fig. 12 is the correlation between the shear index value and the Worst Case Bonding Index. The materials appear to fall into two categories: the ibuprofen lots, which show a strong correlation with Bonding Index, and free flowing excipients such as microcrystalline cellulose, which show no correlation with Bonding Index. Clearly, for the ibuprofen lots, the observed correlation is sensible: poor bonding materials (i.e., low Bonding Index) are less likely to stick together (i.e., bond) and therefore would flow better. Conversely, materials that bond well are more likely to have flow problems, and this is seen for the ibuprofen lots with high Bonding Indices.

A more detailed analysis using multivariable regression of the ibuprofen data demonstrated that a three-parameter model accurately fit the data (Table 7). The Bonding Index and the Heywood shape factor, α, alone explained 86% of the variation, while the best three-variable model, described in what follows, explained 97% of the variation and included the Bonding Index, the Heywood shape factor, and the powder bed density. All three parameters were statistically significant, as seen in Table 7. Furthermore, the coefficients are qualitatively as

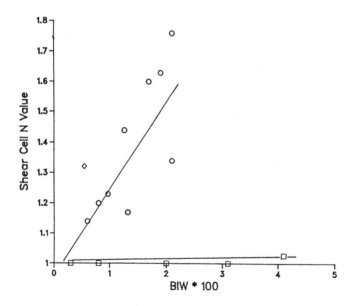

Fig. 12 Plot of the shear index value, N, as determined using simplified shear cell methodology, versus Worst Case Bonding Index, BIW.

Table 7 Ibuprofen Bulk Drug Regression Analysis for the Dependent Variable Shear Index, n

Explained variation	Bonding Index	Heywood shape parameter	Powder bed density	Particle diameter	Permanent deformation pressure
56%		X			
61%	X				
86%	X	X			
97.6%	X	X	X		
98.2%	X	X	X		X
98.4%	X	X	X	X	X

Best three-variable model: $r^2 = 0.976$

Variable	Coefficient	Prob $> F$
Intercept	−1.026	
Shape parameter	0.121	0.0012
Bonding Index	37.3	0.0003
Powder bed density	2.01	0.0044

expected; increasing the Bonding Index, Heywood shape factor (which indicates the particles are becoming less spherical), and powder bed density all resulted in poorer flowing materials (i.e., a higher shear index value, n).

The lack of correlation for the free flowing excipients is interesting, and the reason is unclear. A couple of things can be said, however, that warrant further investigation. For example, the excipients all had high permanent deformation pressures, which may mean that under the experimental conditions used for powder flow, little permanent deformation could occur. Therefore, even though they are good bonding materials, the forces necessary to create significant strength in the powder bed were not reached.

C. Characterization of Tableting Properties

The mechanical properties of materials, though not often studied in detail, can have a profound effect on solids processing. Clearly, tableting properties are influenced by the elastic and plastic deformation properties as well as the viscoelastic properties of a material. As we have pointed out, the powder flow properties are also affected, as well as the tendency of materials to "set up" on storage. Because of the importance of mechanical properties, it is important to be able to quantitatively characterize materials. Reliable mechanical property information can be useful in (1) helping to choose a processing method such as granulation or direct compression, (2) selecting excipients with properties that will mask the poor properties of the drug, and (3) helping to document what went wrong when a tableting process is being scaled up or when a new bulk drug process is being tested.

D. Bulk Drug and Excipient Characterization

To give the reader a feel for the tableting indices, previously published results are shown in Table 8–10 [31]. In general, the indices need to be looked at together to interpret or predict the tableting properties of a material. For example, pure methenamine, erythromycin, and phenacetin are most likely to fracture either on decompression or on ejection. One can see in Table 8, however, that their properties vary significantly. Phenacetin is less brittle than sucrose or ibuprofen, yet the tendency for phenacetin to fracture (a brittle failure) is significantly greater. Note, however, that sucrose and ibuprofen have twice the Bonding Index of phenacetin, which suggests that the poor bond in combination with moderate brittleness leads to phenacetin fracture on uniaxial decompression. Erythromycin and methenamine have very large Bonding Indices but also a high Brittle Fracture Index. Erythromycin also has a high Strain Index and, interestingly, does not fail on decompression but only on ejection. The high Strain Index is consistent with the idea that erythromycin expands substantially on ejection

Table 8 Typical Properties of Some Common Drugs and Excipients Determined at a Compact Solid fraction of 0.8 to 0.9

Material	Bonding Index $\times$ 10^2	Brittle Fracture Index	Strain Index $\times$ 10^2
Methenamine	1.7	0.83	0.62
Erythromycin (anhydrous)	4.0	0.68	4.0
Ibuprofen	1.2	0.40	1.1
Sucrose	1.0	0.35	1.6
Phenacetin	0.5	0.28	1.3
Starch (StaRx 1500)	1.3	0.27	2.3
Spray-dried lactose	0.5	0.18	2.1
Microcrystalline cellulose	4.0	0.04	2.5

Data from Ref. 31.

and stress concentration at the edge of the die on ejection is sufficient to cause fracture. Methenamine, with its lower Bonding Index, fails on decompression just as phenacetin does.

Lot to lot variations of several drugs and excipients are shown in Table 9. Specially crystallized lots of ibuprofen, for example, show substantial changes in Brittle Fracture and Bonding Indices. Phenacetin shows a significant increase in brittleness at higher relative humidity. It was observed by Hiestand and Smith [31] that compacts of dried phenacetin did not fracture, while the lot equilibrated at 40% relative humidity did, consistent with the Brittle Fracture Index change.

The effect of mixing a brittle material like methenamine or sucrose with

Table 9 Lot to Lot Variations of the Mechanical Properties of Some Common Drugs and Excipients Determined at a Compact Solid Fraction of 0.8 to 0.9

Material	Bonding Index $\times 10^2$	Brittle Fracture Index
Methenamine		
Lot A	1.7	0.96
Lot B		0.83
Lot C		0.50
Erythromycin		
Anhydrous	4.0	0.68
Dihydrate	2.0	0.98
Ibuprofen		
Lot A	1.9	0.05
Lot D	1.8	0.57
Lot E	1.2	0.40
Phenacetin		
Dried		0.19
Equil. 40% RH	0.5	0.28
Spray-dried Lactose		
Lot A	0.5	0.18
Lot B	0.6	0.12
Lot C		0.16
Microcrystalline cellulose		
Lot A	4.0	0.04
Lot B		0.09

Data from Ref. 31.

Table 10 Properties of Compacts of Mixtures of
Spray-Dried Lactose and Sucrose

| Mixture | | Brittle | Bonding |
Spray-dried lactose	Sucrose	Fracture Index	Index $\times 10^2$
100%	0%	0.12	0.6
50%	50%	0.09	
25%	75%	0.18	
15%	85%	0.23	0.8
0%	100%	0.42	1.0

Data from Ref. 31.

spray-dried lactose is shown in Tables 10 and 11. Clearly, the brittleness decreased with added lactose until, at a mixture level of about 50% lactose, no further significant decrease in the Brittle Fracture Index was observed. It should be pointed out, however, that while mixture studies of this sort suggest that the properties of mixtures can be predicted by the pure material properties, much has yet to be done. Leuenberger [78] investigated these issues and found that, indeed, there is a need for "interaction terms" to account for unusual behavior. Recent publications by Leuenberger and coworkers [79,80] have speculated on the usefulness of percolation theory to account for some of the observations. Nevertheless, the data demonstrate the utility of the indices in helping to select

Table 11 Properties of Compacts of Mixtures of
Spray-Dried Lactose and Methenamine

| Mixture | | Brittle |
Methenamine	Spray-dried lactose	Fracture Index
0%	100%	0.16
25%	75%	0.15
50%	50%	0.16
65%	35%	0.26
75%	25%	0.27
85%	15%	0.52
90%	10%	0.75
100%	0%	0.96

Data from Ref. 31.

excipients. A brittle material could benefit by the addition of a less brittle material, while the addition of a poor bonding excipient to a poor bonding drug like phenacetin would likely be of little use [31].

The microcrystalline cellulose data in Table 9 is worthy of consideration for a moment. It has a very low Brittle Fracture Index and a very high Bonding Index, two properties that make it an excellent tableting excipient. The low Brittle Fracture Index can successfully mask the brittle properties of drugs, while the high Bonding Index will overcome the poor bond of the active component. The disintegration properties of the material are such that high bond can be overcome. All of these properties make it an excellent direct compaction tableting excipient.

E. Studies of Bulk Drug and Granulation Properties

Hiestand, Amidon, and colleagues [81,82] investigated the effect of varying crystallization procedures on the mechanical properties of ibuprofen. The crystallization procedures are shown in Table 12 and were selected to yield different crystallization rates. The mechanical properties of the bulk drug lots are shown in Table 12. Clearly, the properties differed substantially; the Bonding Index (BI) varied from 0.002 to 0.027, while the Brittle Fracture Index (BFI) varied from 0.05 to 0.57. It is interesting to note that lots A and B were prepared in the same way, except that lot B was a 12-kg lot whereas lot A was approximately 1 kg; yet the mechanical properties differ significantly. This points out the real difficulty of reproducing material with consistent mechanical properties as one scales up the synthesis. A similar situation occurred for lots C and D. A study of the compaction profile of these lots shown in Fig. 13

Table 12 Mechanical Properties of Ibuprofen Bulk Drug and Granulations

	Bulk drug			Granulation		
Lot	BI	BFI	*P*	BI	BFI	*P*
A[a]	2.7	0.05	1.91	—	—	—
AB[b]	1.1	0.16	4.47	1.9	0.18	9.54
B[c]	2.0	0.18	5.10	2.0	0.18	9.36
C[d]	1.9	0.22	7.04	1.6	0.18	12.42
D[e]	0.2	0.57	7.93	—	—	—

[a]Disolve in hexane, seeded at 40°C; slowly cook over 3 days to 15°C. Lot size 1 kg.
[b]Disolve in hexane, chill rapidly to 15°C, and seed, hold at 15°C for 3 days; dry without heat. Lot size 12 kg.
[c]Prepare same as lot A. Lot size 12 kg.
[d]Disolve in hexane, slowly add to cold hexane seeded at –10°C, and dry without heat. Lot size 12 kg.
[e]Prepare same as lot C. Lot size 1 kg.

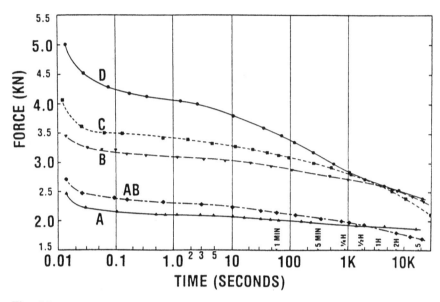

Fig. 13 Compaction profile of five specially crystallized lots of ibuprofen. See Table 12 for description.

demonstrates that the tableting properties of the pure materials differ as well. As expected, lot D with the largest permanent deformation pressure, P, required the largest compaction force to produce a tablet with a solid fraction of 0.87.

The mechanical properties of granulations of lots AB, B, and C were also studied by preparing proprietary granulations of each of the lots. Satisfactory granulations were prepared of all lots (the granulation contained approximately 65% drug and 35% excipient), though processing time and the quantity of water necessary was different for each. These differences could likely be influenced by differences in bulk drug particle size. The mechanical properties of the granulations are shown in Table 12. Clearly, the mechanical properties of the bulk drug were ameliorated by the granulation process: the Bonding Index and Brittle Fracture Index were essentially the same for all three granulations. The permanent deformation pressure for the granulations appears to be influenced by the bulk drug, since the bulk drug with the highest P resulted in the granulation with the highest P. The process of granulation in this case was very beneficial in reducing the varying mechanical properties of the bulk drug.

Table 13 Summary of Tableting Indices for Losulazine HCl and Furegrelate Na as a Solid Fraction of 0.9

Index	Losulazine HCl Lot A	Furegrelate sodium Lot A	Furegrelate sodium Lot B
$BI_w \times 10^2$	2.9	2.6	2.4
BFI	1.63	0.24	0.05
$SI \times 10^2$	3.0	1.8	1.7

Data from Ref. 83.

F. Use of Tableting Indices in Preformulation and Formulation Development

The mechanical properties of two drugs under development at The Upjohn
Company were determined by Middleton, Amidon, and Rowe [83] during
preformulation. The mechanical properties of losulazine hydrochloride bulk drug
are summarized in Table 13. The results for losulazine HCl indicate that the
Strain Index and the Bonding Index are quite high. The Brittle Fracture Index
is extremely high, which indicates that this material is very brittle. Two direct
compression formulations were tested for this drug: one containing a 9.9:1 ratio
of excipients to drug and one containing an 18.8:1 ratio of excipients to drug.
Physical tests of formulated losulazine HCl are shown in Table 14. It appears
that sufficient excipient was added to overcome the brittleness and prevent
capping, since this was not observed, though for the 9.9:1 ratio sticking was
observed around the tablet score. Of particular interest was the poor dissolution
rate, shown in Fig. 14, for the 9.9:1 ratio formulation relative to the 18.8:1

Table 14 Physical Tests of Formulated Losulazine HCl Tablets

Property	18.8 : 1 mixture	9.9 : 1 mixture
Dissolution, % in 15 min	63%	46.8%
Crushing strength SCU	8.5	9.1
Disintegration time, sec	120	170
Sticking/picking	Not seen	Sticking around tablet score

Data from Ref. 83.

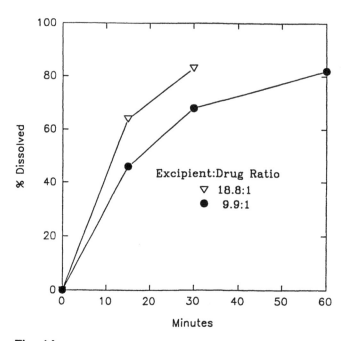

Fig. 14   Dissolution profile of two formulations of losulazine HCl.

formulation. The poor dissolution can be attributed to the good bonding nature of the drug. The additional excipients, which had poor bonding characteristics, improved dissolution and disintegration.

For furegrelate sodium (Table 13), both the Strain Index and the Brittle Fracture Index are relatively low, indicating that there should be little problem with capping or picking. The difference in Brittle Fracture Indices between the two lots may be due, in part, to particle size differences, since lot A was milled and lot B was not, but in either case is considered low from a practical standpoint. The Bonding Index is quite high, indicating excellent bond but a potential for poor disintegration or dissolution. Poor dissolution was a major problem for furegrelate, and high levels of poorly bonding excipients were necessary to achieve acceptable dissolution rates.

VII. CONCLUSIONS

The beauty of science is its ability to predict the future. A fundamental understanding of the properties that affect powder compaction and flow, indeed a fundamental understanding of any aspect of pharmaceutical technology, will make dosage form development more scientific and predictable. This ability to predict the future is what is required of us when we formulate, test, and validate procedures and market drugs. Research on the effects of the physical and mechanical properties of materials, when successful, leads to more enlightened and efficient development. The principles of powder flow and the closely associated aspects of compaction are applicable to all solid dosage forms, tablets and capsules as well as conventional and controlled release dosage forms. A better understanding of the fundamental principles, when combined with reliable and relevant data, results in better formulations and products.

REFERENCES

1. E. Nelson, *J. Am. Pharm. Assoc., Scientific Ed.*, *44*, 435–437 (1955).
2. T. M. Jones and N. Pilpel, *J. Pharm. Pharmacol.*, *18*, 81–93 (1966).
3. C. F. Harwood and N. Pilpel, *Chem. Proc. Engin.*, *49*, 92–96 (1968).
4. F. Q. Danish and E. L. Parrott, *J. Pharm. Sci.*, *60*, 548–554 (1971).
5. N. Pilpel, *J. Pharm. Pharmacol.*, *16*, 705–716 (1964).
6. H. Rumpf, in *Agglomeration* (William A. Knepper, ed.), Wiley Interscience, 1962, 379–418.
7. D. C. H. Cheng, *Chem. Engin. Sci.*, *23*, 1405–1420 (1968).
8. D. C. H. Cheng, R. Farley and F. H. H. Valentin, *I Chem. E Symposium Series* *29*, 14–24 (1968).
9. D. C. H. Cheng, *Proc. Soc. Anal. Chem.*, *10*, 17–19 (1973).
10. R. Rupp, *Boll. Chim. Farm.*, *116*, 251–266 (1977).
11. J. E. Rees, *Boll. Chim. Farm.*, *116*, 125–141 (1977).
12. J. N. Staniforth and J. P. Hart, *Anal. Proc.*, *24*, 7880 (1987).
13. K. Ridgway and R. Rupp, *J. Pharm. Pharmacol.*, *21*, 30S–39S (1969).
14. K. Ridgway and R. Rupp, *Powder Tech.*, *4*, 195–202 (1970/71).
15. C. R. Chang, J. K. Beddow, and A. F. Vetter, *Particulate Sci. Technol.*, *1*, 433–449 (1983).
16. C. F. Harwood and N. Pilpel, *J. Pharm. Pharmacol.*, *21*, 721–730 (1969).
17. K. Ridgway and J. B. Scotton, *J. Pharm. Pharmacol.*, *22*, 24S (1970).
18. R. Rupp, *Boll. Chim. Farm.*, *116*, 251–266 (1977).
19. G. A. Turner and M. Balasubrananian, *Powder Tech.*, *10*, 121–127 (1974).
20. M. C. Coelho and N. Harnby, *Powder Tech.*, *20*, 197–200 (1978).
21. M. C. Coelho and N. Harnby, *Powder Tech.*, *20*, 201–205 (1978).
22. M. J. Kontny, G. P. Gandolfi, and G. Zografi, *Pharm. Res.*, *4*, 104–112 (1987).
23. G. Zografi, M. J. Kontny, A. Y. S. Yang, and G. S. Brenner, *Int. J. Pharm.*, *18*, 99–116 (1984).
24. W. R. Harper, in *Powders in Industry*, Society of Chemical Industry, London, 1961, 115–126.
25. J. N. Staniforth, *Int. J. Pharm.*, *11*, 109–111 (1982).
26. H. Hertz, in Miscellaneous Papers, Macmillan, London, 1896.
27. K. L. Johnson, K. Kendall, and A. D. Roberts, *Proc. Royal Soc. Ser. A*, *324*, 301–313 (1971).
28. D. Tabor, in *The Hardness of Metals*, Clarendon Press, Oxford, 1951.

29. E. N. Hiestand, J. E. Wells, C. B. Peot, and J. F. Ochs, *J. Pharm. Sci.*, *66*, 510 (1977).
30. E. N. Hiestand, J. M. Bane, and E. P. Strzelinski, *J. Pharm. Sci.*, *60*, 758–763 (1971).
31. E. N. Hiestand and D. P. Smith, *Powder Tech.*, *38*, 145–159 (1984).
32. E. N. Hiestand and D. P. Smith, *Int. J. Pharm.*, *67*, 217–229 (1991).
33. E. N. Hiestand and D. P. Smith, *Int. J. Pharm.*, *67*, 231–246 (1991).
34. Z. Chowhan and Y. P. Chow, *Drug Dev. Ind. Pharm.*, *6*, 1–13 (1980).
35. Z. T. Chowhan and Y. P. Chow, *Int. J. Pharm.*, *4*, 317–326 (1980).
36. Z. T. Chowhan and I. C. Yang, *J. Pharm. Sci.*, *70*, 927–930 (1981).
37. E. T. Cole, P. H. Elworthy, and H. Sucker, *J. Pharm. Pharmacol.*, *26*, 57P (1974).
38. A. Gioia, *Pharm. Technol.*, *4*, 65–68 (1980).
39. R. Ho, D. F. Bagster, and M. J. Crooks, *Drug Dev. Ind. Pharm.*, *3*, 475–489 (1977).
40. J. W. Wallace, J. T. Capozzi, and R. F. Shangraw, *Pharm. Technol.*, *7*, 94–104 (1983).
41. D. Train, *J. Pharm. Pharmacol.*, *10*, 127T–135T (1958).
42. R. L. Carr, *Chem. Eng.*, *72*, 69 (1965).
43. R. L. Carr, *Chem. Eng.*, *72*, 163–168 (1965).
44. R. L. Carr, *Brit. Chem. Engin.*, *15*, 1541–1549 (1970).
45. A. W. Jenike, "Storage and Flow of Solids," *Bull. of the Univ. of Utah, Utah Engineering Expt. Station*, *53* (1964).
46. G. Gold, R. N. Duvall, B. T. Palermo, and J. G. Slater, *J. Pharm. Sci.*, *55*, 1291–1295 (1966).
47. H. Nyqvist and A. Brodin, *Acta Pharm. Suec.*, *19*, 81–90 (1982).
48. G. E. Amidon and M. E. Houghton, *Pharm. Manuf.*, *2*, 20 (1985).
49. S. Kocova and N. Pilpel, *J. Pharm. Pharmacol.*, *8*, 33–55 (1973).
50. R. Farley and F. H. H. Valentin, *Powder Tech.*, *1*, 344–354 (1967/68).
51. L. L. Augsburger and R. F. Shangraw, *J. Pharm. Sci.*, *55*, 418–423 (1966).
52. J. F. Bavits and J. B. Schwartz, *Drug and Cosmetic Ind.*, *104*, 44 (1974).
53. G. Gold, N. Duvall, and B. T. Palermo, *J. Pharm. Sci.*, *55*, 1133–1136 (1966).
54. G. Gold, R. N. Duvall, B. T. Palermo, and J. G. Slater, *J. Pharm. Sci.*, *57*, 667–671 (1968).
55. Gold, G., R. N. Duvall, B. T. Palermo, and J. G. Slater, *J. Pharm. Sci.*, *57*, 2153–2157 (1968).
56. R. P. Jordan and C. T. Rhodes, *Drug Dev. Ind. Pharm.*, *5*, 151–167 (1979).
57. L. E. Dahlinder, M. Johansson, and J. Sjogren, *Drug Dev. Ind. Pharm.*, *8*, 455 (1982).
58. N. Pilpel, *Endeavour*, *23*, 73 (1969).
59. M. J. Crooks, R. Ho, and D. F. Bagster, *Drug Dev. Ind. Pharm.*, *3*, 291 (1977).
60. K. Marshall and D. Sixsmith, *J. Pharm. Pharmacol.*, *28*, 770–771 (1976).
61. H. M. Sutton, *Process Engin.* 84–86, August 1972.
62. S. Kocova and N. Pilpel, *Powder Tech.*, *5*, 329 (1971).
63. J. C. Williams and A. H. Birks, *Powder Tech.*, *1*, 199–206 (1967).
64. C. I. Irono and N. Pipel, *J. Am. Pharm. Assoc., Scientific Ed.*, *44*, 435 (1955).
65. H. Nyqvist, A. Brodin, and L. Nilsson, *Acta Pharm. Suec.*, *17*, 215 (1980).

66. K. H. Roscoe, *Geotechnique*, 20, 129–170 (1970).
67. J. H. Nash, G. G. Leiter, and A. P. Johnson, *I&EC Product Res. & Dev.*, 4, 140–145 (1965).
68. E. N. Hiestand and C. J. Wilcox, *J. Pharm. Sci.*, 57(8), 1421–1427 (1968).
69. E. N. Hiestand and C. J. Wilcox, *J. Pharm. Sci.*, 58, 1403–1410 (1969).
70. E. N. Hiestand, S. C. Valvani, C. B. Peot, E. P. Strzelinski, and J. F. Glasscock, *J Pharm. Sci.*, 62, 1513–1517 (1973).
71. M. D. Ashton, R. Farley, and F. H. H. Valentin, *J. Sci. Instrum.*, 41, 763 (1964).
72. P. York, *Int. J. Pharm.*, 6, 89 (1982).
73. G. E. P. Box, W. G. Hunter, and J. S. Hunter, in *Statistics for Experimenters*, John Wiley and Sons, New York, 1978.
74. E. N. Hiestand and J. E. Wells, *Proc. Int. Powder and Bulk Solids Handling and Proc. Conf.*, Rosemont, Ill., May 1977.
75. P. T. Stainforth and R. E. R. Berry, *Powder Tech.*, 8, 243–251 (1973).
76. G. E. Amidon, "Physical and Mechanical Property Effects on Powder Flow," AAPS Symposium, November 1988.
77. G. E. Amidon, "Mixing and Flow," Proceedings of the 31st International Industrial Pharmaceutical Research Conference, University of Wisconsin—Extension, 1989.
78. H. Leuenberger, *Int. J. Pharm.*, 12, 41–55 (1982).
79. L. E. Holman and H. Leuenberger, *Int. J. Pharm.*, 46, 35–44 (1988).
80. D. Blattner, M. Kolb, and H. Leuenberger, *Pharm. Res.*, 7, 113–117 (1990).
81. E. N. Hiestand, G. E. Amidon, D. P. Smith, and B. D. Tiffany, "Variation of Physical Properties with Changes in Crystallization Rate," APhA National Meeting, November 1980.
82. G. E. Amidon, D. P. Smith, E. N. Hiestand, and B. D. Tiffany, "Flowability of Powders Using a Simplified Shear Cell," APhA National Meeting, November 1981.
83. K. R. Middleton, G. E. Amidon, and E. L. Rowe, "Tableting Indices and Tableting Behavior of Losulazine HCl and Furegrelate Na," AAPS National Meeting, November 1986.

11

Solubility of Pharmaceutical Solids

David J. W. Grant
University of Minnesota, Minneapolis, Minnesota

Harry G. Brittain
Ohmeda, Inc., Murray Hill, New Jersey

I. INTRODUCTION

A.   Solubility as an Equilibrium Process

The solubility of a solid in a relevant solvent medium is a crucial characteristic. Solubility is defined as the concentration of the dissolved solid (the solute) in the solvent medium, which becomes the saturated solution and which is in equilibrium with the solid at a defined temperature and pressure. The solubility depends on the physical form of the solid, the nature and composition of the solvent medium, the temperature, and the pressure [1].

The most common solvent media are liquids or liquid mixtures, which give rise to liquid solutions of the solute [1] and which constitute the main subject of this chapter. Less common solvent media are solid phases, which act as the "host" for the solute species that are present as "guest" molecules to form a solid solution of the solute [2,3]. Interest in gases as solvents, often in the supercritical state, has recently been growing because the gaseous solutions that are formed provide useful means for dissolving, crystallizing, chromatographing, separating, transporting, or removing certain solutes [4], as has traditionally been achieved using liquid solutions.

B.   Units Employed for Expressing Solubility

Solubilities may be expressed in any appropriate units of concentration, such as the quantity of the solute dissolved (weight or number of moles) divided by the quantity either of the solvent (weight, volume, or any number of moles) or of the solution (weight, volume, or number of moles). The most convenient expressions of concentration are listed and exemplified in Table 1.

C. Metastable Solubility

As a result of processing operations, such as those listed in Table 2, the individual particles of a pharmaceutical solid may exist in different states of energy and disorder [5–7]. The disorder is represented by the entropy, S. At constant (atmospheric) pressure, the total energy is represented by the enthalpy, H. The ability of the system, that is, the solid, to perform work and to undergo a spontaneous change at constant (atmospheric) pressure is represented by the Gibbs free energy, G, given by

$$G = H - TS \tag{1}$$

Table 1 Expressions for Concentration and Solubility

Symbol	Physical Quantity	Example
M_2	Molecular weight of the solute	$M_2 = 300$ g/mol
W_2	Weight of the solute	$W_2 = 1.0$ g
M_1	Molecular weight of the solvent	$M_1 = 100$ g/mol
W_1	Weight of the solvent	$W_1 = 20$ g
v_1	Volume of the solvent	$v_1 = 25$ ml
ρ_1	Density of the solvent	$\rho_1 = 0.8$ g/ml
ρ	Density of the solution	$\rho = 0.9$ g/ml

Concentration Units and Expressions			Calculated from Above
A (g/100 g solvent)	=	$100\ W_2/W_1$	$A = 5.00$ g/100g
B (g/100 ml solvent	=	$A\rho_1$	$B = 4.00$ g/100ml
	=	$100\ W_2/v_1$	
D (g/100 g solution = % w/w)	=	$\dfrac{100A}{100 + A}$	$D = 4.76$ g/100g
E (g/100 ml solution = % w/v)	=	$D\rho$	$E = 4.29$ g/100 ml
m (mole/kg solvent = molality)	=	$10\ A/M_2$	$m = 0.167$ mol/kg
c (mole/L solution = molarity)	=	$10\ E/M_2$	$c = 0.145$ mol/L
x (mole fraction)	=	$\dfrac{W_2/M_2}{(W_1/M_1) + (W_2/M_2)}$	$x = 0.0164$

Source: Adapted with permission from J. Jacques, A. Collet, and S. H. Wilen, *Enantiomers, Racemates, and Resolutions*, Wiley, New York, 1981, p. 168.

Table 2 Properties of the Crystals or Particles that May Be Modified by the Processing Stresses that Are Imposed During the Manufacture of Solid Dosage Forms

Crystal properties	Processing stresses	Manufacturing procedures
Crystal structure	Temperature	Crystallization
Polymorphism	Pressure	Precipitation
Crystal solvation (solvates)	Mechanical	Milling
Crystal habit (shape)	Radiation	Mixing
Crystallinity (crystal defects)	Exposure to liquids	Drying
Crystal surface constitution	Exposure to gases	Granulation
(polarity, irregularity,	and vapors	Compressing
wettability)		Coating
Expansivity, compressibility,		Storage
viscoelasticity, elasticity,		Transport
plasticity, hardness		Handling
Particle size (distribution)		

Source: Adapted with permission from P. York, *Int. J. Pharm.*, *14*, 1 (1983).

where T is the absolute temperature [8]. When the system undergoes any sort of physical or chemical change, G, H, and S change by amounts represented by ΔG, ΔH, and ΔS, respectively, and thus

$$\Delta G = \Delta H - T \Delta S \qquad (2)$$

ΔG, ΔH, and ΔS are positive quantities for increases in the corresponding property, negative for decreases. There is a natural tendency for the free energy of the system to decrease, so that ΔG is negative for a spontaneous process.

During manufacturing processes, processing stresses are applied to pharmaceutical solids; these invariably cause one or more of the fundamental crystal properties to change [5–7], as indicated in Table 2. The processing stress causes G, H, and S of the solid to usually move to a state of higher Gibbs free energy than that of the mildly stressed material. An example of mild stressing is relatively slow crystallization. The mildly stressed or unstressed material is in a stable equilibrium state in which the molecules, on average, vibrate about a mean portion in a free energy "well," as shown in Fig. 1. When this solid material is equilibrated with a solvent medium, the solubility attains a minimum value representative of a stable equilibrium, as demonstrated by the lower dynamic solubility plot, for untriturated digoxin in water, in Fig. 2 [9].

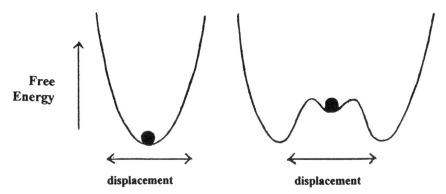

Fig. 1 Stable equilibrium state (left) given by molecules vibrating about their mean position in a free energy well in an unstressed solid. Metastable equilibrium state (right) given by molecules vibrating in an elevated free energy well in a stressed solid (above the stable minimum for an unstressed solid).

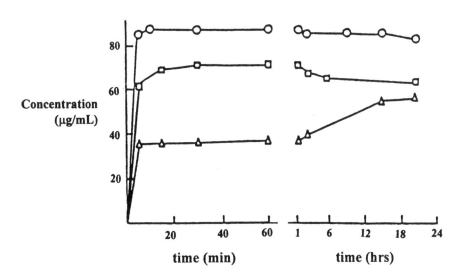

Fig. 2 Dynamic solubility study in water at 37°C for untriturated digoxin powders (△), freshly triturated digoxin (○), and digoxin triturated and stored for 2–5 months (□). (Reproduced from Ref. 9 with permission of the copyright owner, the American Pharmaceutical Association.)

On the other hand, an appreciably stressed solid material is in a metastable equilibrium state in which the molecules, on average, vibrate in an elevated free energy state that is higher in energy than are the stable equilibrium "wells" on either side, as shown in Fig. 1. The higher metastable well is separated from the lower stable wells by activation–free energy barriers, whose height determines the kinetics of the change from the metastable to the stable state by direct analogy with chemical reaction kinetics [10]. When the activation–free energy is so high that the solid changes only slowly (for example, with a half-life measured in many days, weeks, or months) the activated solid is said to achieve a metastable equilibrium with its surroundings, which, for solubility determination, is an appropriate solvent medium. This solid material has a metastable solubility higher than that of the stable solubility of the stable solid. An example of a metastable solubility is provided by the highest dynamic solubility plot in Fig. 2, for freshly triturated digoxin powder [9].

Processing stresses increase the (metastable) solubility by causing relatively small increases, represented by δ, in the Gibbs free energy. This increase in free energy, δG, is related to the measured solubility values, C_s, according to

$$\delta G = RT \ \ln \frac{C_{s,2}}{C_{s,1}} \tag{3}$$

where $C_{s,1}$ is the solubility of the stable solid, $C_{s,2}$ is the solubility of the metastable (stressed) solid, R is the gas constant, and T is the absolute temperature [8]. For example, Fig. 2 shows that $C_{s,2}$ for freshly triturated digoxin is 87 μg/ml, while $C_{s,1}$ for untriturated digoxin is 33 μg/ml at 37°C. Substitution of these values into Eq. (3) enables δG for the trituration of digoxin to be calculated; it is 2470 J/mol (or 590 cal/mol).

The metastable (energized) solid will return to the stable (unenergized) state if sufficient time is allowed to elapse, because there is always a finite probability that the molecules in the solid will rearrange by thermal activation so as to overcome the activation-free energy barriers in Fig. 1 [5,10]. This process is in progress for triturated digoxin that had been stored for 2–5 months [9], as shown in Fig. 2 by the middle dynamic solubility curve, for which $C_{s,2}$ is 70 μg/ml and hence δG has been reduced by 531 J/mol (or 127 cal/mol) to 1939 J/mol (or 463 cal/mol). Given sufficient time, the triturated sample may be expected to give a dynamic solubility that corresponds closely to the lowest curve in Fig. 2. Metastable solids may be expected to return to the stable state, for which $\delta G = 0$, if facilitated by an appropriate kinetic mechanism [5], such as recrystallization through contact with a solvent or through contact with the vapor phase, if the saturated vapor pressure of the solid is appreciable.

The concept of metastable solubility values is recognized by scientists familiar with processed (stressed) solids, but it is not well known to scientists accustomed to highly crystalline materials. Because of the relative sensitivity of pharmaceutical solids (which are largely organic molecular crystals) to processing stresses, some variability of reported solubility values may be seen in the literature [11–18]. This situation is exemplified by the results summarized for griseofulvin in water at 37°C (Table 3). This type of variability can have critical consequences for low-dose drugs, such as digoxin (Fig. 2). For digoxin, unexpectedly high solubility values, and hence abnormally high dissolution rates, have resulted in overdosing of patients before the phenomenon of solid state activation was properly understood and controlled [9]. Because of the critical nature of this phenomenon, its discussion has been placed early in the chapter, before more classical, unactivated solids are considered.

D. Particle Size Effects

For a given solid material, a progressive reduction of particle size corresponds to increases in the surface/volume ratio and the escaping tendency of the molecules until the nature of the surface dominates the properties of the material. Two related thermodynamic consequences of this effect are an increase of solubility in any solvent and an increase of vapor pressure as the size of the particle is reduced. For a spherical particle of radius r, thermodynamic arguments lead to the Thomson–Freundlich equation [19]:

$$\ln \frac{C_{s,r}}{C_{s,\infty}} = \frac{2\gamma V_m}{rRT} \tag{4}$$

where $C_{s,r}$ and $C_{s,\infty}$ are the solubilities of a particle of radius r and of a large particle, respectively, γ is the interfacial tension (i.e., interfacial free energy) between the solid surface and the surrounding medium (the solvent), V_m is the molar volume of the particle (the molecular weight divided by the density), R is the gas constant, and T is the absolute temperature. In the analogous equation for vapor pressure (known as the Kelvin equation [20], which applies to both liquids and solids), $C_{s,r}$ and $C_{s,\infty}$ in Eq. (4) are replaced by P_r and P_∞, respectively. These are the vapor pressures of the substance next to a particle of

Table 3 Reported Values for the Solubility of Griseofulvin in Water at 37°C

Solubility (mg/L)	Reference
12.5	[13]
13.1	[14]
13.2	[15]
15.0	[16]
31.8	[17]

radius r and next to a large particle or flat surface of the substance, respectively. The quantity γ in Eq. (4) now represents the surface free energy of the substance in a gaseous environment.

Equation (4) indicates that the solubility of the solid at a given temperature increases exponentially with decreasing radius. The smaller the radius of the particle, the higher the chemical potential (partial molar free energy) of the molecules in the particle will be, and hence the greater the solubility [19]. The practical significance of this effect can be examined by plotting (as has been done in Fig. 3) the solubility ratio, $C_{s,r}/C_{s,\infty}$, against the radius, r, for a typical pharmaceutical solid having a molecular weight of 200 g/mol, a density of 2 g/cm^3, and an interfacial free energy of 30 mJ/m^2 (dyne/cm) at 298 K (25°C). Figure 3 shows that the effect of particle size reduction on solubility only becomes significant for particles of radius less than about 10 nm (100 Å). The solubility ratio predicted by the Kelvin equation increases steeply as the radius is reduced from 10 to 1 nm. This size range encompasses that of the critical nucleus to that of the unit cell, and radii below 1 nm are approaching molecular dimensions. The minuscule particle sizes at which the increase in solubility becomes effective

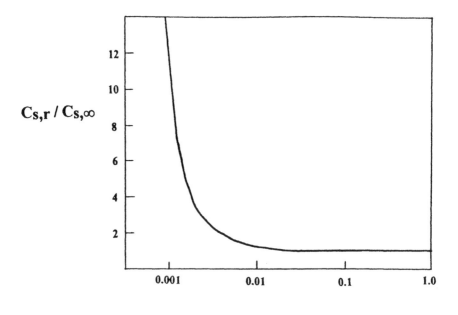

particle radius (μm)

Fig. 3   Solubility $(C_{s,r})$ of small particles of radius, r, compared with the solubility $(C_{s,\infty})$ of large particles. The solid is assumed to have a molecular weight of 200 g/mol, a density of 2 g/cm^3, and an interfacial free energy of 30 mJ/m^2 (dyne/cm) at 298 K (25°C).

greatly limits the practical usefulness of the Kelvin equation for solubility enhancement.

The small but definite increase of solubility with decreasing particle size for microscopic solid particles predicted by Eq. (4) does, however, account for the increase in the average particle size when crystals of various sizes are allowed to age in constant with a saturated solution (the mother liquor). This phenomenon, known as Ostwald ripening, occurs because a smaller particle, having a higher solubility, will dissolve in the unsaturated solution that is saturated with respect to a larger particle of lower solubility. Conversely, a larger particle having a lower solubility will grow in the supersaturated solution that is actually saturated with respect to a smaller particle of higher solubility. Larger particles will therefore grow at the expense of smaller particles, and the concentration of the "saturated" solution will decrease asymptotically. This phenomenon occurs as a precipitate is allowed to age in contact with its mother liquor. Other factors being equal, this process takes place more rapidly at a higher temperature according to the theory of reaction rates.

II. DETERMINATION OF SOLUBILITY

A. Concepts of Solubility Measurement

Methods for the determination of solubility have been thoroughly reviewed [21,22]. Solubility is normally highly dependent on temperature, and so the temperature must be recorded for each solubility measurement. Plots of solubility against temperature, as exemplified by Fig. 4 [23,24], are commonly used for characterizing pharmaceutical solids and have been extensively discussed [1,24]. Frequently (especially over a relatively narrow temperature range), a linear relationship may be given either by a van't Hoff plot according to [23]

$$\ln x_2^{sat} = -\frac{a}{R}\frac{1}{T} + c' \tag{5}$$

or by a Hildebrand plot according to

$$\ln x_2^{sat} = \frac{b}{R}\ln T + c'' \tag{6}$$

where x_2^{sat} is the mole fraction solubility of the solid solute at an absolute temperature T, a is the apparent molar enthalpy of solution, b is the apparent molar entropy of solution, and c' and c'' are constants. Over a wide temperature range (e.g., 0–60°C), however, such plots may not be linear, as has been illustrated in Fig. 4. Under such circumstances, it may be convenient to fit the solubility–temperature data to the following equation:

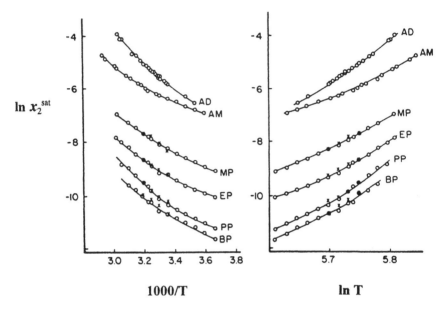

Fig. 4 Van't Hoff plots (left) and Hildebrand plots (right) representing the influence of temperature (absolute) on the aqueous mole fraction solubility of adipic acid (AD), acetaminophen (AM), methyl p-hydroxybenzoate (MP), ethyl p-hydroxybenzoate (EP), propyl p-hydroxybenzoate (PP), and butyl p-hydroxybenzoate (BP). The data represented by "O" were taken from Ref. 24, while the data represented by "×" were obtained from Ref. 23. (Reproduced with permission of the copyright owner, Elsevier Science Publishing Co.)

$$\ln x_2^{\text{sat}} = -\frac{a}{R}\frac{1}{T} + \frac{b}{R}\,\ln T + c \tag{7}$$

where a, b, and c are now adjustable parameters, defined so that a is the apparent enthalpy of solution at $T = 0$ K, b is the difference between the molar heat capacity of the solute in solution and that of the pure solute, and c is a constant [24]. More than three adjustable parameters are rarely necessary and have little practical significance.

Two general methods are available for determining solubility; these can be identified as the analytical method and the synthetic method. In the analytical method, the temperature of equilibration is fixed and the concentration of the solute in the saturated solution is determined at equilibrium by a suitable analytical procedure. In other words, a saturated solution in the presence of an excess of the undissolved solute is prepared at an accurately known temperature. This situation may be achieved by suitable contact between the undissolved solid and the solvent in a thermostat until equilibrium is reached. The analytical method can be subdivided into the traditional common batch agitation method and the more recent flow column method (also known as the generator column method). In the synthetic method, the composition of the solute–solvent system is fixed by appropriate addition and mixing of the solute and solvent, and then the temperature at which the solid solute just dissolves or just crystallizes is carefully bracketed. Each method enables points on the solubility–temperature curve to be determined.

Solubility determination of solids that are moderately soluble, that is, greater than 1 mg/ml (greater than 1 g/L, or greater than 0.1% w/v), normally poses no serious problems. However, the direct determination of solubilities much less than 1 mg/ml, corresponding to very sparingly soluble organic solids, is hindered by problems such as (1) slow equilibrium resulting from a low rate of dissolution during measurement, (2) the influence of impurities, and (3) the apparent heterogeneity in the energy content of the crystalline solid [11], all of which can lead to large discrepancies in reported values. For example, the reported aqueous solubilities of griseofulvin at 37°C range from 12.5 to 31.8 mg/L (Table 3), while those for cholesterol range from 0.025 to 2600 mg/ml [25].

B. Common Batch Agitation Method

In the traditional common batch agitation method, the solvent is agitated or stirred with the solid solute in a suitable vessel. An aliquot of the saturated solution is then removed from the system by centrifugation at the temperature of equilibration by filtration, by straining through a plug of glass wool, or even by decantation, and it is analyzed by a suitable method. Temperature control better than ±1°C is normally essential and should preferably be better than ±0.2°C, especially if the solubility is strongly temperature-dependent.

Equilibration is approached asymptotically at a decreasing rate that depends on the volume of the solvent, on the surface area of the solid solute, and on the nature and extent of the agitation. The method and time of equilibration should be established prior to the solubility determination by measuring the dissolved concentration of the solute as a function of time until no further change is noted. This approach, known as a dynamic solubility study, is illustrated in Fig. 2 for digoxin in water at 37°C. Analysis at times beyond the equilibration time is desirable to verify that a state of true equilibrium has been reached. The equilibrium value of solubility may be confirmed by agitating a supersaturated solution until its concentration falls to a constant value, which should be the same as that obtained on approaching equilibrium from the usual undersaturated side. Supersaturation may be conveniently achieved by equilibrating the solid solute with the solvent at a higher temperature than required for solubility determination. The decreasing concentration versus time plot from a supersaturated solution should converge with the increasing plot obtained from an undersaturated solution.

It is important to ascertain whether the solid phase of the solute changes during equilibration to produce a different polymorph or solvate, by analyzing the solid phase (using either chemical or thermal analysis, or x-ray diffraction). If a solid–solid phase transition occurs during equilibration, the measured equilibrium solubility will be that of the new solid phase of the solute. Methods of circumventing this problem have been proposed and evaluated [26].

Phase separation of the saturated solution from the excess solid solute is a critical process. If a filter is employed, it must be inert to the solvent, it must not release plasticizers, and its pore size must be small enough to retain the smallest particles of the solid solute. Furthermore, steps must be taken to monitor, minimize, and preferably avoid losses of the dissolved solute by adsorption onto the filter material [27–30] and/or onto the vessels, pipettes, and syringes. Typically, the first small volume of filtrate is discarded until the surfaces of the filter and/or vessels are saturated with the adsorbed solute, to ensure that the filtrate analyzed has not suffered significant adsorption losses. Adsorption can be a serious problem for hydrophobic solutes, for which filtration would not be recommended.

If decantation or centrifugation is employed for phase separation of the saturated solution, any disturbance and carryover of the undissolved solid solute (whether precipitated or floating) must be monitored and avoided.

C. Flow Column (Generator Column) Method

This more recent version of the analytical method has been developed by Wasik and coworkers [31,32]. A suitable column, which may be made of glass or stainless steel, is packed with the solid solute [31] or with a suitable supporting material, such as glass beads, onto which the solute has previously been adsorbed by evaporation of a suitable solution [32]. The solvent is pumped through the column or is forced through by applying gas pressure to the closed solvent vessel. The large area of contact between the solid solute and the solvent hastens the attainment of equilibrium, so that the solution emerging from the column is saturated; it is analyzed as just described. For example, by packing the column with solid particles of the solute, the aqueous solubilities of adenine and guanine (8.7 and 39 μmol/L, respectively, at 25°C) were measured. To avoid excessive pressures and to speed up the solvent flow, suitable modification of the packing of the solute into the column may be necessary, as was done for guanine.

The flow column method has useful advantages. Manipulation of the system prior to analysis is minimized, and so problems such as adsorption or evaporation that may arise from separation of the saturated solution and the undissolved solute are reduced. The method is rapid and precise [22,33,34], and it is valuable for sparingly soluble systems, such as hydrophobic solutes in water.

If the solute is adsorbed onto a support material in the flow column method, certain problems may arise. The possibility that the polymorphic form, solvate form, or melting point may change on evaporating the solution for the coating process must be considered. The solid solute that coats the support material should have a defined crystalline form. Furthermore, a strong binding interaction between the adsorbed solute and the support material may reduce the thermodynamic activity of the adsorbed solute below that of the normal crystalline form, so that measured solubilities may be reduced below those determined by the common batch agitation method. This effect has been observed for polynuclear aromatic hydrocarbons when determining their aqueous solubilities [35,36]. Saturation of the support material, as a stationary phase, may be turned to advantage to determine the solubility of solutes in water/cosolvent mixtures [22].

Sources of solubility data are available in the literature [1,22].

D. Synthetic Method

For this method, either a weighed amount of the solute (or a definite amount of the solvent) is placed in a suitable vessel. While agitating the system at constant temperature, known amounts of the solvent (or the solute) are added gradually until the solubility limit is reached. Appropriate checks must be carried out to ensure that the system is very close to equilibrium when the content or temperature of the system is recorded. In this method of temperature variation, attention is usually focused on the last small crystal. The equilibrium temperature is taken as the mean of the two temperatures at which the crystal either slowly grows or slowly dissolves. This procedure may also be carried out at the microscale by examining a small volume of the system under a hot-stage microscope.

E. Purity of the Solute and Analytical Implications

While the high purity required for individual liquid solvents can be obtained relatively easily (therefore presenting few problems), the purity of the solid state may often be a questionable property of the sample under study.

When in solid solution in the solid state, an impurity will alter the crystallinity by introducing impurity defects into the crystal lattice, thereby changing the thermodynamic and other physical properties of the solid, including the solubility and dissolution rate [2,37]. Prolonged equilibration of the solid state with the saturated solution, however, usually leads to recrystallization of the solute and to a consequent return of the crystallinity and the measured solubility of the solid state to that of the pure, highly crystalline solid.

When present as a separate phase, the impurity usually, but not invariably, exerts only a small influence in the solubility of the solid state of interest. However, if the impurity is more soluble than the solute under study, it will tend to concentrate in the solution, changing its solvent properties. To minimize this effect, the amount of solid in excess of that required to saturate the solution should be minimized.

The purity of the solid solute also has fundamental analytical implications. In general, the analytical procedure employed for determining the solubility should be specific for the solute of interest. For this purpose, an analyte-specific chromatographic method (such as high-performance liquid chromatography, HPLC) is preferred. Such a method will also enable the impurities and any possible decomposition products to be identified and quantified.

On the other hand, nonspecific analytical methods (such as liquid scintillation counting, spectrophotometry, and titration procedures) may not distinguish between the solute under study and certain impurities. When using such methods, the solid state of interest should be highly pure and the amount of solid in excess of that required to saturate the solution should be minimized.

III. PHASE SOLUBILITY ANALYSIS

Phase solubility analysis is a technique to determine the purity of a substance based on a careful study of its solubility behavior [38,39]. The method has its theoretical basis in the phase rule, developed by Gibbs, in which the equilibrium existing in a system is defined by the relation between the number of coexisting phases and components. The equilibrium solubility of a material in a particular solvent, although a function of temperature and pressure, is nevertheless an intrinsic property of that material. Any deviation from the solubility exhibited by a pure sample arises from the presence of impurities and/or crystal defects, and so accurate solubility measurements can be used to deduce the purity of the sample.

The experimental procedure for conducting phase solubility analysis is rather simple; it consists of mixing increasing amounts of sample with a fixed volume of solvent and then determining the mass of sample that has dissolved after each addition. It is not necessary to exceed the solubility limit of the analyte species, but attainment of this condition makes it easier to recognize trend within the plots. An experimental protocol for phase solubility analyses is available [39]. The data are most commonly plotted with the *system composition* (total mass of sample added per gram solvent) on the *x* axis, and the *solution composition* (mass of solute actually dissolved per gram of solvent) on the *y* axis.

Schirmer has succinctly summarized the strengths and limitations of phase solubility analysis [40]. The principal advantages are that (1) a reference standard known purity is not required, (2) the number and types of impurities in the sample need not be known, (3) all required solubility information is obtained from the analysis, (4) the technique can be applied to the analysis of any solute that can be dissolved in some solvent, (5) the deduced results are both precise and accurate, (6) only small quantities of recoverable analyte are required, and (7) only simple instrumentation is required for the performance of the work. The main disadvantages are that (1) the method is somewhat slow and tedious, (2) impurities present at less than 0.1% are not normally detected, (3) selection of a suitable solvent for the work involves some trial and error, and (4) interactions between the impurities and the drug substance (such as the formation of a solid solution or molecular adduct) can make interpretation of the results difficult or impossible.

A. Solubility Analysis with Noninteracting Components

A basic exposition of Gibbs phase rule is essential for understanding phase solubility analysis, and detailed presentations of theory are available [41,42]. In a system where none of the chemical species interact with each other, the number of independently variable factors (i.e., the number of degrees of freedom, F) in the system is given by

$$F = C + 2 - P \tag{8}$$

where C is the number of distinct chemical species and P is the number of phases present. Since solubility is normally measured under conditions of fixed temperature and pressure, a reduction in the number of variables results. In that case, the phase rule reduces to

$$F' = C - P \tag{9}$$

where F' (which equals $F - 2$) is the number of compositional degrees of freedom.

When a quantity of pure solid is totally dissolved in a liquid, a single phase is obtained, which consists of the two components. In this system, only one degree of freedom (which is the solute concentration) is possible, and that condition persists as the solute concentration varies from zero to saturation. This behavior is represented by the A–B segment of Fig. 5. When the data are plotted so as to illustrate the dependence of the solution composition on the system composition, one obtains a straight line (the A–B segment) with a slope of unity. Since the saturation limit is defined only with respect to a solid phase, if no undissolved solid is present, the system is undefined.

When the saturation limit is exceeded and excess pure solid remains undissolved and in contact with the solvent, the number of phases present now equals two. However, there are still only two components in the system, leading to the deduction that the number of degrees of freedom is zero. In practical terms, this means that there can be no variation in concentration as more solute is added to the system, and segment B–C of Fig. 5 is obtained. When solubility diagrams are obtained that exactly match the type shown in Fig. 5, it can safely be assumed that the solute under analysis is at least 99.9% pure.

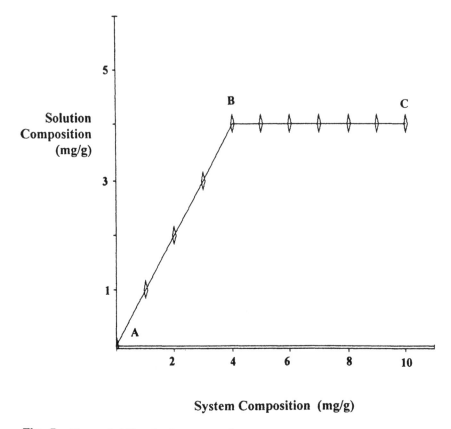

Fig. 5   Phase solubility plot for a pure solute.

When an impurity is present in the solute, the system is unlikely to be saturated with the additional component, and an additional contribution to the overall solubility will result. Application of the phase rule equation indicates that a plot of system composition versus solution composition now consists of three linear segments. When the solvent is not saturated with respect to either the analyte or its impurity, the number of degrees of freedom equals two, indicating that the concentration of each species varies linearly with the amount of solid added. This is illustrated by line segment D–E in Fig. 6. Owing to its large percentage in the solid, the analyte species will first saturate the solvent, and one solid phase will result. This effect decreases the number of degrees of freedom to one, which corresponds solely to the concentration of the impurity. At this point (segment E–F in Fig. 6), there can be no change in the concentration of the analyte as the amount of solid is increased. Finally, if enough solid is

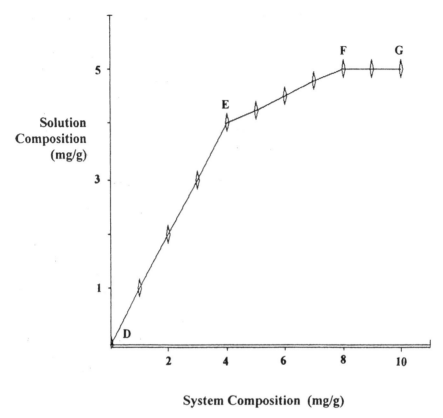

Fig. 6 Phase solubility plot for an impure solute consisting of 75% of the analyte substance and 25% of an impurity.

added, the system becomes saturated with respect to the impurity, and the number of degrees of freedom becomes zero. At this point (segment F–G in Fig. 6), a plateau in solution composition is reached, and no additional solid can be dissolved in the solvent.

When the solubility data are plotted as in Figs. 5 and 6, it is easy to calculate the weight purity of the analyte sample. This information is obtained by computing the slope of segment E–F and using the relation

$$\% \text{ purity} = (1 - \text{slope}_{EF}) \times 100 \tag{10}$$

The presence of more than one impurity in a material will yield additional segments in the plot of system composition versus solution composition.

During the analysis of real samples, the saturation limit of any impurity species is rarely reached. For that reason, the phase solubility curves normally consist of segments D–E and E–F of Fig. 5, with the nonzero slope of segment E–F indicating the purity of the sample. Phase solubility analysis data, summarized in Table 4, were obtained for fluphenazine dihydrochloride in absolute ethanol. As indicated in Table 4, a slope of 0.0073 was obtained, indicating a compound purity of 99.27% (i.e., $(1 - 0.0073) \times 100\%$). The intercept obtained in the analysis was found to be 7.36, permitting the conclusion that the solubility of fluphenazine dihydrochloride in absolute ethanol is 7.36 mg/g. This value is in excellent agreement with an empirically determined equilibrium solubility value of 7.75 mg/g, demonstrating the accuracy of the analysis.

B. Solubility Analysis with Interacting Components

All the methodology just described requires that the solubilities of the analyte and its impurities be totally additive, which implies that the solubility of any given species cannot be affected by the presence of any other dissolved substance. Such a lack of independence is most commonly indicated by the existence of curved lines when the data are plotted in the conventional manner, and it has been discussed in great detail [38]. In the presence of specific molecular interactions, the phase solubility method cannot be used without the benefit of detailed knowledge of the nature and magnitude of the interactions. Such interactions can either increase or decrease the overall solubility, and the outcome is difficult to predict a priori. A thorough investigation is required to deduce the nature of the interactions if the phase solubility method is to be used.

Table 4 Phase Solubility Analysis of Fluphenazine Dihydrochloride in Absolute Ethanol

Point	Mass sample taken (mg)	System composition (mg/g)	Solution composition (mg/g)
1	15.02	4.7846	4.3422
2	26.63	8.4761	7.1348
3	44.45	14.1065	7.3639
4	66.62	21.2364	7.6418
5	85.70	27.2539	7.6628
6	106.01	33.6307	7.4670
7	124.50	39.5219	7.6701
8	144.30	45.8684	7.7048

Linear least squares analysis of points 3–8 yields the following:
slope $= 0.0073$
intercept $= 7.36$

Since the complexation could either increase or decrease the overall solubility, two directions of deviation from the linear curves are possible. As shown in Fig. 7, when the effect of the interaction is to increase the system solubility, the plateau region will consist of an upwardly concave segment. This deviation leads to an apparent increase in the slope of the impurity curve, which in turn indicates that the impurity concentration has been overestimated. When the effect of the interaction decreases the solubility of the system, a plot of the type illustrated in Fig. 8 is obtained. The downward trend in the plateau values serves to underestimate the amount of the impurity.

A practical solution to the problem of intermolecular interactions is to change the solvent system to one for which the solute exhibits a lower solubility, and to work only in systems for which acceptable plots can be obtained. When

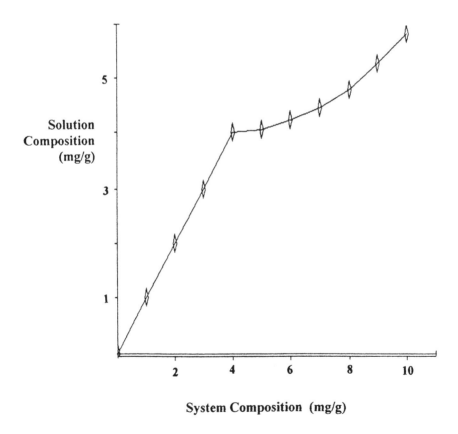

Fig. 7 Phase solubility plot resulting from a positive solute interaction between the analyte substance and an impurity.

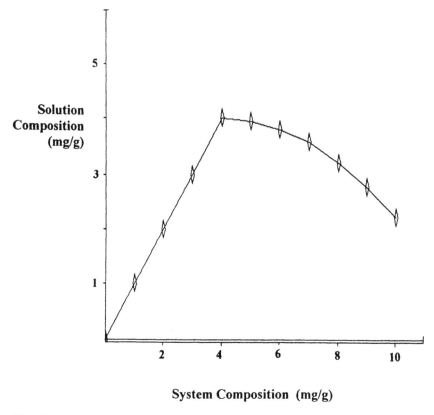

Fig. 8 Phase solubility plot resulting from a negative solute interaction between the analyte substance and an impurity.

an alternative system cannot be identified, the purity of the analyte must be determined by a method other than phase solubility analysis.

IV. SOLUBILIZATION

The solubility of a drug candidate may not always be sufficient for its intended pharmacological purpose, and in such cases the solubility may need to be modified (either increased or decreased). These concerns are especially important for parenterals when the active substance does not exhibit the desired level of solubility. The solubility of a given compound can be profoundly altered in a number of ways, such as through the formation of salt species or drug–substrate complexes, or through the use of surfactants or cosolvents [43]. In the majority of cases, the goal is to solubilize a poorly soluble drug, and to therefore increase the bioavailability of that drug substance.

The solubility of hydrated copper sulfate ($CuSO_4 \cdot 5H_2O$) provides a simple example of how the solubility of a compound can be manipulated. $CuSO_4 \cdot 5H_2O$ itself is very soluble in water, exhibiting an equilibrium solubility of 207 mg/ml at 20°C [44]. This high solubility is due to the dissociation of copper sulfate into its component ions upon dissolution into an aqueous solution:

$$CuSO_4 \rightarrow Cu^{2+} + SO_4^{2-} \tag{11}$$

The slow addition of aqueous ammonia solution to an aqueous copper sulfate solution causes an immediate effect on the solubility, since essentially all the Cu^{2+} ion is consumed by the hydroxide ion of the added reagent, forming in the process a product of very low solubility:

$$Cu^{2+} + 2OH^- \rightarrow Cu(OH)_2 \downarrow \tag{12}$$

Since the solubility of $Cu(OH)_2$ is only 58 μg/ml [45], essentially all the Cu^{2+} ions in the solution immediately precipitate out of the solution. If the addition of aqueous ammonia solution is continued, however, a second reaction takes place. The excess ammonia molecules react with the $Cu(OH)_2$, forming a new complex species that is highly soluble:

$$Cu(OH)_2 + 4NH_3 \rightarrow [Cu(NH_3)_4]^{2+} + 2OH^- \tag{13}$$

The solubility of the $[Cu(NH_3)_4]^{2+}$ complex ion is very high, with the solubility of the sulfate salt exceeding 500 mg/ml [44].

This example of the copper ion–hydroxide–ammonia system clearly demonstrates how complexation effects can either increase or decrease the solubility of a species of interest, and how such effects can be manipulated at the whim of the formulator.

A. Equilibrium Constants and Solubility

Any association between two solute species (molecules and/or ions) in solution can be considered as constituting *complexation*. Whenever a compound is capable of forming an associated species with another dissolved solute molecule, the complexation of the two can profoundly influence the solubility of the system. The simplest type of complex that may be formed between a drug entity (denoted in the following discussion as D) and an interacting substrate or ligand (which will be identified as S) is the species in which they combine in a 1:1 stoichiometric ratio:

$$D + S \rightleftharpoons DS \tag{14}$$

The equilibrium constant, K_1, defining the extent of this particular reaction is defined by

$$K_1 = \frac{[DS]_e}{[D]_e[S]_e} \tag{15}$$

where $[DS]_e$, $[D]_e$, and $[S]_e$, are the concentrations of each species after equilibrium is reached. If we assume that the concentration of D is increased by the addition of S, it must follow that the equilibrium concentration of free D cannot exceed its maximum solubility in the absence of S, or

$$[D]_e = D_0 \tag{16}$$

where D_0 is the equilibrium solubility of the drug. The amount of complexed drug at equilibrium is calculable, since it must be given by the total amount of drug dissolved in the system (identified as D_{total}, whose value is set at the beginning of the determination) less the amount consumed in the complexation reaction:

$$[DS]_e = D_{total} - D_0 \tag{17}$$

The amount of substrate remaining at equilibrium is similarly given by the difference between its total amount (identified as S_{total}, whose value is also set at the beginning of the determination) less the amount consumed in the complexation reaction:

$$[S]_e = S_{total} - [DS]_e \tag{18}$$

The substitution of Eqs. (16–18) into Eq. (15) yields a simple equation that readily permits the calculation of the equilibrium constant, K_1:

$$K_1 = \frac{D_{total} - D_o}{[D]_o \left\{ S_{total} - [D]_{total} + D_o \right\}} \tag{19}$$

One of the key aspects of 1:1 complexation is that the increase in concentration of the drug substance by the substrate is a linear function of the substrate concentration until the limiting solubility of the DS complex is reached. This deduction is more readily evident in the rearrangement of Eq. (19) into the form of a straight line:

$$D_{total} = \frac{K_1 D_o S_{total}}{1 + K_1 D_o} + D_o \tag{20}$$

When the drug entity is capable of forming complexes that have higher stoichiometric ratios than 1:1, the construction of equilibrium constant expressions becomes more difficult. For the general case of *m:n* stoichiometry, as defined by

$$mD + nS \rightleftharpoons D_m S_n \tag{21}$$

the corresponding equilibrium constant is given by

$$K_{mn} = \frac{[D_m S_n]_e}{[D]_e^m [S]_e^n}$$
(22)

and, as before, the concentrations of all species at equilibrium are defined in terms of known quantities:

$$[D]_e = D_0$$
(23)

$$[D_m S_n]_e = \frac{D_{total} - D_o}{m}$$
(24)

$$[S]_e = S_{total} - n[D_m S_n]_e$$
(25)

For the particular case in which $n = 1$, Higuchi and Connors [50] have shown that the substitution of Eqs. (23)–(25) into (22) ultimately yields an equation of the form

$$D_{total} = \frac{mK_m D_o^m S_{total}}{1 + K_m D_o^m} + D_o$$
(26)

in which K_{mn} is replaced by K_m. Equation (26) is in the form of a straight line, so that a plot of D_{total} against S_{total} permits the calculation of K_m from the observed slope.

The computation of formation constants is considered to be the most important aspect of equilibrium theory, since this knowledge permits a full specification of the complexation phenomena. Once this information is in hand, the formulator can literally define the system at a given temperature through the manipulation of solution-phase parameters to obtain the required drug solubility.

B. Formation of Soluble Salts

The solubility of any solid can be either increased or decreased by the addition of an electrolyte to the solvent, a phenomenon known as the salt effect. *Salting-out* describes the situation in which the solubility of the solid is decreased by the salt effect, whereas *salting-in* is the term used when the solubility is increased. Salting-out takes place when the added electrolyte sufficiently modifies the water structure so that the amount of water available for solute dissolution is effectively reduced, and it is a procedure convenient for the isolation of highly soluble substances.

The salting-in effect may be used to increase the solubility of a drug substance through the formation of associated ion pairs, most commonly making use of anionic countering (hydrochloride being the most popular). Detailed reviews of pharmaceutical salts have been published, which contain extensive tables of anions and cations acceptable for pharmaceutical use [44,47]. These articles also describe useful processes for the selection of the most desirable salt form. Recently, Morris et al. have described an integrated approach to the selection of the optimal salt form for a new drug [48].

Since the formation of a salt species is essentially an acid–base neutralization reaction, every compound that contains a protonatable or deprotonatable group can participate in salt formation. The appropriate choice of counterion is dictated by the nature of the functionality, and so acidic solutes will require the use of a cation, whereas basic salts will require an anion. Naturally, the salt form should not be more toxic than the original drug in its free base or free acid form, or else the pharmaceutical advantages of the salt cannot be used. Salt formation represents a means of modifying the physical characteristics of a drug entity without modifying its chemical structure. Nevertheless, the regulatory authorities require every salt form to be treated as a new drug entity.

The pharmaceutical literature contains many examples showing how salt formation can be used to increase the solubility of drug substances. For example, the solubility of nicardipine hydrochloride was found to be greatly affected by the nature of the carboxylate buffer system in which it was dissolved [49]. As evident in Table 5, the degree of solubilization was found to increase with the length of the alkyl chain, although solubility limitations within the various buffer systems limited the scope of the study. Not all buffer systems were found to increase the drug solubility; both phosphate and citrate buffers actually salted-out the compound. Depending on the nature of dosage form being developed, it is clear that the solubility of nicardipine hydrochloride can be adjusted to fit the desired level of activity.

C. Molecular Complexes

The pharmaceutical literature abounds with examples of how poorly soluble drug substances have been solubilized through the formation of molecular complexes.

Table 5 Solubilization of Nicardipine Hydrochloride by Aqueous Carboxylate Buffer Systems

Solubility (mg/ml)	Buffer system
5.1	Unbuffered water
69	5.0 M acetate (ethanoate)
270	5.0 M proprionate (propanoate)
32	1.5 M butyrate (1-butanoate)
5.8	0.1 M valerate (1-pentanonate)

Source: The data were obtained from Ref. 6.

These studies were initiated by Higuchi, who studied the effect of complexation on the solubility of numerous drug compounds. The results obtained for more than 500 systems were cited in the primary review of this work [50].

One of the classic examples in this series is the solubilization of *p*-aminobenzoic acid (PABA) by caffeine [51], for which the essential data are summarized in Fig. 9. The solubility of PABA in the absence of caffeine was reported to be 6.2 mg/ml, which could be increased to 7.7 mg/ml by the addition of at least 2 mg/ml of caffeine. The linear increase in PABA concentration as a function of caffeine concentration is consistent with the formation of a 1:1 stoichiometric complex. From the data obtained in the linear concentration region, a value of 48 L/mol was calculated for K_1. Further increases in the caffeine concentration up to 6 mg/ml had no effect on the PABA solubility. However, larger concentrations of caffeine led to a reduction in the dissolved

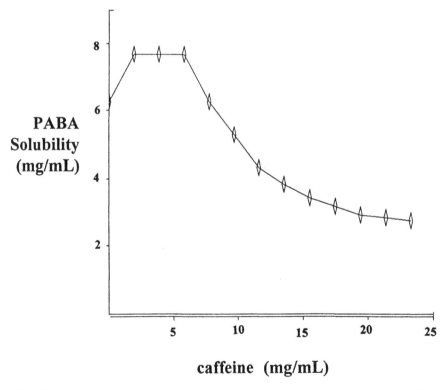

Fig. 9 Phase solubility diagram showing the changes in the apparent aqueous solubility of *p*-aminobenzoic acid (PABA) brought about by the addition of the complexing agent, caffeine, at 30°C. (The data are adapted from Ref. 51.)

PABA concentration (Higuchi type B$_s$), undoubtedly due to the formation of complexes of higher stoichiometry.

The pH of the medium can exert an influence, not only on the degree of association between drug and substrate, but also on the stoichiometry of the complexes. It was found that mg/ml concentrations of the normally insoluble moricizine could be obtained upon complexation with nicotinamide, and that higher degrees of solubilization could be reached at pH 6 rather than at pH 7 [52]. Through a detailed analysis of the data, it was deduced that both 1:1 and 1:2 moricizine–nicotinamide complexes could be formed, and that the K_{mn} constants varied with solution pH. At pH 6 (where moricizine is 70% protonated), it was calculated that $K_1 = 16.6$ L/mol and $K_2 = 0.93$ L^2/mol^2. Upon raising the pH to 7 (at which moricizine is only 20% protonated), it was deduced that $K_1 = 7.7$ L/mol and $K_2 = 5.41$ L^2/mol^2. These findings were interpreted as indicating that protonated moricizine favored a complex of 1:1 stoichiometry, while the free base preferred to form a complex of 1:2 stoichiometry.

One other example will serve to illustrate how complexation may be used to modify the solubility of a drug entity. In an attempt to overcome the variable bioavailability of nifedipine through increasing its solubility, a series of complexes were formed with substituted phenolic ligands [53]. Through systematic variation in substrate functionality, it was possible to deduce the essential structural requirements for the interaction. A linear free-energy relationship was used to relate the observed K_1 values with Hammett's sigma (σ) and fractional partition coefficient (π). A statistical analysis of the data showed that electronic factors were more important in determining the extent of the interaction than were hydrophobic interactions. Substrates having electron-withdrawing abilities were found to yield stronger complexes, and the presence of a lipophilic group further increased the degree of interaction.

D. Cyclodextrin Inclusion Complexes

Cyclodextrins are torus-shaped, cyclic oligosaccharides consisting of either six (denoted as α-cyclodextrin), seven (β-cyclodextrin), or eight (γ-cyclodextrin) D-glucose units. Owing to their hydrophobic interior, the various cyclodextrins are capable of including a variety of solutes within the inner cavity. These association complexes are normally of 1:1 stoichiometry, but other ratios of host and guest are known [54–57]. The predominant forces responsible for the formation of the host–guest complexes are hydrogen bonding [58] and van der Waals forces [59]. Since most inclusion complexes are amorphous, the solubility enhancement plots are essentially linear in cyclodextrin concentration (Higuchi type A), and the calculation of K_1 formation constants by Eq. (20) is quite straightforward.

For most compounds of pharmaceutical interest, the cavity provided by

β-cyclodextrin appears to be optimal for complex formation. In one study, the interaction of four barbiturates with β-cyclodextrin was studied by solubility analysis and other methods [60]. It was found that the solubility of the barbiturates increased significantly upon formation of the inclusion complexes, with the relative strength of interaction with β-cyclodextrin being phenobarbital > pentobarbital > amobarbital > barbital. In another work, it was found that formation of the β-cyclodextrin inclusion complex improved the solubility of retinoic acid from less than 0.5 mg/ml to 160 mg/ml, eliminating the formulation problems associated with solubility limitations [61].

The magnitude of association between a drug compound and various cyclodextrins depends critically on the details of the fit of the substrate into the cyclodextrin cavity. As shown in Table 6, the experimental compound RS-82856 forms the strongest complexes with β-cyclodextrin, while maximal solubility is reached with γ-cyclodextrin [62]. Formation of the β-cyclodextrin complex dramatically increased the dissolution rate of the compound as well. For RS-82856 itself, 20% dissolved within 20 minutes, while more than 80% of the drug–β-cyclodextrin complex was found to be dissolved at the same time point.

Unfortunately, the relatively low solubility of β-cyclodextrin has precluded its widespread use as a solubilizing agent. To enhance the solubility of β-cyclodextrin, much work has been carried out on possible derivatizations of the β-cyclodextrin skeleton. 2-Hydroxypropyl-β-cyclodextrin has been found to exhibit the desired enhanced solubility, while still retaining the optimal binding characteristics of underivatized β-cyclodextrin. In addition, this derivative is relatively nontoxic and may even be used in parenteral applications [63].

The number of studies describing the solubilization of drug compounds by cyclodextrins is extraordinarily large, and the recent state of the field has been summarized [64,65].

Table 6 Formation Constants, K_1, and Maximum Solubilities for the Various Cyclodextrin Complexes with the Experimental Compound RS-82856

Complex	K_1 (L/mole)	Maximum solubility (μg/ml)
α-cyclodextrin	136.5	62.5
β-cyclodextrin	370.4	130.7
γ-cyclodextrin	64.7	240.0

The data were all obtained at ambient temperature [62].

E. Solubilization by Surfactants

Molecules that possess both hydrophilic and hydrophobic structures may associate in aqueous media to form dynamic aggregates, commonly known as *micelles*. The properties of micellar structures have been discussed in great detail [66–69], but their main pharmaceutical application lies in their ability to provide enhanced solubility to compounds lacking sufficient aqueous solubility [70]. The ability of a micelle to solubilize compounds of limited aqueous solubility can be understood from consideration of the schematic drawing of Fig. 10a. Above the critical micelle concentration, these molecules orient themselves with the polar ends in interfacing with the aqueous solution and the nonpolar ends at the interior. A hydrophobic core is formed at the interior of the micelle, and hydrophobic solute molecules enter and occupy this region.

Other surface-active compounds self-assemble into bilayer structures (schematically illustrated in Fig. 10b), which normally spherilize into structures termed *vesicles*. When vesicles are formed from phospholipids, the term *liposome* is used to identify the structures, which also provide useful drug delivery systems [71]. Solutes may be dispersed into the lipid bilayer or into the aqueous interior, to be subsequently delivered through a variety of mechanisms. Liposomes have shown particular promise in their ability to act as modifiers for sustained or controlled release.

(b)

(a)

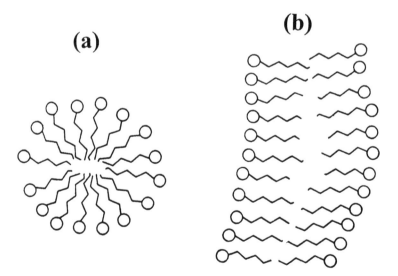

Fig. 10 Cross sections of (a) spherical and cylindrical micelles, and (b) cylindrical and lamellar vesicles in aqueous solution. Each surfactant molecule making up the structures has a polar head-group, depicted as a circle, and a nonpolar, hydrophobic chain, depicted as a zigzag.

In a series of studies, Thoma and coworkers have used micelles to solubilize a wide variety of compounds under investigation as local anesthetic agents [72,73]. Some of the compounds themselves were found to be surface-active, and also capable of self-associating into assembled species. It was found that the bioavailability of the agents could be increased by decreasing the self-association through dispersion in micellar media, and that an increase in efficacy also resulted. These findings were demonstrated using several in vitro models of efficacy. In addition, the ability of the various agents to be solubilized in the micellar systems could be correlated with the octanol–water partition coefficients when measured at pH 5.7.

Certain compounds are known to achieve higher absorption rates from the GI tract if they are taken with food, and this observation has been linked to their solubilization by bile salts [74]. Bile salts, especially those of cholic and deoxycholic acids, have been used to solubilize steroid hormones [75], antibiotics [76], and nonsteroidal antiinflammatory drugs [77]. For example, amphotericin B (an antifungal agent) has been solubilized for parenteral use in micelles composed of sodium desoxycholate [78]. As illustrated in Fig. 11, the degree of solubilization of carbamazepine by sodium desoxycholate is minimal below the critical micelle concentration but increases rapidly above this value [79]. At sufficiently high concentrations, when the micelles become saturated in carbamezepine, the apparent solubility reaches a limiting value approximately seven times the true aqueous solubility in the absence of desoxycholate.

F. Solubilization by Cosolvents

When the aqueous solubility of a drug candidate is insufficient for its intended purpose, the use of organic cosolvents to aid the solubilization is a common practice. The solvents most frequently utilized for this purpose are propylene glycol, glycerol, ethanol, and polyethylene glycol (PEG) 400 [80]. Solvents are characterized by their polarity, dielectric constant, surface tension, and partition coefficient, and these quantities are useful in the prediction of the solubilizing power associated with a particular cosolvent [81]. The degree of solubilization may often reach several orders of magnitude, and it is frequently sufficiently large to permit the development of an acceptable parenteral product.

Yalkowsky has shown that the solubility of a compound in a cosolvent mixture (S_m) can be estimated through a log-linear solubility relation:

$$\log \frac{S_m}{S_w} = f\sigma \tag{27}$$

where S_w is the solubility of the compound in water, f is the volume fraction of the cosolvent, and σ is the slope of the f vs. log (S_m/S_w) plot [82]. This relation has been explored for phenytoin [83,85], as well as for diazepam and benzocaine [84,85]. As shown in Fig. 12, an essentially linear plot was obtained for

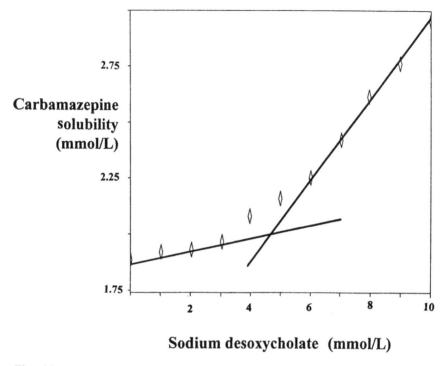

Sodium desoxycholate (mmol/L)

Fig. 11   Solubilization of carbamazepine in aqueous solution by addition of sodium desoxycholate at 25°C. (The data are adapted from Ref. 79.)

hydrocortisone (with only a slight deviation at high percentages of cosolvent), but a greater departure from linearity was noted for triazolam [82]. Yalkowsky has suggested that deviations from the linear relation are associated with subtle changes in solvent structure taking place as the composition is varied from one extreme to another [86].

One potential risk that formulators run when using cosolvents as drug solubilizers is the possibility of vehicle toxicity. Each cosolvent is characterized by an acceptable concentration range, which cannot be exceeded without incurring biological damage. To avoid the requirement for in vivo testing, several in vitro models have been advanced to evaluate the relative safety of cosolvent excipients. The most useful in vitro procedure follows the hemolysis of red blood cells, which has been correlated with in vivo animal tests [87,88].

V. DISSOLUTION RATE

Evaluation of the dissolution rates of solid drugs is extremely important in the development, formulation, and quality control of solid pharmaceuticals. The

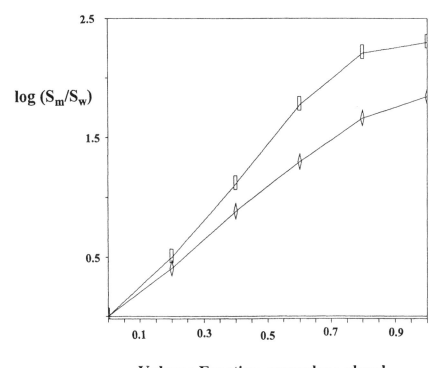

Volume Fraction, propylene glycol

Fig. 12 Solubilization of hydrocortisone (◇) and triazolam (□) by propylene glycol, when mixed with water at 25°C. (The data are adapted from Ref. 82.)

rates of dissolution of solids may be measured experimentally, modeled theoretically, and sometimes predicted from the physicochemical properties of the system.

The wide variety of methods for determining the dissolution rates of solids may be categorized either as batch methods (Fig. 13A) or as continuous-flow methods (Fig. 13B). The common batch-type dissolution methods are derived from the beaker–stirrer method of Levy and Hayes [89] and include a number of thoroughly standardized procedures, especially those defined by the U.S. Pharmacopoeia [90].

The dissolution rate of a solid may be defined as dm/dt, where m is the mass of solid dissolved at time t. In a batch dissolution method, the analyzed concentration, c_b, in the solution (if well stirred) is representative of the entire volume, V, of the dissolution medium, so that

$$m = Vc_b \tag{28}$$

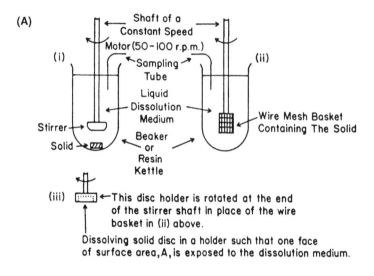

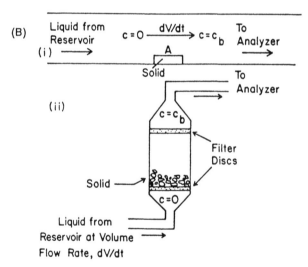

Fig. 13 Schematic diagrams of various types of apparatus for measuring the dissolution rates of solids. (A) Batch-type dissolution apparatus: (i) USP 23 (1995) Dissolution Apparatus 2; (ii) USP 23 (1995) Dissolution Apparatus 1; (iii) rotating disc method [104,111]. (B) Continuous-flow dissolution apparatus: (i) schematic representation; (ii) column-type flow-through dissolution cell [92]. Each filter disc may be replaced by a stainless-steel sieve (US 40 mesh) on top of which glass beads (1 mm diameter) are placed. (Reproduced with permission of the copyright owner, John Wiley and Sons, Inc., from Ref. 1, p. 475.)

and

$$\frac{dm}{dt} = V \frac{dc_b}{dt} \qquad (29)$$

If the dissolved solute is premitted to accumulate and if sufficient solid solute is placed in the dissolution medium, the dissolution curve will resemble Fig. 14A. Eventually a saturated solution will be formed.

While batch dissolution methods are simple to set up and to operate, are widely used, and may be carefully and reproducibly standardized, they suffer from the following disadvantages: (1) the hydrodynamics are usually poorly characterized, with the notable exception of the rotating disc method, (2) a small change in dissolution rate will often create an undetectable and therefore an immeasurable perturbation in the dissolution time curve, and (3) the solute concentration c_b may not be uniform throughout the solution volume V.

By rotating sticks of benzoic acid or lead chloride (examples of sparingly

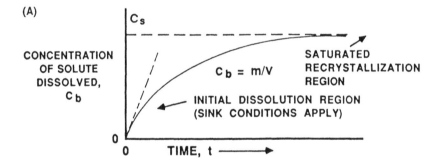

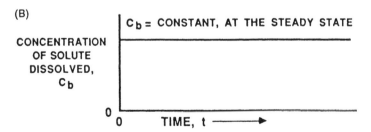

Fig. 14 Dissolution-time profiles for a batch-type dissolution apparatus (A), and a continuous-flow dissolution apparatus (B). (Reproduced with permission of the copyright owner, John Wiley and Sons, Inc., from Ref. 1, p. 476.)

soluble compounds) in water within a batch-type dissolution apparatus, Noyes and Whitney [91] found that the dissolved concentration c_b of each solute increased with time according to first-order kinetics:

$$\frac{dc_b}{dt} = k_{nw} (c_s - c_b) \tag{30}$$

where k_{nw} is the first-order rate constant and c_s is the equilibrium solubility of the solute. This well-known Noyes–Whitney relationship (30) is represented by Fig. 14A. When combined, Eqs. (29) and (30) lead to

$$\frac{dm}{dt} = k_{nw}V (c_s - c_b) \tag{31}$$

From the onset of dissolution at $t = 0$ until c_b attains a value of about 5–10% of c_s, the assumption $c_b \ll c_s$ is often valid. In that instance "sink conditions" are said to apply. From Eqs. (29) and (31), the following expressions for the initial dissolution rate are obtained:

$$\left(\frac{dm}{dt}\right)_{t \to 0} = \left(\frac{dc_b}{dt}\right)_{t \to 0} V = k_{nw}V c_s \tag{32}$$

Continuous-flow (i.e., column dissolution) methods, depicted schematically in Fig. 13B, have limited, but growing, application [92–94]. The volume flow rate, dV/dt, of the dissolution medium must remain constant to achieve the steady state shown in Fig. 14B, so that

$$\frac{dm}{dt} = \left(\frac{dV}{dt}\right) c_b \tag{33}$$

Continuous-flow methods have the following advantages: sink conditions may be easily achieved by adjusting dV/dt so that $c_b \ll c_s$; and a change in dissolution rate is reflected in a change in c_b. Continuous-flow methods suffer from the disadvantages that a high flow rate will require a large volume of dissolution medium; and if the solid has a low solubility and dissolution rate, c_b will be small, requiring a very sensitive analytical method.

Since the rate of absorption of many drugs from the gastrointestinal tract is controlled by their dissolution rate, this becomes the rate-limiting step. Accurate and reliable measurements of dissolution rate are therefore required in the pharmaceutical sciences. The measurement and interpretation of the dissolution rates of solid drugs in the pure state or from formulations, such as tablets, capsules, and suppositories, has an extensive pharmaceutical literature [95–100]. Moreover, the design, operation, and interpretation of dissolution rate measurements on pharmaceutical solids have been the subject of considerable scientific study, technical development, and debate.

The dissolution rate of a given solid is usually strictly proportional to the wetted surface area A of the dissolving solid, shown in Fig. 13; thus,

$$\frac{dm}{dt} \propto A \tag{34}$$

The dissolution rate per unit surface area is the mass flux, J, usually termed the "intrinsic dissolution rate" in the pharmaceutical sciences and is given by

$$J = \left(\frac{dm}{dt}\right) \frac{1}{A} \tag{35}$$

From Eqs. (31) and (35) under constant defined conditions,

$$J = \left(\frac{dm}{dt}\right) \frac{1}{A} = k_1 (c_s - c_b) \tag{36}$$

where k_1 is the mass transfer coefficient, which has the dimensions of length/time and which is related to k_{nw}. Thus,

$$k_{nw} = k_1 \frac{A}{V} \tag{37}$$

Under sink conditions, $c_b \ll c_s$, so

$$J_{t \to 0} = \left(\frac{dm}{dt}\right)_{t \to 0} \left(\frac{1}{A}\right) = k_1 c_s \tag{38}$$

Equations (36) and (38) are generally applicable, independent of the mechanism of dissolution.

A. Theories and Mechanisms of Dissolution

Among the various mechanisms that have been proposed for the dissolution of solids [101,102], two of the simplest are depicted in Fig. 15. The common features of these are that an infinitesimally thin film of saturated solution of concentration c_s (the solubility) is formed at the solid–liquid interface; and that in the well-mixed bulk of solution, the concentration of the dissolving solid at any given time is c_b.

The dissolution process, in general, consists of the following chemical reaction at the solid–liquid interface:

$$\text{solid} + \text{solvent} = \text{solution} \tag{39}$$

for which the reaction rate constant is k_R. This is followed by transport of the solute away from the interface into the bulk solution, for which the transport rate constant is k_T. For the overall process, the observed rate constant, k_1, in Eqs. (36) and (38) is given by

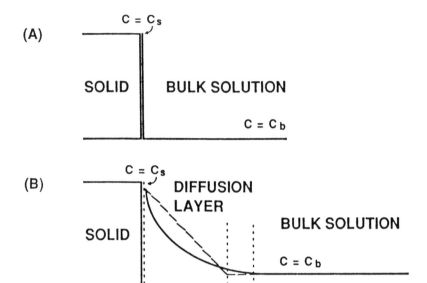

Fig. 15 Two of the simplest theories for the dissolution of solids: (A) the interfacial barrier model, and (B) the diffusion layer model, in the simple form of Nernst [105] and Brunner [106] (dashed trace); and in the more exact form of Levich [104] (solid trace). c is the concentration of the dissolving solid, c_s is the solubility, c_b is the concentration in the bulk solution, and x is the distance from the solid–liquid interface of thickness h or δ, depending on how it is defined. (Reproduced with permission of the copyright owner, John Wiley and Sons, Inc., from Ref. 1, p. 478.)

$$\frac{1}{k_1} = \frac{1}{k_R} + \frac{1}{k_T} \qquad\qquad (40)$$

According to the diffusion layer theory, for which the transport process is rate-limiting, $k_T \ll k_R$, so that $k_1 = k_T$. According to the interfacial barrier theory, for which the surface reaction is rate-limiting, $k_R \ll k_T$, so that $k_1 = k_R$.

The interfacial barrier theory is illustrated in Fig. 15A. Since transport does not control the dissolution rate, the solute concentration falls precipitously from the surface value, c_s, to the bulk value, c_b, over an infinitesimal distance. The interfacial barrier model is probably applicable when the dissolution rate is limited by a condensed film absorbed at the solid–liquid interface; this gives rise to a high activation energy barrier to the surface reaction, so that $k_R \ll k_T$. Reaction-controlled dissolution is somewhat rare for organic compounds. Examples include the dissolution of gallstones, which consist mostly of cholesterol, in bile acid media. Furthermore, for the dissolution of certain solid carbon acids, such as deuterated phenylbutazone, in an ionizing medium, such as buffered water, under rapid agitation, $k_R \ll k_T$ and the rate-limiting step is the slow ionization of the carbon acid [103].

The diffusion layer theory, illustrated in Fig. 15B, is the most useful and best-known model for transport-controlled dissolution. The dissolution rate here is controlled by the rate of diffusion of solute molecules across a diffusion layer of thickness h, so that $k_T \ll k_R$ in Eq. (40), which simplifies to $k_1 = k_T$. With increasing distance, x, from the surface of the solid, the concentration, c, decreases from c_s at $x = 0$ to c_b at $x = h$. In general, c is a nonlinear function of x, and the concentration gradient dc/dx becomes less steep as x increases. The hyrodynamics of the dissolution process has been fully discussed by Levich [104]. In a stirred solution, the flow velocity of the liquid dissolution medium increases from zero at $x = 0$ to the bulk value at $x = h$.

The original and simplest form of the diffusion layer theory was developed by Nernst [105] and Brunner [106], who assumed that the mass flux is given by Fick's first law of diffusion. In that case,

$$J = \left(\frac{dm}{dt}\right)\left(\frac{1}{A}\right) = -D\frac{dc}{dx} \tag{41}$$

where dc/dx is the concentration gradient and D is the diffusivity (diffusion coefficient) of the dissolving solute. They also assumed that dc/dx is constant within the diffusion layer of thickness h in Fig. 15B; thus,

$$\frac{dc}{dx} = \frac{c_b - c_s}{h} \tag{42}$$

Eliminating dc/dx from the previous two equations yields

$$J = \left(\frac{dm}{dt}\right)\left(\frac{1}{A}\right) = \frac{D}{h}(c_s - c_b) \tag{43}$$

Equation (43) describes the transport-controlled dissolution rate of a solid according to the diffusion layer theory in its simplest form. The mass transfer coefficient here is given by $k_1 = k_T = D/h$.

The diffusion layer model satisfactorily accounts for the dissolution rates of most pharmaceutical solids. Equation (43) has even been used to predict the dissolution rates of drugs in powder form by assuming approximate values of D (e.g., 10^{-5} cm^2/sec), and h (e.g., 50 μm) and by deriving a mean value of A from the mean particle size of the powder [107,108]. However, as the particles dissolve, the wetted surface area, A, decreases in proportion to the 2/3 power of the volume of the powder. With this assumption, integration of Eq. (38) leads to the following relation, known as the Hixon–Crowell [109] cube root law:

$$M^{1/3} - M_0^{1/3} = k_{HC}t \tag{44}$$

where M is the mass of the powder undissolved at time t, M_0 is the original mass of powder at time $t = 0$, and k_{HC} is the Hixon–Crowell dissolution rate constant. For spherical particles of equal size,

$$k_{HC} = \frac{2Dc_sM_0^{1/3}}{hd} \tag{45}$$

where d is the diameter of a particle at time $t = 0$ and the other quantities already have been defined. The Hixon–Crowell equation has proved useful for quantifying the dissolution of pharmaceutical powders.

B. Intrinsic Dissolution Rates

The intrinsic dissolution rates of pharmaceutical solids may be calculated from the dissolution rate and wetted surface area using Eq. (36) or (37). For powdered solids, two common methods are available: the "powder" intrinsic dissolution rate method, and the "disc" intrinsic dissolution rate method. In the former method, the initial dissolution rate of one gram of powder is determined by a batch-type procedure as illustrated in Fig. 13A. The initial wetted surface area of one gram of powder is assumed to equal the specific surface area determined by an established "dry" procedure, such as monolayer gas adsorption by the Brunauer, Emmett, and Teller (BET) procedure [110].

In the "disc" method, the powder is compressed by a punch in a die to produce a compacted disc, or tablet. The disc, with one face exposed, is then rotated at a constant speed without wobble in the dissolution medium. For this purpose the disc may be placed in a holder, such as the Wood et al. [111] apparatus, or may be left in the die [112]. The dissolution rate, dm/dt, is determined as in a batch method, while the wetted surface area is simply the area of the disc exposed to the dissolution medium. The powder x-ray diffraction patterns of the solid after compaction and of the residual solid after dissolution should be compared with that of the original powder to test for possible phase changes during compaction or dissolution. Such phase changes would include polymorphism, solvate formation, or crystallization of an amorphous solid [113].

The dissolution rate of a solid from a rotating disc is governed by the controlled hydrodynamics of the system, and it has been treated theoretically by Levich [104]. This theory considers only forced convection due to rotation and ignores natural convection, which may occur at low speeds of rotation. Figure 16 shows the solvent flow held near the surface of the rotating disc. The apparent thickness, h, of the diffusion layer next to the surface of the disc is given by

$$h = 1.612(cm)\, D^{1/3}\nu^{1/6}\omega^{-1/2} \tag{46}$$

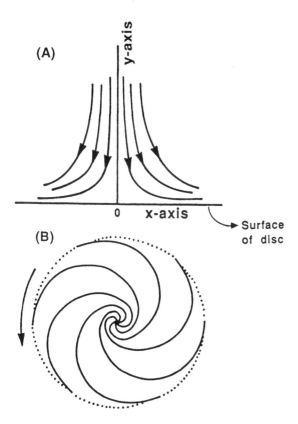

Fig. 16 Lines of flow of liquid in the Levich rotating disc method of determining dissolution rates. There is a transition from (A) flow essentially normal to the surface to (B) flow parallel with the surface, pointing to the existence of a viscous boundary layer. (Reproduced with permission of the copyright owner, the Royal Society of Chemistry, from Ref. 101.)

where D is the diffusivity of the dissolved solute, ω is the angular velocity of the disc in radians per second, and ν is the kinematic viscosity of the fluid, which is given by

$$\nu = \frac{\eta}{\rho} \tag{47}$$

where η and ρ are the dynamic viscosity and density of the fluid, respectively. If the rotation speed of the disc is expressed as W revolutions per second (Hz),

$$\omega = 2\pi W \tag{48}$$

then

$$h = 0.643 \text{ (cm) } D^{1/3} \nu^{1/6} W^{-1/2} \tag{49}$$

The actual thickness, δ, of the diffusion layer corresponds to the nonlinear decrease in c with increasing x in Fig. 15B and is proportional to the apparent thickness, h; thus,

$$h = 0.893\delta \tag{50}$$

Elimination of h from Eqs. (43) and (44) or (49) gives the following expressions for the intrinsic dissolution rate:

$$J = 0.620 D^{2/3} \nu^{-1/6} (c_s - c_b) \omega^{1/2} \tag{51}$$

$$J = 1.555 D^{2/3} \nu^{-1/6} (c_s - c_b) W^{1/2} \tag{52}$$

The dependence of J on $\omega^{1/2}$ predicted by Eq. (51) is illustrated by Fig. 17 for 2-naphthoic acid dissolving in an aqueous solution having an ionic strength of

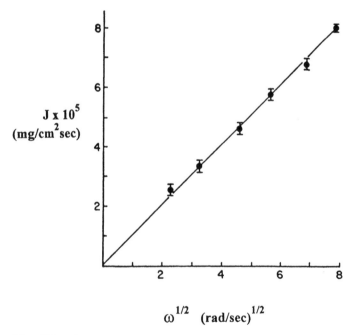

Fig. 17 Plot of the flux, J, of 2-naphthoic acid as a function of the square root of the rotation speed, ω, in 0.01 M HC, at an ionic strength $\mu = 0.5$ M (potassium chloride) at 25°C. The error bars represent the standard deviation for each point. (Reproduced with permission of the copyright owner, the American Pharmaceutical Association, from Ref. 114.)

0.5 M at 25°C [114]. Equation (51) or (52) enables the diffusivity of a solute to be measured. For example, from the slope of the line in Fig. 17 under sink conditions, D is calculated to be 6.1×10^{-6} cm^2/sec for 2-naphthoic acid. At low rotational speeds, the dissolved solute may not be uniformly distributed throughout the volume of the dissolution medium, and/or natural convection may become significant. The former effect may complicate the analytical procedure, while the latter effect will cause positive deviations of J values from Eqs. (51) and (52). At high rotational speeds, turbulence may disturb the flow pattern in Fig. 16, causing other deviations [101,104].

Equations (46), (49), (51), and (52) assume the presence of only one diffusing species. For several species having differing diffusivities, significant errors may result. However, as mentioned, many small organic molecules have similar diffusivities (of the order 10^{-5} cm^2/sec in water at 25°C), and J depends on the 2/3 power of D. Consequently, the errors arising from several diffusing species only become significant if one or more species exhibit abnormal diffusivities. Examples of high diffusivity are hydroxonium or hydroxyl ions, whereas macromolecules would represent an example of abnormally low diffusivities.

The diffusivity, D, of a particle is inversely proportional to its radius, r, according to the Stokes–Einstein equation,

$$D = \frac{kT}{6\pi\eta r} \tag{53}$$

where k is the Boltzmann constant, T is the absolute temperature, and η is the dynamic viscosity. For approximately spherical particles or molecules of volume v, r is proportional to $v^{1/3}$, which is also proportional to $\overline{V}^{1/3}$ ($\overline{V}$ is the partial molar volume), and which is also proportional to $M^{1/3}$ (M is the molecular weight). Finally, it is deduced that D is proportional to $\overline{V}^{-1/3}$, and in turn proportional to $M^{-1/3}$ [115]. For approximately planar particles or molecules of surface area a, r is proportional to $a^{1/2}$, which is proportional to $v^{1/2}$, in turn proportional to $\overline{V}^{1/2}$ and proportional to $M^{1/2}$, so that D is finally deduced to be proportional to $M^{-1/2}$ [116]. These proportionalities show that the diffusivity is only weakly dependent on the molecular weight, and they are useful for estimating the diffusivities of solute molecules from that of a standard molecule of known diffusivity under the same conditions in solution. In most cases, predictions agree quite well with experiment [116].

C. Factors that Influence Dissolution Rates

It has been shown that the dissolution rates of solids are determined or influenced by a number of factors, 10 of which have been outlined in the preceding sections. These may be summarized as follows:

1. Solubility of the solid, c_s, and the temperature.
2. Concentration in the bulk solution, c_b, if not under sink conditions (Eq. 31).
3. Volume, V, of the dissolution medium in a batch-type apparatus (Eq. 31), or volume flow rate, dV/dt, in a continuous-flow apparatus (Eq. 33).
4. Wetted surface area, A, which consequently is normalized in measurements of intrinsic dissolution rate, J (Eq. 35).
5. Conditions in the dissolution medium, which, together with the nature of the dissolving solid, determine the dissolution mechanism (see Theories and Mechanisms of Dissolution, pp. 355–358).

The conditions in the dissolution medium that may influence the dissolution rate are as follows:

6. The rate of agitation, stirring, or flow of solvent, if the dissolution is transport-controlled, but not when the dissolution is reaction-controlled. Increasing the agitation rate corresponds to an increased hydrodynamic flow rate and to an increased Reynolds number [104, 117] and results in a reduction in the thickness of the diffusion layer in Eqs. (43), (45), (46), (49), and (50) for transport control. Therefore, an increased agitation rate will increase the dissolution rate, if the dissolution is transport-controlled (Eqs. (41–46,49,51,52), but will have no effect if the dissolution is reaction-controlled. Turbulent flow (which occurs at Reynolds numbers exceeding 1000 to 2000 and which is a chaotic phenomenon) may cause irreproducible and/or unpredictable dissolution rates [104,117] and should therefore be avoided.
7. The diffusivity, D, of the dissolved solute, if dissolution is transport-controlled (Eqs. 41–46,49,51,52). The dissolution rate of a reaction-controlled system will be independent of D.
8. The viscosity (dynamic, η, or kinematic, ν) and density, ρ (Eq. 47), influence the dissolution rate if the dissolution is transport-controlled, but not if the dissolution is reaction-controlled. In transport-controlled dissolution, increasing η or ν will decrease D (Eq. 53), will increase h (Eqs. 46 and 49) and will reduce J (Eqs. 51 and 52). These effects are complex. For example, if an additional solute (such as a macromolecule) is added to the dissolution medium to increase η, it may also change ρ and D. The ratio of $\eta/\rho = \nu$ (Eq. 47) and D directly influence h and J in the rotating disc technique, while ν directly influences the Reynolds number (and hence J) for transport-controlled dissolution in general [104].
9. The pH and buffer concentration (if the dissolving solid is acidic or basic), and the pK_a values of the dissolving solid and of the buffer

[103,114,118–120]. The influence of these factors has also been summarized [1,120].

10. Complexation between the dissolving solute and an interactive ligand [116], or solubilization of the dissolving solute by a surface-active agent in solution [121]. Each of these phenomena tends to increase the dissolution rate.

VI. APPLICATION: POLYMORPHISM, SOLVATES, AND SOLUBILITY

A large number of compounds of pharmaceutical interest are capable of being crystallized in either more than one crystal lattice structure (polymorphs), with solvent molecules included in the crystal lattice (solvates), or in crystal lattices that combine the two characteristics (polymorphic solvates) [122,123]. A wide variety of structural explanations can account for the range of observed phenomena, as has been discussed in detail [124,125]. The pharmaceutical implications of polymorphism and solvate formation have been recognized for some time, with solubility, melting point, density, hardness, crystal shape, optical and electrical properties, vapor pressure, and virtually all the thermodynamic properties being known to vary with the differences in physical form [126].

That the crystal structure can have a direct effect on the solubility of a solid can be understood using a simple model. For a solid to dissolve, the disruptive force of the solvent molecules must overcome the attractive forces holding the solid intact. In other words, the solvation free energy released upon dissolution must exceed the lattice free energy of the solid for the process to proceed spontaneously. The equilibrium solubility of the solid in question (which represents the free-energy change of the system) will be determined by the relative balancing of the attractive and disruptive forces. The balance of these forces is determined by the enthalpy change, and the increase in disorder of the system (i.e., the entropy change). Since different crystal structures are characterized by different lattice energies (and enthalpies), it follows that the solubility of different crystal polymorphs (or solvate species) must differ as well.

It should be emphasized that the solubility differences between polymorphs or solvates will be maintained only when a less stable form cannot convert to the most stable form. When such conversion can take place, the equilibrium solubility of all forms will approach a common value, namely that of the most stable form at room temperature.

The effect of polymorphism becomes especially critical on solubility since the rate of compound dissolution must also be dictated by the balance of attractive and disruptive forces existing at the crystal–solvent interface. A solid having a higher lattice free energy (i.e., a less stable polymorph) will tend to dissolve faster, since the release of a higher amount of stored lattice free energy will increase the solubility and hence the driving force for dissolution. At the same time, each species would liberate (or consume) the same amount of solvation energy, since all dissolved species (of the same chemical identity) must be thermodynamically equivalent. The varying dissolution rates possible for different structures of the same drug entity can in turn lead to varying degrees of bioavailability for different polymorphs or solvates. To achieve bioequivalence for a given drug compound usually requires equivalent crystal structures in the drug substance, although exceptions are known to exist.

A. Equilibrium Solubility Studies

Phenylbutazone has been found capable of existing in five different polymorphic structures, each of which exhibits a characteristic x-ray powder diffraction pattern and melting point [127]. The equilibrium solubilities of all five polymorphs of phenylbutazone in three different solvent systems are summarized in Table 7. Form I exhibits the highest melting point (implying the highest value for lattice energy at the elevated temperature), and its solubility is the lowest in each of the three solvent systems studied. This finding demonstrates that this particular crystal form is thermodynamically the most stable polymorph both at room temperature and at the melting point (105°C). Identifying the sequence of stability for the other forms is not quite as simple. Following the accepted convention, the polymorphs had been named in the order of decreasing melting points, but the solubility data of Table 7 do not follow this order. This finding implies that the order of stability at room temperature is not equivalent to that at 105°C. In fact, only measurements of solubility can be used to deduce the stability order.

The effect of solvent composition on the solubility of polymorphs is illustrated by cimetidine [128]. The onset of melting of the two forms is essentially indistinguishable, making it impossible to apply the conventional nomenclature to the labeling of the polymorphs. Form B was found to be less soluble than form A, identifying it as the more stable polymorph at room

Table 7 Equilibrium Solubilities, in mg/ml, of Phenylbutazone Polymorphs at Ambient Temperature in Different Solvent Systems

Solvent system	Polymorphis in order of increasing free energy at ambient temperature				
	I	II	IV	V	III
pH 7.5 phosphate buffer	4.80	5.10	5.15	5.35	5.90
Above buffer with 0.05% Tween 80	4.50	4.85	4.95	5.10	5.52
Above buffer with 2.25% PEG 300	3.52	5.77	5.85	6.15	6.72

temperature. As evident in Fig. 18, the two forms were more soluble in mixed water–isopropanol solvents than in either of the pure solvents, which reflects the balance between solvation and lattice energies, and the entropy change in the various solvent systems. At constant temperature, the ΔG term and the solubility ratio are constant, independent of the solvent system:

$$\Delta G \ (\text{II} \rightarrow \text{I}) = -RT \ln \frac{S_{\text{II}}}{S_{\text{I}}} \tag{54}$$

where S_{II} and S_{I} are the solubilities of forms II and I, respectively.

Amiloride hydrochloride can be obtained in two polymorphic dihydrate forms [129]. No distinction can be made between the two solvates, since each dehydrates around 115–120°C, and the resulting anhydrous solids melt at the

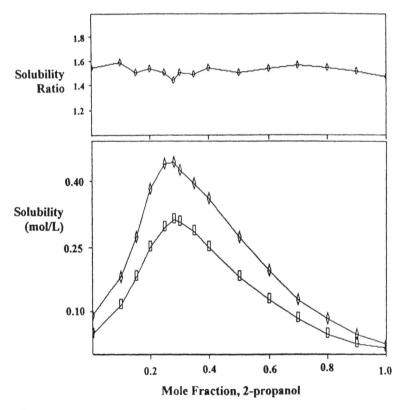

Fig. 18 Solubilities at 20°C of the two polymorphs of cimetidine, form A ($\Leftrightarrow$) and form B ($\boxminus$), as a function of the mole fraction of 2-propanol in the 2-propanol/water system. Also shown are the solubility ratios (form A/form B) calculated at each solvent composition. (The data are adapted from Ref. 128.)

same temperature. However, form B was found to be slightly less soluble than form A between temperatures of 5 and 45°C, suggesting it to be the thermodynamically stable form at room temperature. The temperature dependencies of the solubility data were processed by the van't Hoff equation to yield apparent enthalpies of solution for the two polymorphic dihydrates.

B. Intrinsic Dissolution Studies

A very powerful method for the evaluation of solubility differences between polymorphs or solvates is that of intrinsic dissolution, which entails measurements of the rates of solution. One method for this work is to simply pour loose powder into a dissolution vessel, and to monitor the concentration of dissolved solute as a function of time. However, data obtained by this method are not readily interpretable unless they are corrected by factors relating to the surface area or particle size distribution of the powder. In the other approach, the material to be studied is filled into the cavity of a circular dissolution die, compressed until it exhibits the effective planar surface area of the circular disc, and then the dissolution rate is monitored off the surface of the rotating disc in the die [130].

The types of intrinsic dissolution profiles obtainable through the loose powder and constant surface area methods are shown in Fig. 19. Oxyphenbutazone was obtained as the crystalline anhydrate and monohydrate forms, with the monohydrate being the less soluble [129]. The loose powder dissolution profiles consisted of sharp initial increases, which gradually leveled off as the equilibrium solubility was reached. In the absence of supporting information, the solubility difference between the two species cannot be adequately understood until equilibrium solubility conditions are reached. In addition, the shape of the data curves is not amenable to quantitative mathematical manipulation. The advantage of the constant surface area method is evident in that its dissolution profiles are linear with time, and more easily compared. Additional information about the relative surface areas or particle size distributions of the two materials is not required, since these differences were eliminated when the analyte disc was prepared.

Sulfathiazole has been found to crystallize in three distinct polymorphic forms, all of which are kinetically stable in the solid state but two of which are unstable in contact with water [130]. As evident in Fig. 20, the initial intrinsic dissolution rates are different, but as forms I and II convert into form III, the dissolved concentrations converge. Only the dissolution rate of form III was constant during the studies, indicating it to be the thermodynamically stable form at room temperature. Aqueous suspensions of forms I or II were all found to convert into form III over time, supporting the finding of the dissolution studies. Interestingly, around the melting points of the three polymorphs, form I exhibited

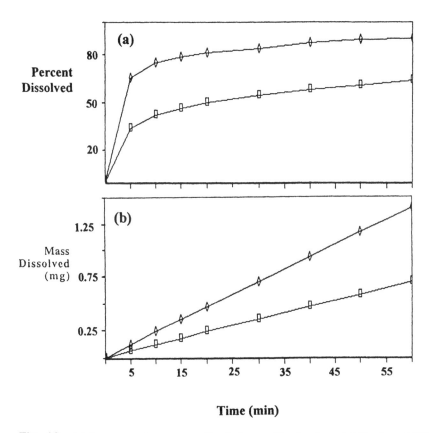

Fig. 19 (a) Loose powder aqueous dissolution profiles for the solubilization at 37°C of oxyphenylbutazone anhydrate (◇) and monohydrate (⊟). Also shown (b) are the aqueous intrinsic dissolution profiles obtained from compacted discs prepared using the same pseudopolymorphs. (The data are adapted from Ref. 131.)

the highest melting point and was thus assigned as being the most stable at the elevated temperature. This behavior would indicate an enantiotropic relationship between forms I and III.

When drug polymorphs cannot interconvert as a result of being suspended in aqueous solution, a different bioavailability of the two forms usually results [126]. For instance, the peak concentration of chloramphenicol in blood serum was found to be roughly proportional to the percentage of the B-polymorph of chloramphenicol palmitate present in a matrix of the A-polymorph [133]. The same concept has been found to apply to hydrate species, where the higher solubility and dissolution rate of the anhydrous phase relative to the trihydrate phase resulted in measurably higher blood levels when using the anhydrate as

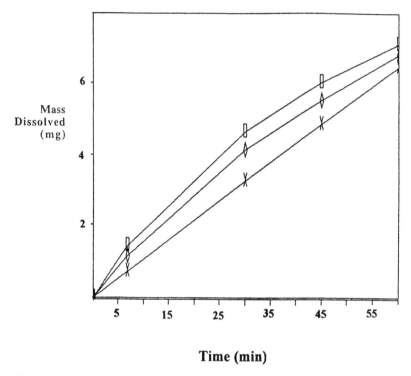

Fig. 20 Intrinsic aqueous dissolution profiles obtained at 37°C for sulfathiazole, form I (◇), form II (⊟), form III (✕). (The data are adapted from Ref. 132.)

the drug, ampicillin [134]. However, when drug polymorphs can undergo phase transformations, no real difference in bioavailability parameters can be detected when administering the different polymorphic modifications. This situation has been realized wth carbamazepine, where the rapid transformation of the anhydrate phase into the dihydrate ensures bioequivalence of the two forms [135].

C. Solution Calorimetry Studies

One of the most powerful techniques available for the evaluation of the solubility attributes of polymorphic or solvate species is solution calorimetry. Each distinct crystal phase is characterized by a well-defined heat of solution as it dissolves into a given solvent system, and the difference between the heats of solution of each phase in the same defined solvent system equals the heat of transition between them at the temperature of measurement. Solution calorimetry has been used to complement or to investigate a wide variety of crystal properties, dissolution phenomena, complexation reactions, and compound stabilities [136, 137].

The polymorphism of auranofin has been investigated by solution calorimetry [138], and the key results of this study are collected in Table 8. As would be expected for chemically identical forms differing only in their crystal lattice energies, and by analogy with the free energy differences, the heats of transition computed from the solution calorimetry data are independent of the solvents used. In this same work, thermal analysis was also used to deduce the heat of transition around the melting point, and a value of 3.20 kcal/mole was obtained. This value is in good agreement with that obtained from solution calorimetry, indicating that heat capacity corrections to the transition enthalpy are relatively small. It was also recognized in this study that forms A and B must be enantiotropic in their relationship, since the lower melting form (less stable around the melting point) was thermodynamically more stable at room temperature.

In some instances, distinct polymorphic forms can be isolated that do not interconvert when suspended in a solvent system, but that also do not exhibit differences in intrinsic dissolution rates. One such example is enalapril maleate, which exists in two bioequivalent polymorphic forms of equal dissolution rate [139], and therefore of equal free energy. When solution calorimetry was used to study the system, it was found that the enthalpy difference between the two forms was very small. The difference in heats of solution of the two polymorphic forms obtained in methanol was found to be 0.51 kcal/mol, while the analogous difference obtained in acetone was 0.69 kcal/mol. These results obtained in two different solvent systems are probably equal to within experimental error. It may be concluded that the small difference in lattice enthalpies (ΔH) between the two forms is compensated by an almost equal and opposite small difference in the entropy term ($-T \Delta S$), so that the difference in free energy (ΔG) is not sufficient to lead to observable differences in either dissolution rate or equilibrium solubility. The bioequivalence of the two polymorphs of enalapril maleate is therefore easily explained thermodynamically.

Table 8 Heats of Solution Measured for Auranofin Polymorphs in Different Solvent Systems at Ambient Temperature

	95% Ethanol	Dimethylformamide
ΔH_{sol} (form A)	12.42 kcal/mol	5.57 kcal/mol
ΔH_{sol} (form B)	9.52 kcal/mol	2.72 kcal/mol
$\Delta(\Delta H_{sol})$ (form A → form B)	2.90 kcal/mol	2.85 kcal/mol
= ΔH_{trans} (form A → form B)		

When one polymorph can be thermally converted to another, differential scanning calorimetry (DSC) analysis cannot be used to deduce the heat of transition between the two forms, and so solution calorimetry represents an alternative methodology. This situation was encountered when evaluating the polymorphs of losartan [140]. Enthalpies of transition were obtained in water ($\Delta(\Delta H_{sol})$ = 1.723 kcal/mol) and in N,N-dimethylformamide ($\Delta(\Delta H_{sol})$ = 1.757 kcal/mol), with the equivalence in results demonstrating the quality of the results. Although enthalpy does not indicate stability, the authors deduced from solution calorimetry that form I was more stable than form II at ambient temperature.

VII. APPLICATION: CHIRAL SOLUTES, ENANTIOMERS, AND RACEMATES

Dissymmetric molecules are distinctly characterized by the nonsuperimposability of their mirror images. Many drugs are inherently chiral, such as ephedrine, pseudoephedrine, various antibiotics (e.g., ampicillin, griseofulvin), various alkaloids (e.g., morphine), and various nonsteroidal antiinflammatory drugs based on 2-arylpropanoic acid (e.g., ibuprofen and naproxen). Most synthetic chiral drugs are commercially available as a mixture of the two enantiomers, whereas most naturally occurring or semisynthetic drugs are commercially available only as one enantiomer. The majority of excipients as obtained from natural sources are chiral, with sugars, cyclodextrins, and all cellulose derivatives serving as examples. The number of chiral drugs is progressively growing as drugs of increasing molecular complexity receive approval. Regulatory agencies are now treating the enantiomers of a compound as separate chemical entities during the approval process, and they are emphasizing the importance of enantiomeric purity.

The enantiomeric purity, p, of a partially resolved mixture of two opposite enantiomers is defined in terms of the mole fraction, x, of the predominant enantiomer. The mole fraction of the contaminating enantiomer is therefore $1 - x$, and so p is defined by

$$p = \frac{x - (1 - x)}{x + (1 - x)} = 2x - 1 \tag{55}$$

As x varies from 0.5 to 1, p varies from 0 to 1.

The solubility behavior of enantiomers and racemates has been reviewed by Jacques et al. [141]. Other solid state properties of chiral drugs have been reviewed by Brittain [142].

A. Ternary Phase Solubility Diagrams

A system consisting of the solid enantiomers, D and L, in the presence of solvent, S, is a ternary system [141]. The variations of the solubilities of both D and L and of any other solid phase(s) formed by them are represented by a triangular prism whose vertical axes denote the equilibrium temperature of the system (Fig. 21). The three faces of the prism constitute the binary phase diagrams D+L, D+S, and L+S. A horizontal slice through the prism perpendicular to the vertical axes represents the phases of the system D+L+S at constant temperature T_0. This isothermal slice is normally presented as an equilateral triangle, although other planar representations have sometimes been used. This section considers the influence of composition on the solubility behavior by focusing on the isothermal slices, and the influence of temperature will not be considered in detail. Increasing temperature usually increases the solubility of each solid phase, as discussed in Section II. The effects of temperature on the D+L+S system may be quite complex [141], and they may even correspond to a change from one type of phase diagram to another in accordance with the nature of the isothermal slices, as may be seen in Fig. 21.

To construct a triangular diagram consisting of two enantiomers and the solvent at constant temperature requires the determination of the concentration of the saturated solutions as a function of the total composition of the system, and the number and nature of the solid phases in equilibrium with the saturated solution. The determination of the solubility of mixtures of enantiomers is

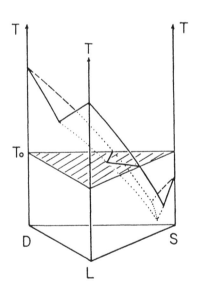

Fig. 21 Three-dimensional representation of a ternary system of two enantiomers in a solvent, S. One of the faces of the prism (at left) corresponds to the binary diagram of D and L (here a conglomerate). Shaded area: isothermal section representing the solubility diagram at temperature T_0. (Reproduced with permission of the copyright owner, John Wiley and Sons, Inc., New York, from Ref. 141, p. 169.)

experimentally similar to the common batch agitation method described in Section V. Similar precautions should be taken, including the use of well-stoppered flasks and verification of equilibrium, and the temperature control over the system should be to at least ±0.1°C.

Since the D and L enantiomers are mirror images of each other, their crystal structures and physical properties must be identical, apart from those properties associated with their chirality [142]. The latter properties include the behavior of the D and L molecules in a chiral environment, typified by polarized light or a chiral stationary phase in chromatography. The similarities between the enantiomers indicate that several mixtures covering the full enantiomeric composition range (that is, from 100 to 0% enantiomeric purity) may be sufficient to construct the phase diagram. This phase diagram must necessarily be symmetrical about the racemic mixture. The differences associated with dissymmetry indicate that chiral technology (i.e., optical rotation or chiral chromatography) must be employed to differentiate between the enantiomers. While chiral HPLC can be conveniently used to determine the concentration of each enantiomer present in a solution, optical rotation measurements must be supplemented by an achiral analytical method that determines the sum of the enantiomer concentrations. These measurements will provide the relative amounts of D, L, and S in each saturated solution, which is represented by point M on the solubility curve and which occurs at the intersection of lines ss' and Sm in Fig. 22.

The saturated solution, M, is in equilibrium with a solid of composition represented by n in Fig. 22. These two phases in equilibrium are joined by the "tie line" Mn. While it is possible to analyze the dry solid, n, itself, it is often more

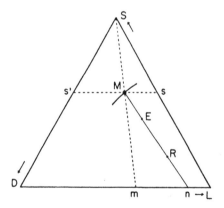

Fig. 22   Construction of a triangular solubility diagram. (Reproduced with permission of the copyright owner, John Wiley and Sons, Inc., New York, from Ref. 141, p. 176.)

convenient to determine points E or R on the tie line and to then extrapolate ME or MR to n. Point R represents the composition of the "wet residue," which is determined by analyzing the solid sample with some saturated solution clinging to it [143]. In an alternative method, the composition of the entire system (which may be known beforehand or determined analytically) is plotted to give point E [144]. The method of "wet residues" provides a shorter extrapolation to n and is therefore more precise. A bundle of tie lines converging at a certain point indicates a definite phase, as shown in Fig. 23. The phase rule of Gibbs [145,146] is frequently used to calculate the number of degrees of freedom or variance of the system.

B. Racemic Conglomerates and Eutectics

Figure 24 shows the ternary phase diagram (solubility isotherm) of an unsolvated conglomerate that consists of physical mixtures of the two enantiomers that are capable of forming a racemic eutectic mixture. It corresponds to an isothermal (horizontal) cross section of the three-dimensional diagram shown in Fig. 21. Examples include *N*-acetyl-leucine in acetone, adrenaline in water, and methadone in water (each at 25°C) [141].

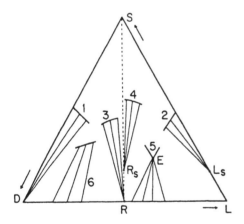

Fig. 23 Tie lines associated with different systems: (1) a solid phase D (pure enantiomer) in the presence of mother liquor of variable composition, (2) a solid phase L_S (solvated enantiomer) in mother liquor of variable composition, (3) a solid phase R (pure racemic compound) in mother liquor of variable composition, (4) a solid phase R_S (solvated racemic compound) in mother liquor of variable composition, (5) two solid phases, one enantiomer and the racemic compound (or two enantiomers if E is on SR, i.e., for a conglomerate) in mother liquor of fixed composition E (eutectic), and (6) the tie lines do not converge; one solid phase is present (solid solution of D and L) in mother liquor of variable composition. (Reproduced with permission of the copyright owner, John Wiley and Sons, Inc., New York, from Ref. 141, p. 177.)

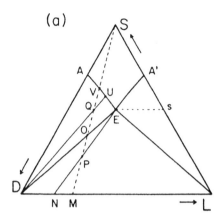

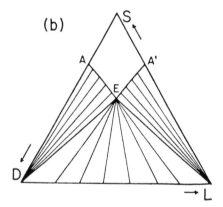

Fig. 24   Isothermal solubility diagrams for a racemic conglomerate, i.e., a eutectic system of the two opposite enantiomers. The appearance of the tie lines is shown in (b). Symbols are defined in the text. (Reproduced with permission of the copyright owner, John Wiley and Sons, Inc., New York, from Ref. 141, p. 178.)

In Fig. 24 points A and A' represent the equal solubilities at the temperature T_0 of the pure enantiomers, while E represents the solubility of the eutectic mixture. AEA' is the solubility–composition curve, above which undersaturated solutions exist and below which a saturated solution is in equilibrium with the two solid phases, D and L. Figure 21 shows that, with an increase in temperature, A and A' move toward the pure enantiomers D and L, while E moves toward the pure eutectic. All of these trends indicate an increase in solubility upon increasing the temperature.

On gradual addition of solvent, S, to a solid mixture (conglomerate) of the two enantiomers of composition M, several changes occur. The solvent becomes saturated with equal amounts of the two enantiomers, and the composition of the residual solid becomes richer in D. For example, when the system reaches point P, the composition of the saturated solution is given by E at the upper end of the tie line, while the composition of the residual solid is given by N at the lower end of the tie line. On addition of additional solvent, more L dissolves, and the enantiomeric purity of D increases until point O is reached. At this point, the residual solid phase is pure D in equilibrium with E. Addition of more solvent causes the composition of the saturated solution to become richer in D, and to move along the curve EA as D dissolves. When the composition of the system corresponds to point Q, the composition of the saturated solution is represented by point U, which is still in equilibrium with pure solid D. When point V is reached on the solubility curve, all the solid D has dissolved. Addition of further solvent produces unsaturated solutions along VS. An analogous series of changes occurs for solids richer in L and D. The variance of the system can be calculated using the Gibbs phase rule [145,146]. If the enantiomers are solvated, the tie lines focus at the appropriate stoichiometry between D and S and between L and S.

If the molecular species of the solute present in solution is the same as those present in the crystals (as would be the case for nonelectrolytes), then to a first approximation, the solubility of each enantiomer in a conglomerate is unaffected by the presence of the other enantiomer. If the solutions are not dilute, however, the presence of one enantiomer will influence the activity coefficient of the other and thereby affect its solubility to some extent. Thus, the solubility of a racemic conglomerate is equal to twice that of the individual enantiomer. This relation is known as Meyerhoffer's double solubility rule [147]. If the solubilities are expressed as mole fractions, then the solubility curves are straight lines, parallel to sides SD and SL of the triangle in Fig. 24.

If a solute of the general formula AX_n (A is the chiral ion and X is an achiral ion) dissociates completely into ions once dissolved, then the solubility of the racemic conglomerate, S_R, is equal to $\sqrt[n+1]{2} \cdot S_A$ (where S_A is concentration of A in a solution saturated with AX_n). If the solute is of the type AX, then $S_R = \sqrt{2} \cdot S_A$. The subscript n refers to the achiral ion and may be fractional, and so A_2X must be represented by $AX_{1/2}$. If dissociation of AX_n is incomplete, S_A lies between $\sqrt[n+1]{2} \cdot S_A$ and $2S_A$. For weakly dissociated electrolytes (such as carboxylic acids), S_R is approximately $2S_A$.

C. Racemic Compounds

Figure 25 shows the ternary phase diagram (solubility isotherm) for an unsolvated racemic compound. Examples of this type include benzylidenecamphor in methanol, or *N*-acetylvaline in acetone [141]. In Fig. 25, A and A' represent the

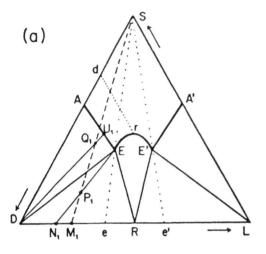

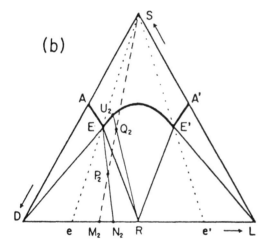

Fig. 25 Solubility isotherm for an unsolvated racemic compound. (Reproduced with permission of the copyright owner, John Wiley and Sons, Inc., New York, from Ref. 141, p. 193.)

saturated solutions of the pure enantiomers, for which the solubilities are equal to SA. R represents the solid racemic compound, DL, and r represents its saturated solution, for which the solubility is equal to 2Sd. There are two eutectics, E between enantiomer D and R, and E' between enantiomer L and R. The compositions of these eutectics are exactly the reverse of one another.

Point M_1 in Fig. 25a represents a physical mixture of solid D and solid R. Gradual addition of solvent S to this system will produce changes analogous to those described for M in Fig. 24; the eutectic material gradually dissolves, leaving a solid that becomes increasingly rich in D until it also dissolves upon crossing the solubility curve AE. Thus, when the enantiomeric purity of the original solid is greater than that of the eutectic, crystallization will increase the enantiomeric purity.

On the other hand, if the enantiomeric purity of the original solid is less than that of the eutectic (as in the case of M_2 in Fig. 25b), crystallization results in a decrease in enantiomeric purity. For example, when sufficient solvent has been added to correspond to point P_2, the tie line shows that the solid N_2 contains less of the predominant enantiomer D than M_2 and is in equilibrium with E, which corresponds to a saturated solution of the eutectic solid, e. When the system reaches the composition represented by point Q_2, the solid that crystallizes out is the racemic compound, R, which is in equilibrium with the saturated solution, U_2, containing the racemic compound and enantiomer D.

If solvent is added to either of the solid eutectics represented by e or e' in Fig. 25a or b, the undissolved solid retains this composition while the saturated solution maintains the composition E or E', respectively. Again, Gibbs phase rule [145,146] can provide further insight into these systems. If the solid enantiomers are solvated, the compositions of the equilibrium solids are displaced symmetrically along the DS or LS axes to an extent determined by the stoichiometry of the solvates. Similarly, if the racemic compound is solvated, the stoichiometry of the equilibrium solid is displaced from R along the line RS to an extent determined by the stoichiometry of the solvate.

D. Pseudoracemates and Solid Solutions

Certain pairs of enantiomers, such as 25°C solutions of camphor in aqueous ethanol (Fig. 26a), 2,2,5,5-tetramethyl-l-pyrrolidinoxy-3-carboxylic acid in chloroform (Fig. 26b), or carvoxime in hexane (Fig. 26c), cocrystallize throughout the entire range of mole fractions. The crystals that are in equilibrium with the saturated liquid solution are solid solutions (mixed crystals), since they constitute a single phase. The racemic solid solution is termed a "pseudoracemate" [141].

Figure 26 shows the ternary phase diagrams (solubility isotherms) for three types of solid solution. The solubilities of the pure enantiomers are equal to SA, and the solid–liquid equilibria are represented by the curves ArA'. The point r represents the equilibrium for the pseudoracemate, R, whose solubility is equal to 2Sd. In Fig. 26a the pseudoracemate has the same solubility as the enantiomers, that is, 2Sd = SA, and the solubility curve AA' is a straight line parallel to the base of the triangle. In Figs. 26b and c, the solid solutions including the pseudoracemate are, respectively, more and less soluble than the enantiomers.

(a)

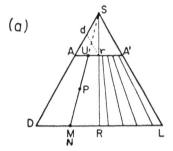

(b)

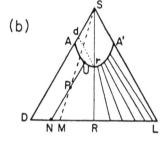

(c)

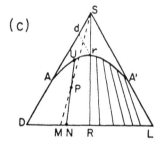

Fig. 26   Isothermal solubility diagrams of the three types of pseudoracemates (see text). The appearance of the tie lines is shown in the right half of the triangles. (Reproduced with permission of the copyright owner, John Wiley and Sons, Inc., New York, from Ref. 141, p. 197.)

In the region below the solid–liquid equilibrium curves in Fig. 21, the tie lines do not focus to a point within the region. In Figs. 26a–c the slopes of the tie lines, such as UN, are respectively equal to, less than, or greater than those of the lines passing through S, such as SPM. Examination of the tie lines shows that crystallization does not change the enantiomeric purity of the solid solutions whose solubility curve is represented by a horizontal line (Fig. 26a) but leads to an increase in enantiomeric purity for solid solutions having a minimum for the pseudoracemate (Fig. 26b) and to a decrease in enantiomeric purity for solid solutions having a maximum for the pseudoracemate (Fig. 26c).

Gibbs phase rule [145,146] shows that the compositions of the liquid solution and of the solid solution are mutually interdependent. Thus, the establishment of a real solubility equilibrium would require that any changes in the composition of the liquid phase be matched by corresponding changes in the composition of the entire solid phase. Such changes in the solid phase are subject to high activation energy barriers and are therefore slow. Consequently, experiments for obtaining ternary phase diagrams of pseudoracemates should employ as large a ratio of liquid solution to solid solution (crystals) as possible, so that the change in concentration of the solution during equilibration is negligible and the crystals are likely to be homogeneous. Solubility diagrams for pseudoracemates are rare, at present, and are often complicated by polymorphism and solvate formation [141].

This discussion has considered the influence of a continuous range of solid solutions on solubility. In addition, the formation of terminal solid solutions of L in crystals of D and vice versa appears to be an inevitable result of imperfect resolution followed by crystallization. These terminal solid solutions may have very limited compositions, corresponding to a solid solubility not exceeding a few mole percent in a eutectic system. The consequent increase in the lattice free energy of the host crystals then corresponds to an increase in the metastable solubility, as discussed in Section I, and hence to an increase in the intrinsic dissolution rate, as discussed in Section V. This behavior has been demonstrated for (SS)-$(+)$-pseudoephedrinium salicylate doped with trace quantities of its opposite enantiomer, which, in equimolar proportions, would give rise to a eutectic (i.e., a racemic conglomerate) [148,149]. Similar behavior has been described for $(RS)(-)$-ephedrinium 2-naphthalenesulfonate doped with trace quantities of its opposite enantiomer, which, in equimolar quantities, would give rise to a racemic compound [150]. In addition, similar behavior has been observed when this racemic compound is doped with trace quantities of either enantiomer, corresponding to an enantiomeric excess in solid solution in $(\pm)$-ephedrinium 2-naphthalenesulfonate [151]. In all three examples, solid solubility is limited to a few mole percent, but the increase in intrinsic dissolution rate may approach 12, 14.5, and 27%, respectively, emphasizing the physical pharmaceutical consequences of chiral impurity content.

E. Diastereomers

Two compounds are diastereomers when they contain more than one chiral center. If the number of dissymmetric centers is given by N, then the number of possible diastereomers is given by 2^N. Of these 2^N diastereomers, each will be characterized by its mirror image, so that the number of enantiomers is given by $2^N/2$. Whereas the physical properties of enantiomers in an achiral environment are necessarily identical, the physical properties (including solubility) of diastereomers are normally different. The differences arise since there is no structural requirement that the crystal lattices of different diastereomers be the same. For instance, the solubility of an (SS)-diastereomer could differ substantially from that of the (RS)-diastereomer. However, it should be remembered that the solubility of the (SS)-diastereomer must be exactly identical to that of the (RR)-diastereomer, since these compounds are enantiomers of each other. At the same time, the solubilities of the (SR)-diastereomer and the (RS)-diastereomer must also be identical.

A racemic conglomerate, being a eutectic, may be resolved into its enantiomers by crystallization [141,142], but a true racemate cannot be so resolved. A common means to resolve a racemate, DL, is to react it to form an association complex with a resolving agent, A. The resolving agent is itself chiral, and it is reacted in its enantiomerically pure form. As a result of the complexation, a pair of diastereomers, DA and LA, are produced, and these can usually be separated by exploiting differences in their physical properties. This could involve chromatography with an achiral stationary phase, but preferential crystallization is most commonly exploited. In this process, the less soluble diastereomer is precipitated, while the more soluble diastereomer concentrates in the solution. For example, a racemic acid may be resolved by a chiral, resolved base such as an amine [142], thus:

Racemic compound (or mixture)	Chiral resolving agent	Diastereomeric salts	Diastereomer designation
(+)-R-COOH		$(+)\text{-R-COO}^{-+}\text{NH}_3\text{-R}'(+)$	p salt
	$+\ 2\ \text{NH}_2\text{-R}'(+) \rightarrow$		
(−)-R-COOH		$(-)\text{-R-COO}^{-+}\text{NH}_3\text{-R}'(+)$	n salt

The (RR) and (SS) diastereomers (or the $(++)$ and $(--)$ diastereomers) are termed the p salts, while the (RS) and (SR) diastereomers (the $(+-)$ and $(-+)$ diastereomers) are termed the n salts [152]. Because the diastereomeric salts exhibit different physical properties [152], they may be separated by physical means, such as fractional crystallization. The greater the solubility difference, the more efficient will be the resolution [152]. After the diastereomers are separated, the resolving agent is dissociated from the salt and recovered for further use.

The use of dissociable diastereomers for enantiomer resolution may be illustrated by the case where racemic mandelic acid is resolved using enantiomerically pure α-methylbenzylamine. The n and p salts of α-methylbenzylamine mandelate have aqueous solubilities of 49.1 and 180 g/L, respectively, at 25°C [153]. A more recent example, which focuses on the crystallographic origin of the solubility differences, is provided by the resolution of ($\pm$)-mandelic acid with (−)-ephedrine in water or methanol solution [154]. In general, the relative solubilities of the n and p salt pairs are strongly influenced by the choice of solvent medium and temperature, which provide considerable flexiblity in optimizing the crystallization conditions and the efficiency of resolution. This process may be facilitated by the development of a full solubility phase diagram.

REFERENCES

1. D. J. W. Grant, and T. Higuchi, *Solubility Behavior of Organic Compounds*, Vol. 21, Techniques of Chemistry (W. H. Saunders Jr., series ed.), John Wiley and Sons, New York, 1990.
2. A. H.-L. Chow, P. K. K. Chow, W. Zhongshan, and D. J. W. Grant, *Int. J. Pharm.*, *24*, 239 (1985).
3. M. J. Pikal and D. J. W. Grant, *Int. J. Pharm.*, *39*, 243 (1987).
4. E. M. Phillips and V. Stella, *Int. J. Pharm.*, *94*, 1 (1993).
5. R. Hüttenrauch, *Acta Pharm. Technol.*, *6*, Suppl., 55 (1978).
6. P. York, *Int. J. Pharm. 14*, 1 (1983).
7. P. York, and D. J. W. Grant, "Crystal Modifications and Powder Compaction," Proceedings of the Pharmaceutical Technology Conference '86, Cherry Hill, New Jersey, 1986, pp. 163–173.
8. G. N. Lewis, and M. Randall, *Thermodynamics*, 2nd ed., revised by K. S. Pitzer and L. Brewer, McGraw-Hill, New York, 1961.
9. W. L. Chiou and L. E. Kyle, *J. Pharm. Sci.*, *68*, 1224 (1979).
10. R. Hüttenrauch, S. Fricke, and P. Zielke, *Pharm. Res.*, *2*, 302 (1985).
11. T. Higuchi, F.-M. L. Shih, T. Kimura, and J. H. Rytting, *J. Pharm. Sci.*, *68*, 1267 (1979).
12. S. Venkataram, Ph.D. thesis, University of Alberta, Edmonton, Alberta, Canada, 1986, p. 81.
13. A. H. Goldberg, M. Gibaldi, and J. L. Kanig, *J. Pharm. Sci.*, *55*, 487 (1966).
14. D. T. Cook, Ph.D. thesis, University of Nottingham, UK, 1978, p. 49.
15. R. Kaur, Ph.D. thesis, University of Nottingham, UK, 1980, p. 21.
16. B. Katchen and S. Symchowicz, *J. Pharm. Sci.*, *56*, 1108 (1967).
17. N. Aoyagi, H. Ogata, N. Kaniwa, M, Koibuchi, T. Shibazaki, and A. Ejima, *J. Pharm. Sci.*, *71*, 1165 (1982).
18. A. A. Elamin, C. Ahlneck, G. Alderborn, and C. Nyström, *Int. J. Pharm.*, *111*, 159 (1994).
19. R. A. Swalin, *Thermodynamics of Solids*, 2nd ed., John Wiley and Sons, New York, 1972, p. 183.

20. D. J. Shaw, *Introduction to Colloid and Surface Chemistry*, 2nd ed., Butterworth, London, 1970, pp. 57, 107.
21. W. J. Mader, R. D. Vold, and M. J. Vold, in *Physical Methods of Organic Chemistry*, 3rd ed., Vol. 1, Part I (A. Weissberger, ed.), Interscience Publishers, New York, 1959, pp. 655–688.
22. S. H. Yalkowsky and S. Banerjee, *Aqueous Solubility Methods of Estimation for Organic Compounds*, Marcel Dekker, New York, 1992, pp. 149–154.
23. K. S. Alexander, B. Laprade, J. W. Mauger, and A. N. Paruta, *J. Pharm. Sci.*, *67*, 624 (1978).
24. D. J. W. Grant, M. Mehdizadeh, A. H.-L. Chow, and J. E. Fairbrother, *Int. J. Pharm.*, *18*, 25 (1984).
25. E. B. Bagley and J. M. Scigliano, in *Solutions and Solubilities*, Part II, Techniques of Chemistry, Vol. VIII (M. R. J. Dack, ed.), John Wiley, New York, 1976, pp. 437–485.
26. S. Ghosh and D. J. W. Grant, *Int. J. Pharm.*, *114*, 185 (1995).
27. J. Reilly and W. N. Rae in *Physico-Chemical Methods*, Vol. 1, 3rd ed., Van Nostrand, Princeton New Jersey, 1939, p. 589.
28. M. J. Mader and L. T. Grady, in *Physical Methods of Chemistry, Part V, Determination of Thermodynamic and Surface Properties*, (A. Weissberger and B. W. Rossiter, eds.), John Wiley and Sons, New York, 1971, p. 257.
29. S. A. Butter, *J. Chem. Educ.*, *51* (1), 70 (1974).
30. D. K. Wyatt and L. T. Grady, in *Physical Methods of Chemistry, 2nd Ed., Vol. VI, Determination of Thermodynamic Properties* (B. W. Rossiter and R. C. Baetzold, eds.), John Wiley and Sons, New York, 1992, p. 283.
31. H. De Voe and S. P. Wasik, *J. Soln. Chem.*, *13*, 51 (1984).
32. W. E. May, S. P. Wasik, and D. H. Freeman, *Anal. Chem.*, *50*, 175, 997 (1978).
33. K. J. Friesen, L. P. Sarna, and G. R. B. Webster, *Chemosphere*, *14*, 1267 (1985).
34. R. M. Dickhut, A. W. Andren, and D. E. Armstrong, *Environ. Sci. Technol.*, *20*, 807–810 (1986).
35. R. S. Pearlman, S. H. Yalkowsky, and S. Banerjee, *J. Phys. Chem. Ref. Data*, *13*, 555–562 (1984).
36. J. W. Billington, G.-L. Huang, F. Szeto, W. Y. Shiu, and D. Mackay, *Environ. Toxicol. Chem.*, *7*, 117–124 (1988).
37. S. P. Duddu and D. J. W. Grant, *Thermochim. Acta*, *248*, 131–145 (1995).
38. T. Higuchi and K. A. Connors, *Adv. Anal. Chem. Instrum.*, *4*, 117 (1965).
39. A. J. Mader, *Crit. Rev. Anal. Chem.*, *1*, 193 (1970).
40. R. E. Schirmer, in *Modern Methods of Pharmaceutical Analysis*, Volume III, CRC Press, Boca Raton, Florida, 1982, pp. 157–169.
41. R. M. Herriot, *Chem. Rev.*, *30*, 413 (1942).
42. T. J. Webb, *Anal. Chem.*, *20*, 100 (1948).
43. S. H. Yalkowsky, *Techniques of Solubilization of Drugs*, Marcel Dekker, New York, 1981.
44. J. A. Dean, *Lange's Handbook of Chemistry*, 11th ed., McGraw-Hill, New York, 1973, pp. 10–12.
45. W. H. Hall, *Textbook of Quantitative Analysis*, 2nd ed., John Wiley and Sons, New York, 1935, pp. 148–149.

46. S. M. Berge, L. D. Bighley, and D. C. Monkhouse, *J. Pharm. Sci.*, *66* 1 (1977).
47. P. L. Gould, *Int. J. Pharm.*, *33*, 201 (1986).
48. K. R. Morris, M. G. Fakas, A. B. Thakur, A. W. Newman, A. K. Singh, J. J. Venit, C. S. Spagnuolo, and A. T. M. Serajuddin, *Int. J. Pharm.*, *105*, 209 (1994).
49. M. B. Maurin, S. M. Rowe, C. A. Koval, and M. A. Hussain, *J. Pharm. Sci.*, *83*, 1418 (1994).
50. T. Higuchi and K. A. Connors, *Adv. Anal. Chem. Instrum.*, *4*, 117 (1965).
51. T. Higuchi and J. Lach, *J. Am. Pharm. Assoc. Sci. Ed.*, *43*, 525 (1954).
52. M. A. Hussain, R. C. DiLuccio, and M. B. Maurin, *J. Pharm. Sci.*, *82*, 77 (1993).
53. K. M. Boje, M. Sak, and H.-L. Fung, *Pharm. Res.*, *5*, 655 (1988).
54. M. L. Bender and M. Komiyama, *Cyclodextrin Chemistry*, Springer-Verlag, New York, 1978.
55. J. Szejtli, *Cyclodextrins and Their Inclusion Complexes*, Akademiai Kiado, Budapest, 1982.
56. S. P. Jones, D. J. W. Grant, J. Hadgraft, and G. D. Parr, *Acta Pharm. Technol.*, *30*, 213, 263 (1984).
57. J. Szejtli, *Cyclodextrin Technology*, Kluwer Academic Publisher, Dordrecht, 1988.
58. F. Cramer, *Rev. Pure Appl. Chem.*, *5*, 143 (1975).
59. R. L. VanEtten, J. F. Sabastian, G. A. Clowes, and M. L. Bender, *J. Am. Chem. Soc.*, *89*, 3253 (1967).
60. A. L. Thakkar, P. B. Kuehn, J. H. Perrin, and W. L. Wilham, *J. Pharm. Sci.*, *61*, 1841 (1972).
61. D. Amdidouche, H. Darrouzet, D. Duchêne, and M.-C. Poelman, *Int. J. Pharm.*, *54*, 175 (1989).
62. C. D. Yu, S. A. Sweetana, N. I. Chu, G. J. L. Lee, and I. J. Massey, *Drug Dev. Indust. Pharm.*, *15*, 609 (1989).
63. M. E. Brewster, J. W. Simpkins, M. S. Hora, W. C. Stern, and N. Bodor, *J. Paren. Sci. Tech.*, *43*, 231 (1989).
64. D. Duchêne and D. Wouessidjewe, *Drug Dev. Indust. Pharm.*, *16*, 2487 (1990).
65. O. Bekers, E. V. Uijtendaal, J. H. Beijnen A. Bult, and W. J. M. Underberg, *Drug Dev. Indust. Pharm.*, *17*, 1503 (1991).
66. M. J. Rosen, *Surfactants and Interfacial Phenomena*, John Wiley and Sons, New York, 1978.
67. K. L. Mittel, *Micellization, Solubilization, and Microemulsions*, Plenum Press, New York, 1977.
68. K. L. Mittel and J. H. Fendler, *Solution Behavior of Surfactants*, Plenum Press, New York, 1982.
69. K. L. Mittel and B. Lindman, *Surfactants in Solution*, Plenum Press, New York, 1984.
70. A. T. Florence, "Drug Solubilization in Surfactant Systems," in *Techniques of Solubilization of Drugs* (S. H. Yalkowsky, ed.), Marcel Dekker, New York, 1981, pp. 15–89.
71. N. Weiner, F. Martin, and M. Riaz, *Drug Dev. Indust. Pharm.*, *15*, 1523 (1989).
72. K. Thoma and C.-D. Herzfeldt, *Pharm. Acta Helv.*, *63*, 34, 40, 66, 77, 98 (1989).
73. K. Thoma and F. R. Kasper, *Pharm. Acta Helv.*, *63*, 102, 155, 160, 178, 181, 210, 215 (1989).

74. R. H. Levy, W. H. Pitlic, A. S. Troupin, M. D. Green, and J. M. Neal, *Clin. Pharmacol. Ther.*, *17*, 657 (1975).

75. A. L. Thakkar, *J. Pharm. Sci.*, *59*, 1499 (1970).

76. T. R. Bates, M. Gibaldi, and J. L. Kanig, *J. Pharm. Sci.*, *55*, 191 (1966).

77. S. Miyazaki, H. Inoue, T. Yamahira, and T. Nadai, *Chem. Pharm. Bull.*, *27*, 2468 (1979).

78. I. M. Asher and S. Schwartzman, "Amphotericin B," in *Analytical Profiles of Drug Substances*, Volume 6 (K. Florey, ed.), Academic Press, New York, 1977.

79. M. W. Samaha and M. A. F. Gadalla, *Drug Dev. Indust. Pharm.*, *13*, 93 (1987).

80. S. H. Yalkowsky and T. J. Roseman, "Solubilization of Drugs by Cosolvents," in *Techniques of Solubilization of Drugs* (S. H. Yalkowsky, ed.), Marcel Dekker, New York, 1981, pp. 91–134.

81. J. T. Rubino and S. H. Yalkowsky, *Pharm. Res.*, *4*, 220 (1987).

82. S. H. Yalkowsky and J. T. Rubino, *J. Pharm. Sci.*, *74*, 416 (1985).

83. J. T. Rubino, J. Blanchard, and S. H. Yalkowsky, *J. Paren. Sci. Tech.*, *38*, 215 (1984).

84. J. T. Rubino and S. H. Yalkowsky, *J. Paren. Sci. Tech.*, *39*, 106 (1985).

85. J. T. Rubino, J. Blanchard, and S. H. Yalkowsky, *J. Paren. Sci. Tech.*, *41*, 172 (1987).

86. J. T. Rubino and S. H. Yalkowsky, *Pharm. Res.*, *4*, 231 (1987).

87. F. L. Fort, I. A. Heyman, and J. W. Kesterson, *J. Paren. Sci. Tech.*, *38*, 82 (1984).

88. R. C.-C. Fu, D. M. Lidgate, J. L. Whatley, and T. McCullough, *J. Paren. Sci. Tech.*, *41*, 164 (1987).

89. G. Levy and B. A. Hayes, *New Eng. J. Med.*, *262*, 1053 (1960).

90. *United States Pharmacopeia 23 and National Formulary 18*, United States Pharmacopeial Convention, Inc., Rockville, MD, 1994, pp. 1791–1793.

91. A. A. Noyes and W. R. Whitney, *J. Am. Chem. Soc.*, *19*, 930 (1897).

92. F. Langenbucher, *J. Pharm. Sci.*, *58*, 1265–1272 (1969).

93. M. J. Groves and M. H. Alkan, *Manuf. Chemist Aerosol News*, *46*(5), 37 (1975).

94. K. G. Nelson and A. C. Shah, *J. Pharm. Sci.*, *64*, 610, 1518 (1975).

95. J. Swarbrick, ed., *Current Concepts in the Pharmaceutical Sciences: Biopharmaceutics*, Lea and Febiger, Philadelphia, 1970.

96. J. G. Wagner, *Biopharmaceutics and Relevant Pharmacokinetics*, Hamilton Press, Hamilton, Illinois, 1971.

97. L. J. Leeson and J. T. Carstensen, *Dissolution Technology*, American Pharmaceutical Association, Washington, D.C., 1974.

98. J. T. Carstensen, *Solid Pharmaceutics: Mechanical Properties and Rate Phenomena*, Academic Press, New York, 1980.

99. W. A. Hanson, *Handbook of Dissolution Testing*, Pharmaceutical Technology Publication, Springfield, Oregon, 1982.

100. H. M. Abdou, *Dissolution, Bioavailability, and Bioequivalence*, Mack, Easton, Pennsylvania, 1989.

101. L. L. Birkumshaw and A. C. Riddiford, *Quart. Rev. Chem. Soc.*, *6*, 157 (1952).

102. W. I. Higuchi, *J. Pharm. Sci.*, *56*, 315 (1967).

103. K. G. Mooney, M, Rodriguez-Gaxiola, M. Mintun, K. J. Himmelstein, and V. J. Stella, *J. Pharm. Sci.*, *70*, 1358 (1981).

104. V. G. Levich, *Physicochemical Hydrodynamics*, Prentice-Hall, Englewood Cliffs, New Jersey, 1962.
105. W. Nernst, *Z. Phys. Chem.*, *47*, 52 (1904).
106. E. Brunner, *Z. Phys. Chem.*, *47*, 56 (1904).
107. A. Hussain, *J. Pharm. Sci.*, *61*, 811 (1972).
108. J. A. Hersey, *J. Pharm. Sci.*, *62*, 514 (1973).
109. A. W. Hixon and J. Crowell, *Ind. Eng. Chem.*, *23*, 923–926 (1931).
110. S. Lowell and J. E. Shields, *Powder Surface Area and Porosity*, 2nd ed., Chapman and Hall, London and New York, 1984.
111. J. W. Wood, J. E. Syarto, and H. Letterman, *J. Pharm. Sci.*, *54*, 1068 (1965).
112. R. N. Jashnani, P. R. Byron, and R. N. Dalby, *J. Pharm. Sci.*, *82*, 670 (1993).
113. J. K. Haleblian, *J. Pharm. Sci.*, *64*, 1269 (1975).
114. K. G. Mooney, M. A. Mintun, K. J. Himmelstein, and V. J. Stella, *J. Pharm. Sci.*, *70*, 13 (1981).
115. G. L. Flynn, S. H. Yalkowsky and T. J. Roseman, *J. Pharm. Sci.*, *63*, 479 (1974).
116. T. Higuchi, S. Dayal, and I. H. Pitman, *J. Pharm. Sci.*, *61*, 695 (1972).
117. J. V. Fee, D. J. W. Grant, and J. M. Newton, *J. Pharm. Sci.*, *65*, 48 (1976).
118. K. G. Mooney, M. A. Mintun, K. J. Himmelstein, and V. J. Stella, *J. Pharm. Sci.*, *70*, 22 (1981).
119. T. Higuchi, H. K. Lee, and I. H. Pitman, *Farm. Aikak.*, *80*, 55 (1971).
120. S. S. Ozturk, B. O. Palsson, and J. B. Dressmann, *Pharm. Res.*, *5*, 272 (1988).
121. D. Attwood and A. T. Florence, *Surfactant Systems: Their Chemistry, Pharmacy, and Biology*, Chapman and Hall, London and New York, 1983.
122. L. Borka and J. K. Habelian, *Acta Pharm. Jugosl.*, *40*, 71 (1990).
123. L. Borka, *Pharm. Acta Helv.*, *66*, 16 (1991).
124. S. R. Byrn, *Solid-State Chemistry of Drugs*, Academic Press, New York, 1982, pp. 79–148.
125. S. R. Byrn, R. R. Pfeiffer, G. Stephenson, D. J. W. Grant, and W. B. Gleason, *Chem. Mater.*, *6*, 1148 (1994).
126. J. K. Habelian and W. McCrone, *J. Pharm. Sci.*, *58*, 911 (1969).
127. M. D. Tuladhar, J. E. Carless, and M. P. Summers, *J. Pharm. Pharmacol.*, *35*, 208 (1983).
128. S. Sudo, K. Sato, and Y. Harano, *J. Chem. Eng. Japan.*, *24*, 237 (1991).
129. M. J. Jozwiakowski, S. O. Williams, and R. D. Hathaway, *Int. J. Pharm.*, *91*, 195 (1993).
130. R. N. Jashnani, P. R. Byron, and R. N. Dalby, *J. Pharm. Sci.*, *82*, 670 (1993).
131. M. Stoltz, M. R. Caira, A. P. Lotter, and J. G. van der Watt, *J. Pharm. Sci.*, *78*, 758 (1989).
132. M. Lagas and C. F. Lerk, *Int. J. Pharm.*, *8*, 11 (1981).
133. A. J. Aguiar, J. Krc Jr., A. W. Kinkel, and J. C. Samyn, *J. Pharm. Sci.*, *56*, 847 (1967).
134. J. W. Poole, G. Owen, J. Silverio, J. N. Freyhof, and S. B. Rosenman, *Current Therap. Res.*, *10*, 292 (1968).
135. P. Kahela, R. Aaltonen, E. Lewing, M. Anttila, and E. Kristoffersson, *Int. J. Pharm.*, *14*, 103 (1983).
136. R. M. Izatt, E. H. Redd, and J. J. Christensen, *Thermochim. Acta*, *64*, 355 (1983).

137. G. Buckton and A. E. Beezer, *Int. J. Pharm.*, *72*, 181 (1991).

138. S. Lindenbaum, E. S. Rattie, G. E. Zuber, M. E. Miller, and L. J. Ravin, *Int. J. Pharm.*, *26*, 123 (1985).

139. D. P. Ip, G. S. Brenner, J. M. Stevenson, S. Lindenbaum, A. W. Douglas, S. D. Klein, and J. A. McCauley, *Int. J. Pharm.*, *28*, 183 (1986).

140. L.-S. Wu, C. Gerard, and M. A. Hussain, *Pharm. Res.*, *10*, 1793 (1993).

141. J. Jacques, Collet, and S. H. Wilen, *Enantiomers, Racemates, and Resolutions*, John Wiley and Sons, New York, 1981, pp. 167–213.

142. H. G. Brittain, *Pharm. Res.*, *7*, 683 (1990).

143. F. A. H. Schreinemakers, *Z. Phys. Chem.*, *11*, 75 (1893).

144. A. E. Hill and J. E. Ricci, *J. Am. Chem. Soc.*, *53*, 4305 (1931).

145. S. T. Bowden, *The Phase Rule and Phase Reactions, Theoretical and Practical*, Macmillan, London, UK, 1938.

146. A. Findlay, A. N. Campbell, and N. O. Smith, *The Phase Rule and Its Applications*, 9th ed., Dover, New York, 1951.

147. W. Meyerhoffer, *Ber.*, *37*, 2604 (1904).

148. S. P. Duddu, F. K.-Y. Fung, and D. J. W. Grant, *Pharm. Res.*, *10*, S-151 (1993).

149. S. P. Duddu, F. K.-Y. Fung, and D. J. W. Grant, *Int. J. Pharm.*, in press.

150. S. P. Duddu, F. K.-Y. Fung, and D. J. W. Grant, *Int. J. Pharm.*, *94*, 171 (1993).

151. Z. J. Li and D. J. W. Grant, *Pharm. Res.*, *11*, S-243 (1994).

152. I. Ugi, *Z. Naturforsch*, *20B*, 405 (1965).

153. A. W. Ingersoll, S. H. Babcock, and F. B. Burns, *J. Am. Chem. Soc.*, *55*, 411 (1933).

154. E. J. Valente, J. Zubrowski, and D. S. Eggleston, *Chirality*, *4*, 494 (1992).

12

Sorption of Water by Solids

Mark J. Kontny
Boehringer Ingelheim Pharmaceuticals, Inc., Ridgefield, Connecticut

George Zografi
University of Wisconsin—Madison, Madison, Wisconsin

I. INTRODUCTION

The physical and chemical properties of pharmaceutical solids are critically dependent on the presence of moisture. Pharmaceutical scientists can cite numerous examples of desirable and undesirable properties that result from varied levels of moisture associated with a particular solid or formulations consisting of mixtures of solids. Flow, compaction, caking, disintegration, dissolution, hardness, and chemical stability are just some of the properties influenced by moisture. Since water is present in bulk liquid form or as vapor at some relative humidity in virtually all stages of solid manufacture (active ingredient and excipients), storage, processing into formulations, and final product packaging, a fundamental understanding of the role of water in affecting solid properties (and vice versa) is necessary.

Though the properties of individual solids and the performance of solid dosage forms are dependent on moisture, characterization of the underlying water–solid interaction is often nebulous. For example, many solids are described as "hygroscopic" without further reference to whether and how this relates the rate and amount of moisture uptake as a function of relative humidity and temperature [1]. To illustrate this ambiguity, consider that water-soluble, nonhydrating crystalline substances such as sodium chloride sorb very low levels of moisture (e.g., less than 0.1%) below their critical relative humidities, yet sorb significant quantities of moisture above their critical relative humidities, where the solid actually dissolves in the sorbed moisture. On the other hand, some typical excipient materials used in solid dosage forms, such as starches, celluloses, and gelatin capsules, sorb significant quantities of moisture (e.g., 25–50%), and even though they do not dissolve, they do undergo significant morphological change at high relative humidities (i.e., swelling). For these substances, moisture uptake rate depends on the relative humidity of the environment and the time-dependent moisture content of the solid. On the other hand, sodium chloride will have a very low moisture uptake/loss rate that will decrease to zero if the environmental relative humidity is kept below its critical relative humidity, while the uptake rate will be much greater (and continuous) until all the solid has dissolved if the relative humidity is above the critical value. For situations in which the environmental relative humidity is significantly different from the relative humidities at which the starch, cellulose, or gelatin were previously equilibrated, the initial uptake/loss rate will be significant, but it will approach zero over time (i.e., a constant amount of sorbed moisture will be attained at a given relative humidity). Obviously, very different mechanisms of water sorption/desorption occur for the different samples. In this light, describing sodium chloride and/or starch as "hygroscopic" offers very little toward understanding the water–solid interactions that might affect their physical-chemical properties. These examples illustrate the need to understand the underlying mechanism(s) of uptake for a particular solid. In this regard, therefore, addressing the following questions provides a basis for studying the various mechanisms of water–solid interaction:

1. How much water is present and what is the corresponding water activity (approximated by relative pressure or percent relative humidity/100)?
2. What are the kinetics of moisture uptake or loss, and is the rate constant or changing over time?
3. Where is the water located (i.e., adsorbed to the external surface of crystals, absorbed into crystals as specific or nonspecific water of hydration, absorbed into amorphous regions, condensed into pores, etc.)?
4. What is the state of the moisture associated with the solid (i.e., bulk water, water of hydration, physisorbed water, etc.)?
5. What form of the solid is present (i.e., particle morphology, polymorphic species, degree of crystallinity, (an)hydrate), and is this form thermodynamically stable over the temperature and relative humidity range that the solid is expected to encounter?

It is the objective of this chapter to discuss the various mechanisms whereby water can interact with solid substances, present methodologies that can be used to obtain the necessary data, and then discuss moisture uptake for nonhydrating and hydrating crystalline solids below and above their critical relative humidities, for amorphous solids and for pharmaceutically processed substances. Finally, transfer of moisture from one substance to another will be discussed.

II. THE WATER SORPTION ISOTHERM

The most fundamental manner of demonstrating the relationship between sorbed water vapor and a solid is the water sorption–desorption isotherm. The water sorption–desorption isotherm describes the relationship between the equilibrium amount of water vapor sorbed to a solid (usually expressed as amount per unit mass or per unit surface area of solid) and the thermodynamic quantity, water activity (a_w), at constant temperature and pressure. At equilibrium the chemical potential of water sorbed to the solid must equal the chemical potential of water in the vapor phase. Water activity in the vapor phase is related to chemical potential by

$$\mu = \mu^0 + RT \ln a_w \qquad (1)$$

where μ is the chemical potential of water in the system at equilibrium, μ^0 is the standard chemical potential of water at a specific reference temperature and pressure, R is the gas constant, and T is absolute temperature. Lewis et al. [2] defined the relative activity of any pure substance or component (such as water) as a ratio of fugacities:

$$a_w = \frac{f_w}{f_w^0} \qquad (2)$$

where f_w is the fugacity of water in the system at equilibrium and f_w^0 is the fugacity of pure water at a standard temperature and pressure. For all practical purposes, the fugacity (or "escaping tendency") of water vapor can be approximated by the water vapor pressure in the system. This assumption is valid as long as the water vapor behaves as an ideal gas. For the water pressure range of usual interest at temperatures less than 50°C, this approximation is excellent (<0.2% relative error) [3]. Thus, the relative pressure of water vapor, P/P^0, is usually employed as an estimate of the relative water activity in the system:

$$a_w \sim \frac{P}{P^0} \qquad (3)$$

where P is the water vapor pressure in the system and P^0 is the vapor pressure above pure water at the temperature of interest. Relative humidity (RH) is defined as the relative pressure expressed on a percentage basis:

$$RH = 100 \times \frac{P}{P^0} \qquad (4)$$

The sorption branch of the isotherm is obtained experimentally by measuring the equilibrium amount of water sorbed to a solid at known relative pressure, beginning with a known mass of absolutely dry solid and then progressively increasing the relative pressure in the system. Drying the solid sample under heat, possibly using vacuum to facilitate the removal of desorbed water vapor, is usually necessary to eliminate residual moisture. One must be aware, however, of the effects of such conditions on the chemical and physical stability of the solid. The desorption portion of the isotherm is obtained by progressively decreasing the relative pressure in the system from a relative pressure of approximately unity, again monitoring the equilibrium amount of moisture sorbed at each relative pressure. Generation of water sorption–desorption isotherms for a particular solid can lend considerable insight into the nature of the water–solid interaction, as well as the surface characteristics of the solid. This information is readily obtained from the amounts of moisture sorbed at lower relative humidities in comparison with the specific surface area of the sample; from the general shape of the isotherm; from whether or not water uptake is a completely reversible process (i.e., whether hysteresis is observed between sorption and desorption); and from the shape of the hysteresis loop if it is present. With knowledge of the aforementioned, one can usually obtain an indication of the mechanism of moisture sorption for the material of interest. For example, a material that exhibits sorption at lower relative humidities in much greater amounts than one might expect based on the specific surface area of the sample, and that exhibits hysteresis over the complete range of relative humidities, is most likely absorbing water into its internal structure. On the other hand, a material exhibiting a closed hysteresis loop over the higher relative humidity range while sorbing moisture over the lower relative humidity range, similar to what might be expected based on its specific surface area, is probably quite porous in nature and is most likely sorbing water via capillary condensation over the higher relative humidity range.

III. MODELS DESCRIBING VAPOR ADSORPTION

A. Brunauer, Emmett, and Teller Equation

The model most commonly referred to in the literature describing vapor adsorption onto solid surfaces was put forth in 1938 by Brunauer, Emmett, and Teller [4]. The so-called BET model was originally derived using kinetic arguments in a manner very similar to that used by Langmuir [5]. The BET model has since also been derived using statistical mechanics [6–8]. The BET model assumes that vapor molecules, behaving as an ideal gas, exist in a state of equilibrium with a solid that consists of identical, homogeneous adsorption sites. The first vapor molecule adsorbed to an adsorption site on the solid is proposed to be bound, whereas molecules adsorbing beyond the first layer are assumed to have the properties of bulk liquid. Furthermore, adsorption is proposed to occur such that the adsorbed molecules do not interact laterally. The linear form of the BET equation is

$$\frac{1}{W[(P^0/P) - 1]} = \frac{(C_B - 1)\ (P/P^0)}{W_m C_B} + \frac{1}{W_m C_B} \tag{5}$$

where W is the mass of vapor adsorbed per gram of solid at a particular relative pressure, P/P^0, W_m is the theoretical quantity of vapor adsorbed when each adsorption site has one vapor molecule adsorbed to it, and

$$C_B = k\ \exp\frac{H_1 - H_L}{RT} \tag{6}$$

where H_1 is the heat of adsorption of the first vapor molecule adsorbed to a site, H_L is the heat of condensation of bulk adsorbate, R is the universal gas constant, T is absolute temperature, and k is a constant, usually assumed to be close to unity. The two BET constants, W_m and C_B, can easily be obtained from the linear plotting form of the BET equation given in Eq. (5). Plotting the quantity $1/[W\{(P^0/P) - 1\}]$ versus P/P^0 gives a slope equal to $(C_B - 1)/\ W_m C_B$ and an intercept equal to $1/W_m C_B$. Algebraic manipulation gives

$$W_m = \frac{1}{\text{slope} + \text{intercept}} \tag{7}$$

and

$$C_B = 1 + \frac{\text{slope}}{\text{intercept}} \tag{8}$$

In general, the BET equation fits adsorption data quite well over the relative pressure range 0.05–0.35, but it predicts considerably more adsorption at higher relative pressures than is experimentally observed. This is consistent with an assumption built into the BET derivation that an infinite number of layers are adsorbed at a relative pressure of unity. Application of the BET equation to nonpolar gas adsorption results is carried out quite frequently to obtain estimates of the specific surface area of solid samples. By assuming a cross-sectional area for the adsorbate molecule, one can use W_m to calculate specific surface area by the following relationship:

$$S = \frac{W_m X N_{Av}}{M\Sigma} \tag{9}$$

where S is specific surface area in m^2/g, W_m is the mass of adsorbate adsorbed at monolayer coverage, X is the cross-sectional area of an adsorbed adsorbate molecule (assumed to be 19.5 $Å^2$ for krypton, 16.2 $Å^2$ for nitrogen, and 12.5 $Å^2$ for water [9,10]), N_{Av} is Avogadro's number of molecules, M is the molecular weight of adsorbate, and Σ is the mass of sample. Obviously, calculating surface areas from moisture uptake data that does not lead to monolayer coverage at W_m

(either incomplete coverage [Section V.B] or absorption into the solid [Section VI.C]) will result in (incorrect) values that have no physical meaning.

B. Guggenheim and deBoer Equation

Many attempts to modify the BET adsorption theory have been made since its original derivation. Its simplicity and ability to fit adsorption data extremely well at lower relative pressures, however, have made it the model of choice for estimating surface areas from nonpolar gas adsorption. Most modifications of the BET model, developed to analyze data over the entire range of relative pressures, usually add at least one fitting parameter to the equation. This makes computer fitting a necessity, since only two measurable parameters, W and P/P^0, are available. From a modeling perspective, additional fitting parameters of unknown or undefined physical meaning that arise from such approaches are often a deterrent to the use of multiparameter models because of the consequent difficulty in interpreting results. In this regard, therefore, only a single modification of the BET model, which has been shown to extend the relative pressure range over which vapor adsorption data are able to be fit, will be considered here. This extension of the BET model, independently derived by Guggenheim [11] and deBoer [12], accounts for the adsorption of an intermediate state of vapor between the tightly bound first molecule adsorbing to an adsorption site and the condensed molecules adsorbed at very high relative pressures. Molecules adsorbed in the intermediate range can be considered to interact with the solid, but the interaction is assumed to be considerably less than that of the first molecule sorbed at an adsorption site. This equation is given as

$$W = \frac{W_m C_G K (P/P^0)}{[1 - K(P/P^0)][1 - K(P/P^0) + C_G K(P/P^0)]} \tag{10}$$

where P, P^0, H_L, W, and W_m are identical to the parameters used in the BET equation, and

$$K = B \exp \frac{H_L - H_m}{RT} \tag{11}$$

where B is a constant and H_m is the heat of adsorption of vapor adsorbed in the intermediate layer. The constant C_G is defined as

$$C_G = D \exp \frac{H_1 - H_m}{RT} \tag{12}$$

where D is a constant, H_1 is the heat of adsorption of the first molecule adsorbed at a site, and H_m is the heat of adsorption of the intermediately bound molecule.

C. Water Vapor Absorption by Amorphous Solids

Although water vapor is absorbed into amorphous solids and not simply adsorbed on the surface, it still has been found that such absorption isotherms can be fit to the BET equation up to a P/P^0 of about 0.40 as with vapor adsorption, and over the entire range of P/P^0 using its extension, Eq. (10). Since this was first reported by Anderson [13] to be the case for water absorption, Eq. (10), when applied to water vapor absorption it is often called the GAB equation for Guggenheim, Anderson, and deBoer [14]. Since the theoretical basis for the derivation of the original equation does not translate directly to the absorption process, which involves dissolution of water in the amorphous solid, the significance of fit to the GAB equation is somewhat limited. It is, however, a very useful equation since it does allow one to describe the entire isotherm and to draw out some useful parameters (to be discussed in what follows).

Since water vapor dissolves in the solid during absorption, several models based on solution theory, proposing that the sorbate is taken up into the solid as a solid solution, have been derived and used to describe water sorption on polymers (e.g., Flory–Huggins [15], Hailwood–Horrobin [16]). The development of these sorption theories is based on meaningful physical-chemical principles. As with the many modifications of the BET adsorption model, however, the physical significance of the constants, and the meaning of the values obtained (from computer fitting) from such analyses are often of limited utility in helping to gain a basic understanding of the mechanisms of sorption from a molecular viewpoint. From this perspective, other models based on entirely different theoretical concepts will not be considered in this chapter. For further reference, the reader is directed to several excellent literature reviews of the many sorption theories that have been proposed [17,18].

D. Capillary Condensation

Vapor sorption onto porous solids differs from vapor uptake onto the surfaces of flat materials in that a vapor (in the case of interest, water) will condense to a liquid in a pore structure at a vapor pressure, P_r, below the vapor pressure, P^0, where condensation occurs on flat surfaces. This is generally attributed to the increased attractive forces between adsorbate molecules that occur as surfaces become highly curved, such as in a pore or capillary. This phenomenon is referred to as capillary condensation and is described by the Kelvin equation [19]:

$$\ln \frac{P_r}{P^0} = -\frac{2\gamma V_m}{rRT} \tag{13}$$

where γ is the surface tension of the adsorbed film (assumed equal to that of the bulk liquid), V_m is the molar volume of the liquid, r is the pore radius, R is the gas constant, and T is temperature. The Kelvin equation has been shown to be applicable to pore radii as low as 5 nm for water adsorption onto mica [20,21]. As mentioned in Section II, capillary condensation will result in closed hysteresis loops in the adsorption/desorption isotherms of porous materials. Calculating P_r/P^0 by assuming a surface tension for water of 72.8 ergs/cm^2 and a density of 0.998 g/cm^3 at 293°K shows that condensation is predicted at relative pressures of 0.998, 0.989, 0.898, and 0.340 for pore radii of 1000, 100, 10, and 1 nm, respectively. In this regard, it is clear that capillary condensation need only be considered for very small pore dimensions. In practical terms, one should be concerned about this mechanism of water uptake for microporous pharmaceutical powders that exhibit a relatively large specific surface area (i.e., >100 m^2/g), as determined from nonpolar gas adsorption studies.

IV. METHODOLOGY

A. Control of Relative Humidity

Maintenance of constant relative humidity environments is essential for studying water–solid interactions. There are primarily four techniques that are frequently employed to maintain constant relative humidity:

1. Saturated salt solutions
2. Sulfuric acid solutions
3. Temperature modification of an aqueous solution
4. Mixing wet and dry air streams

Saturated salt solutions and sulfuric acid solutions establish relative humidity by reducing the vapor pressure above an aqueous solution (a colligative effect). Saturated salt solutions at controlled temperature maintain a constant relative humidity as long as there is excess salt and bulk water present. As water is added or removed from the solution, moisture from the headspace will condense/evaporate, with subsequent dissolution/precipitation of salt to maintain the equilibrium vapor pressure. Since the degree of vapor pressure depression is dependent on the number of species in solution and, further, since the solubility of most salts is somewhat dependent on temperature, the relative humidity generated is also temperature-dependent. Hence, use of the same salt at different temperatures can result in different relative humidities. References 22–26 can be consulted for specific saturated salt solutions that result in defined relative humidities as a function of temperature. Since relative humidity is dependent on the number of dissolved species, it is essential that saturation be attained prior to beginning experimentation. In this regard, preparing the salt solutions several days before beginning a sorption study is recommended.

Sulfuric acid solutions of varying concentration [26] are also used to establish relative humidity. Addition or removal of water from the solution by desorption or sorption of water to the solid, however, will alter the concentration of sulfuric acid (and water) in solution, and thus change the relative humidity of the headspace. This technique for controlling relative humidity in the headspace is practically more useful when small amounts of water are sorbed/desorbed from the solid.

Temperature modification of an aqueous solution can also be used to maintain constant relative humidity in the headspace [14]. This technique maintains the solid at one temperature and an aqueous solution connected to the system at another temperature. Due to the strong vapor pressure dependence on temperature, very tight temperature control of the aqueous solution and the solid are required to maintain constant relative humidity in the vicinity of the solid.

Mixing dry and water vapor–saturated air in defined proportions also can be used to generate constant relative humidity. Control of flow rates and the water vapor content of the dry and saturated air are essential [27,28].

B. Measurement of Relative Humidity

Measurement of relative humidity depends on the system used. Systems employing vacuum are usually evacuated prior to introduction of water vapor [29]. For cases in which there is not a gas-forming reaction occurring, measurement of total pressure in the system can be used as a measure of water vapor pressure. Systems in which air is not evacuated require specific measurement of water vapor pressure. (For the latter type of system, caution should be taken to assure that the relative humidity source is in close proximity to the solid, since the diffusion of water vapor through air to the solid is required to maintain a constant relative humidity in the immediate vicinity of the solid.) A wide variety of pressure measuring instrumentation is commercially available with varying accuracy, precision, and cost.

C. Measurement of the Critical Relative Humidity, RH_0

The relative humidity at which a solid begins to deliquesce, RH_0, can be determined in two ways: directly, by measuring the relative humidity above a saturated solution of the substance; or indirectly, by measuring the steady state moisture uptake rate at relative humidities above RH_0 and then extrapolating to the relative humidity at which the moisture uptake rate is zero [1,30,31].

Although other techniques can be used to measure the relative humidity above a saturated solution, one relatively simple procedure is to utilize a vacuum system to remove air from the headspace (by vapor phase expansions) and then, with the vacuum pumps isolated and the saturated solution maintained at a constant temperature, measure water vapor pressure. Water vapor pressure can then be converted to relative humidity by dividing by P^0, the vapor pressure above pure water at the temperature of interest [32].

D. Measurement of Moisture Uptake (Kinetics of Deliquescence)

The rate of moisture uptake above RH_0 requires maintenance and measurement of a range of relative humidities, and the capability of measuring the moisture content of the solid over time. Use of a vacuum system can minimize vapor diffusion through the headspace, thus maintaining constant relative humidity in the vicinity of the sample. Also, since the most reliable estimate of the steady state moisture uptake rate is when the integrity of the solid is intact and the film of sorbed moisture is thin (and saturation most likely), it is advisable to determine the moisture uptake rate at early time periods. In this regard, it is also helpful to be able to view the solid during the experiment to verify that integrity is maintained and excess solid remains [31,33].

E. Measurement of Equilibrium Moisture Sorption

Generation of water sorption/desorption isotherms in a controlled relative humidity environment can be carried out either gravimetrically or volumetrically. Gravimetric methods require

1. A dry sample weight,
2. Constant temperature of the sample,
3. Maintaining predetermined constant relative humidities in the head-space, and
4. Attaining and measuring an equilibrium weight of sorbed water vapor.

Gravimetric measurement of moisture uptake can occur continuously or discontinuously. Continuous measurement usually involves placing a sample on a balance in a temperature- and relative humidity–controlled environment. A Cahn electrobalance in a glass bell jar has been used successfully for this purpose [29,31]. Discontinuous procedures store the samples in a constant relative humidity environment (e.g., in desiccators containing saturated salt solutions) and require periodic sample removal for weighings. The advantages of continuous procedures are that it is obvious when equilibrium is attained; and that improved accuracy and precision are possible since "equilibrated" samples do not have to be removed and subjected to ambient relative humidity, where subsequent sorption or desorption of water vapor can occur. Only a single sample at one relative humidity can be run at a time, however, whereas discontinuous techniques allow many samples to be equilibrated simultaneously in different chambers. Weighing at multiple time points after attaining equilibrium, however, is required to verify that the sample has indeed reached equilibrium.

Volumetric methods require

1. A dry sample weight,
2. Constant temperature of the sample,
3. Water vapor pressure measurement in a dosing volume and, later, in the headspace above the equilibrated sample, and
4. Measuring dead volumes of the individual chambers, including the sample chamber.

In essence, volumetric methods equilibrate a known headspace dosing volume at a given (measured) water vapor pressure, and then they expose the pre-equilibrated sample to this water vapor, with subsequent measurement of the water vapor pressure after equilibration. The mass of water sorbed, Δn (in moles), at the final pressure in the system, P_f, is obtained from the difference, ΔP, between P_f^{calc}, the calculated water vapor pressure at equilibrium, and P_f^{meas}, the final measured water vapor pressure:

$$\Delta n = \frac{\Delta P V}{RT} \tag{14}$$

where V is the final volume, R is the gas constant, and T is absolute temperature [29].

V. WATER SORPTION BY CRYSTALLINE SOLIDS

A. General Model

Figure 1 schematically describes the important steps in the uptake of water vapor by crystalline water-soluble solids. At low relative humidities, water is adsorbed to the surface of a nonhydrate-forming solid. As the relative humidity is increased, some tendency for multilayer sorption is expected. At some relative humidity (characteristic for a given substance), the solid will begin to dissolve in the sorbed film of water. A saturated solution of solute will most likely exist, and this will cause the vapor pressure over the sorbed film of water to be depressed relative to pure water and to be constant and equal to that above a saturated solution for the substance. This vapor pressure may be expressed as the critical relative humidity, RH_0. If the relative humidity in the atmosphere is greater than that over the saturated solution (RH_0), water will spontaneously condense on the aqueous film. This will dilute the film, allowing more solid to dissolve, which, in turn, will maintain the pressure gradient. The process of water vapor uptake will continue until all the solid has dissolved and further solution dilution has occurred. Only when the relative humidity above the solution is elevated to that of the atmosphere will this process terminate. This phenomenon is called deliquescence. Although hydrates undergo solid state transitions in transforming

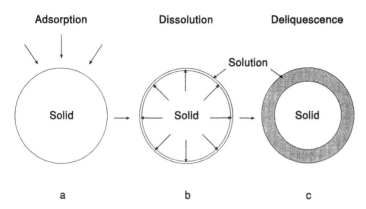

Fig. 1 Water vapor adsorption and deliquescence of a water-soluble solid: (a) Atmospheric relative humidity, $RH_i < RH_0$; (b) $RH_i = RH_0$; and (c) $RH_i > RH_0$.

from the anhydrate to hydrate, as well as from one hydrate species to another, behavior similar to that we have described for nonhydrates is also noted at and above RH_0 for hydrates. In pharmaceutical systems, water-soluble species are frequently encountered in solid dosage forms. Thus, it is important to understand the conditions responsible for deliquescence and the molecular events occurring at relative humidities below the deliquescence point.

B. Water Sorption onto Nonhydrates Below RH_0

The sorption of water vapor onto nonhydrating crystalline solids below RH_0 will depend on the polarity of the surface(s) and will be proportional to surface area. For example, water exhibits little tendency to sorb to nonpolar solids like carbon or polytetrafluorethylene (Teflon) [21], but it sorbs to a greater extent to more polar materials such as alkali halides [34–37] and organic salts like sodium salicylate [37]. Since water is only sorbed to the external surface of these substances, relatively small amounts (i.e., typically less than 1 mg/g) of water are sorbed compared with hydrates and amorphous materials that absorb water into their internal structures.

Unfortunately, the literature is relatively sparse with examples showing the water uptake profile onto crystalline, nonhydrating substances below RH_0. This is most likely due to the difficulty in accurately measuring the small amounts of water that are sorbed. Alkali halides are an exception, however, likely due to their well-characterized particle morphologies [34–37]. Figure 2 shows a water uptake isotherm onto recrystallized sodium chloride [37]. Note that the amount of water sorbed as a function of relative humidity is normalized to the specific surface area of the sample. Since water is sorbed only to the external surface of

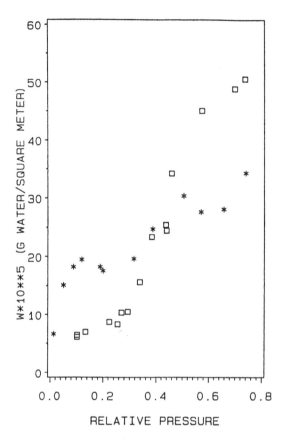

Fig. 2 Water vapor sorption for recrystallized (□) and ground (*) sodium chloride at 20°C. (From Ref. 37.)

this material, this allows comparison of water uptake data from different lots of material, whereas plotting this data on a "per gram" basis would have little or no meaning. For the sodium chloride sample in Fig. 2 (specific surface area = 0.0875 m^2/g, from krypton adsorption studies), only 5×10^{-4} g water/m^2 of sodium chloride is sorbed, even up to 70% relative humidity. Also, note the apparent steplike nature of the isotherm. From BET analysis of the sorption data at the lower relative humidities, a W_m value of 7.6×10^{-5} g/m^2 is obtained. This value is only about 0.32 that of the predicted value for monolayer coverage assuming an area per water molecule of 12.5 Å^2. This suggests that it is quite meaningless to speak of the number of layers of sorbed water as multiples of W_m, except as a point of reference. Interestingly, the second step plateau in Fig. 2 occurs at about three times the moisture content corresponding to W_m, suggesting that for sodium chloride the monolayer is actually completed during the second step of the isotherm. Isosteric heat of sorption results for sodium chloride from Barraclough and Hall [34] suggest that the heat of sorption of water up to W_m is invariant, whereas the heat of sorption decreases and becomes constant at about two times W_m. Considering the experimental error involved in obtaining W_m and the isosteric heats of sorption, this suggests that water is sorbed with a homogeneous binding energy up to W_m and then is bound with a decreasing extent until the monolayer is complete.

As shown in Fig. 2 [37], and also in the work of Barraclough and Hall [34], moisture uptake onto sodium chloride as a function of relative humidity is reversible as long as RH_0 is not attained. This is evidence that actual dissolution of water-soluble crystalline substances does not occur below RH_0. This is consistent with thermodynamic rationale that dissolution below RH_0 would require a supersaturated solution (i.e., an increased number of species in solution would be necessary to induce dissolution at a relative humidity below that of the saturated solution, RH_0). In this regard, one should only need to consider the solid state properties of a purely crystalline material below RH_0. As will be described, other considerations are warranted for a substance that contains amorphous material.

C. Water Sorption onto Hydrates Below RH_0

Solids that form specific crystal hydrates sorb small amounts of water to their external surface below a characteristic relative humidity, when initially dried to an anhydrous state. Below this characteristic relative humidity, these materials behave similarly to nonhydrates. Once the characteristic relative humidity is attained, addition of more water to the system will not result in a further increase in relative humidity. Rather, this water will be sorbed so that the anhydrate crystal will be converted to the hydrate. The strength of the water–solid interaction depends on the level of hydrogen bonding possible within the lattice [21,38]. In some hydrates (e.g., caffeine and theophylline) where hydrogen bonding is relatively weak, water molecules can aid in hydrate stabilization primarily due to their space-filling role [21,38].

Since water molecules occupy regular positions within the lattice of a hydrate with a specific stoichiometry (e.g., 1:1 monohydrate, 2:1 dihydrate, 5:1 pentahydrate) to the solid, relatively large quantities of water are sorbed. Figure 3 shows a moisture uptake isotherm for ipratropium bromide [39]. This substance undergoes an apparent hydration of the crystal between 63% and 75% relative humidity. Above 75% relative humidity, approximately 4.6% water is sorbed (theoretical monohydrate is 4.4 g/g). Interestingly, as anhydrous ipratropium bromide is equilibrated for extended time periods (e.g., 2 months and 5 months respectively, as shown in Fig. (3), hydration of the crystal appears to occur at

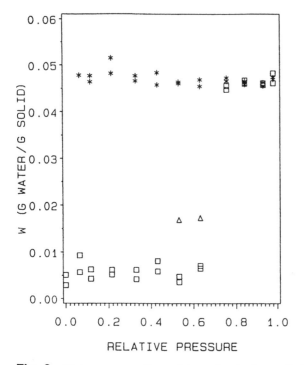

Fig. 3   Water vapor sorption and desorption isotherms for ipratropium bromide at 20°C
(□), 2 month sorption; (Δ), 5 month sorption; (*), 2 and 5 month desorption [Note: All
2 month sorption results, except at 53% and 63% relative humidity, were verified at 5
months.]

53% and 63% relative humidities. This example clearly shows that time periods
of many months are required to attain a reliable estimate of the equilibrium uptake
at select relative humidities. Characteristic of many hydrates, ipratropium
bromide exhibits significant hysteresis between the sorption and desorption
isotherms. This is attributed to the degree of binding and the physical fit of water
in the hydrated lattice.

   Nonspecific hydration, or hydration of the lattice without first-order phase
transitions, also must be considered. Cox et al. [40] reported the moisture
uptake profile of cromolyn sodium, and the related effects on the physical
properties of this substance. Although up to nine molecules of water per
molecule of cromolyn sodium are sorbed into the crystalline lattice at 90%
relative humidity, the sorption profile does not show any sharp plateaus
corresponding to fixed hydrates. Rather, the uptake profile exhibits a gradual
increase in moisture content as relative humidity increases, which results in

marked changes in x-ray diffraction patterns, density, and other physical properties. For this example, moisture uptake onto cromolyn sodium was correlated with expansion of the lattice in the *b* crystallographic direction, which was shown to be reversible on dehydration.

A thorough understanding of the hydration profile for a solid forming a crystal hydrate is important for several reasons. First, since an anhydrate and hydrate(s) are distinct thermodynamic species, they will have different physical-chemical properties (e.g., solubility) that may affect bioavailability. Second, a desired hydrate species can be formed and used (and retained) simply by controlling the desired, established environmental conditions. Third, since significant quantities of water can be sorbed/liberated as a hydrate becomes hydrated/dehydrated, the physical-chemical properties of the immediate system (including other nearby solids) can be markedly affected.

D. The Critical Relative Humidity, RH_0

Knowledge of RH_0 for each component in a formulation, and for the entire system, is extremely important for predicting relative humidities where gross physical changes of the system are expected due to dissolution of the water-soluble components. The value of RH_0, as a colligative property, is determined by the number of species in solution. As a rule of thumb, two general comments can be made. First, compounds exhibiting poor water solubility have RH_0 values in the high 90% range. Second, as solubility increases, RH_0 decreases. Since nonidealities are introduced as solutions become more and more concentrated, however, it is not usually possible to use dilute solution models (e.g., Raoult's law) to predict the expected RH_0 for a solute of significant aqueous solubility. Hence, RH_0 should be measured for individual solids. Examples of RH_0 values for single-component systems are shown in Table 1.

Values of RH_0 for mixtures, on the other hand, can be calculated from the RH_0 values of single components using an equation developed by Ross [42]:

$$\frac{(RH_0)_{mix}}{100} = \frac{(RH_0)_1}{100} \cdot \frac{(RH_0)_2}{100} \cdot \frac{(RH_0)_3}{100} \cdot \cdots \quad (15)$$

where $(RH_0)_{mix}$ is the relative humidity above a saturated solution of the mixture and $(RH_0)_i$ represents the relative humidities of the individual saturated salt solutions. The Ross equation was derived assuming dilute solutions and negligible interaction between components in solution. The results presented in Table 2 compare RH_0 values obtained by calculating RH_0 values for mixtures from the Ross equation and those obtained experimentally. Agreement is very good, especially considering the high levels of dissolved solute(s) that are attained (i.e., estimated as high as 50 molal for the choline bromide/tetrabutylammonium bromide system) [33].

Table 1 RH_0 Values for Single-Component Systems at 25°C

Compound	RH_0	Reference
Potassium chloride	84	[31]
Potassium bromide	81	[31]
Potassium iodide	68	[31]
Sodium chloride	75	[31]
Choline iodide	72	[31]
Choline bromide	41	[31]
Choline chloride	23	[31]
Tetrabutylammonium bromide	61	[31]
Potassium acetate	23	[22]
Potassium carbonate	43	[22]
Sucrose	84	[31]
Fructose	64	[31]
Glucose	87	[39]
Sodium salicylate	79	[29]
Sodium benzoate	88	[29]
Salicylic acid	>99	[41]
Benzoic acid	>99	[41]
Malic acid	78	[41]
Tartaric acid	93	[41]
Fumaric acid	98	[41]
Succinic acid	95	[41]

Table 2 Comparison of Calculated and Experimentally Determined Values of RH_0 for Mixtures of Substances

Mixture	RH_0	
	Calculated	Experimental
Sodium chloride–potassium bromide	61	64
Potassium chloride–sodium chloride	64	67
Potassium chloride–potassium bromide	68	73
Sucrose–potassium bromide	68	66
Sucrose–dextrose monohydrate	69	68
Sucrose–sodium chloride–potassium bromide	51	57
Choline bromide–potassium bromide	33	40
Tetrabutylammonium bromide–potassium bromide	49	57
Tetrabutylammonium bromide–choline bromide	25	34

E. The Kinetics of Deliquescence Above RH$_0$

Initial work by Edgar and Swan [43], Adams and Merz [44], Prideaux [45], Markowitz and Boryta [46], and Carstensen [1] suggested that the rate of moisture uptake onto water-soluble solids above RH$_0$ should depend on the difference between the partial pressure of water in the environment and that of the partial pressure of water above a saturated solution of a water-soluble substance, temperature, the exposed surface area of the solid, the velocity of movement of the moist air, and a specific reaction constant that is characteristic of the individual solid.

Van Campen et al. [31] developed models describing the rate of moisture uptake above RH$_0$ that consider both the mass transport of water to the solid substance and the heat transfer away from the surface. For the special case of an environment consisting of pure water vapor (i.e., initial vacuum conditions), the Van Campen et al. model is greatly simplified since vapor diffusion need not be considered. Here, only the rate at which heat is transported away from the surface is assumed to be an important factor in limiting the sorption rate, W'. For this special case, an expression was derived to express the rate of moisture uptake solely as a function of RH$_i$, the relative humidity of the environment, and RH$_0$.

This model was shown to be applicable for describing moisture uptake kinetics (in vacuum) above RH$_0$ for single-component systems of alkali halides, sugars, and choline salts [31]. The model later was extended to consider the moisture uptake kinetics above RH$_0$ for multicomponent systems of these substances [33].

VI. WATER SORPTION BY AMORPHOUS SOLIDS

A. Isotherm Analyses at Ambient Temperatures

The amount of moisture sorbed by amorphous solids is typically much greater than that sorbed by nonhydrating crystalline substances below their critical relative humidities. Typical substances of pharmaceutical interest in this class of solids include celluloses, starches, poly(vinylpyrrolidone), gelatin, and some lyophilized proteins. Though some of these substances exhibit partially crystalline character, they generally contain significant fractions of amorphous material and, thus, fall into this class of solids. A typical isotherm for microcrystalline cellulose is shown in Fig. 4. Note the significant amounts of water that are sorbed over the entire relative humidity range and that both the sorption and desorption isotherms are characterized by the classical sigmoidal shape often observed with the physical adsorption of gases. Also apparent is the hysteresis between the sorption and desorption portions of the isotherm (i.e., the amount of water associated with the solid is greater for the desorption isotherm than the sorption isotherm for a given

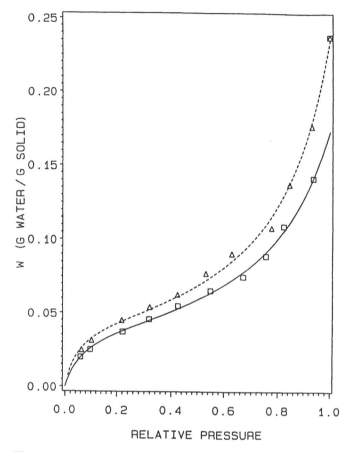

Fig. 4 Water vapor sorption (□) and desorption (Δ) isotherms for microcrystalline cellulose at 20°C. ———, GAB fit to sorption data; – – –, GAB fit to desorption data.

relative pressure). This is typical for these types of materials and is generally attributed to either kinetic effects or to a change in the polymer-chain conformation caused by the plasticization effects of sorbed water [14,47–49].

Figure 4 also shows the excellent fit to the GAB equation (Eq. 10) of the sorption and desorption isotherms for microcrystalline cellulose. In this regard, this equation offers considerable practical utility in fitting isotherms for these types of materials over the entire relative humidity range, especially in contrast to the BET equation, which usually only fits uptake data up to about 40% relative humidity. As we have mentioned, however, this does not in itself confirm the validity of the GAB model for describing moisture sorption data on these materials. Rather, independent confirmation of the physical meaning is necessary.

Considerable physical insight has been gained into the primary binding mechanism of water onto starches and celluloses from isotherm analyses that yield values for W_m (Eqs. 5, 7, 10). This is illustrated in Table 3, which gives W_m values for three types of starches [49]. Table 3 shows that despite significant morphological differences between the various starches, the values of W_m are quite constant. One value of W_m, taken from a fit of desorption data, appears to be slightly higher than values obtained from sorption data. This might be expected if the availability of primary sorption sites had been increased by previous exposure to elevated relative humidities, with subsequent increased levels of water sorption. As shown by Van den Berg et al. [14,47,48], these values of W_m are all close to the value of 0.11 g of water per g of starch, calculated by assuming that one water molecule sorbs per anhydroglucose unit. Since this calculation assumes that all anhydroglucose units are available for primary binding, and since this is not likely to be precisely the case, it is not surprising that the values measured for W_m are slightly less than 0.11 g/g.

Zografi et al. [49,51] have extended this analysis to the sorption of water vapor by various celluloses. For celluloses, corrections are necessary because only the amorphous regions of cellulose take up water vapor. Table 4 shows the W_m values obtained from isotherm analyses of several cellulosic materials before and after accounting for the degree of crystallinity. As expected, celluloses with different degrees of crystallinity exhibit different values of W_m without correction for crystallinity, and all are considerably less than that for the starches. When corrected, however, for the degree of crystallinity, all of the values are in reasonable agreement with each other and with the W_m values obtained for the starches. Especially interesting are the results in Table 4 for microcrystalline cellulose samples having different degrees of crystallinity due to grinding [53]. These results suggest that a similar mechanism of water uptake is occurring in starches and the noncrystalline regions of celluloses.

Similar analyses of moisture uptake data available in the literature for other

Table 3 W_m Values for Various Starches Obtained from BET Analysis of Moisture Uptake Isotherms

Starch	W_m (g/g)	Reference
Corn	0.095^a	[29]
Corn	0.083	[50]
Potato	0.085	[14]
Wheat	0.080	[14]

^{a}This value is taken from the desorption isotherm. Others are from sorption isotherms.
Source: Ref. 49.

Table 4 W_m Values for Various Celluloses Obtained from BET Analysis of Moisture Uptake Isotherms Corrected for Degree of Crystallinity

Cellulose	% Crystallinity	W_m Corr. (g/g)	Reference
Cotton	70	0.093	[52]
Cellophane	40	0.098	[52]
MCC	63	0.095	[53]
MCC[a]	49	0.076	[53]
MCC[a]	38	0.107	[53]
MCC[a]	0	0.086	[53]

[a]MCC ground in a ball mill [53].
Source: Ref. 49.

cellulose and starch derivatives used as pharmaceutical excipients are presented in Table 5. Considering the uncertainties associated with estimating the moisture uptake values from published graphs, the values of W_m are all quite consistent with each other and with a stoichiometry of one water molecule per anhydroglucose unit. It is interesting to note that the two samples derived from cellulose, sodium carboxymethylcellulose and sodium croscarmellose, did not require any correction for degree of crystallinity to conform to close to a 1:1 stoichiometry. It appears quite likely, therefore, that the processing of these materials essentially eliminates the crystallinity of cellulose.

The preceding analysis suggests that water, indeed, penetrates throughout the amorphous regions of these materials and undergoes a specific interaction with available sorption sites, most likely the available hydroxyl groups on the anhydroglucose units. Differential heat of sorption results for various starches [14,50] and celluloses [10,55] support this model. Figure 5 is an example of a

Table 5 W_m Values for Various Pharmaceutical Excipients Obtained from BET Analysis of Moisture Uptake Isotherms

Excipient	W_m (g/g)	Reference
Starch 1500	0.074	[50]
Sodium starch glycolate (Explotab®)	0.081	[54]
Sodium starch glycolate (Primogel®)	0.092	[54]
Crosslinked dextrose (CLD-2®)	0.098	[55]
Croscarmellose, sodium (Ac-Di-Sol®)	0.094	[55]
Sodium carboxymethylcellulose	0.103	[56]

Source: Ref. 49.

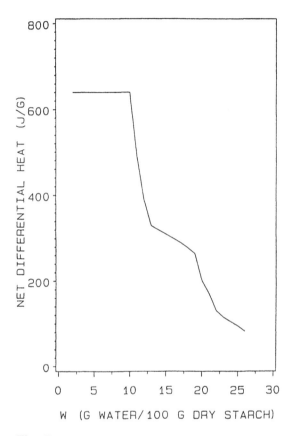

Fig. 5 Net differential heat of sorption for water vapor by native potato starch at 20°C as a function of amount of water sorbed per mass of dry starch. (From Ref. 14.)

pattern that is obtained for all these materials and that supports a model where there is a specific water–solid interaction out to a moisture content of at least the equivalent of 3 times W_m. This means that water is in a more structured state (i.e., reduced mobility) than bulk water in this range. Interestingly, the heats of sorption exhibit discreet breaks corresponding to stoichiometries of one and two water molecules per anhydroglucose unit. Some differential heat of sorption results are nearly constant over the W_m range, suggesting that binding is homogeneous over this range [10,14]. This is, however, not always the case.

Other supportive evidence for a specific water–solid interaction is available from thermal studies showing the amount of nonfreezable water [57–59], nuclear magnetic resonance [29,60–66], and diffusion studies [67,68]. The evidence is less clear, however, concerning whether there is distinct binding of water to sorption sites with discrete energy levels or whether there is a continuum of states where water interacts to a lesser extent with increasing amount sorbed [49,69].

B. Isotherm Analyses as a Function of Temperature

Generally speaking, the absorption of water vapor into amorphous solids as a function of relative humidity decreases as the temperature increases, reflective of an overall exothermic process, normally expected with vapor adsorption processes. Such behavior has been observed with cellulose [70], starch [14], poly(vinylpyrrolidone) [71], and poly(methyl methacrylate) [72]. In such cases it is often assumed that the dominant factor is the negative heat of absorption arising from the change in extent of water binding. The process, however, is made much more complex than this because of the changing morphology of the solid and, hence, an entropy change as well. The complexity of the effects of temperature on water vapor absorption and the possible links to the plasticizing effects of water may be observed in the recent work of Oksanen and Zografi [71], who have reported that the W_m values for poly(vinylpyrrolidone) over the temperature range $-40°C$ to $60°C$ decrease by a factor of three, suggesting that W_m does not reflect the absolute number of available binding sites on the polymer for directly "bound" water. Rather, W_m appears to be related to $W(T_g = T)$, the amount of water sorbed that will reduce the glass transition temperature, T_g to the temperature of the sample, as the ratio of $W(T_g = T)/W_m$ remains nearly constant at 3.0 over the entire temperature range.

In summary, it is clear that water absorbs into amorphous polymers to a significant extent. Interaction of water molecules with "available" sorption sites likely occurs via hydrogen bonding such that the mobility of the sorbed water is reduced and the thermodynamic state of this water is significantly altered relative to bulk water. Yet accessibility of the water to all potential sorption sites appears to be dependent on the previous history and physical-chemical properties of the solid. In this regard, the water–solid interaction in amorphous polymer systems is a dynamic relationship depending quite strongly on water activity and temperature.

C. The Meaning of Specific Surface Areas Calculated from Water Absorption Studies

Simply calculating specific surface areas from the W_m values in Tables 3–5 leads to "apparent" specific surface areas of approximately 400–500 m^2/g [49,51]. Specific surface areas obtained from similar analyses of nonpolar gas (nitrogen or krypton) adsorption studies, however, are typically in the range of 1 m^2/g, independent of sample pretreatment.

Interestingly, the ball-milling studies of microcrystalline cellulose by Nakai (Table 4 [53]) have shown that the W_m values obtained from water sorption studies increase to a much greater extent than the increase in surface area due to comminution of the sample. In fact, as discussed earlier, moisture sorption was shown to be proportional to the amount of amorphous character, suggesting that water is being absorbed throughout the amorphous regions of this substance. In this regard, artifactual specific surface areas are obtained if calculated from water absorption data [51] for these types of substances.

D. The Role of Water as a Plasticizer

Absorption of significant amounts of water into the internal structure of a solid has been shown to influence the properties of the solid. This is apparent, for example, in the hysteresis observed between the sorption and desorption isotherms in Fig. 4. This phenomenon becomes exaggerated to a greater extent for materials that consist of higher proportions of amorphous material. Levine and Slade [73,74], have demonstrated that water, with a very low glass transition temperature, can act as a plasticizer, thereby lowering the glass transition temperature, T_g, of amorphous polymers. Recognizing that the viscoelastic properties of the solid are altered significantly above (rubbery state) T_g, relative to below (glass or vitreous state) T_g, it is likely that the solid will undergo changes of its physical properties at distinct moisture contents and defined temperatures as a result of this phenomenon [71,73]. Oksanen and Zografi [71] have shown with poly(vinylpyrrolidone) that the moisture content at which the moisture sorption isotherm begins to increase significantly correlates very well with the moisture content that will reduce T_g to the temperature of the isotherm. This is illustrated in Fig. 6, which shows water absorption isotherms for poly(vinylpyrrolidone) over the temperature range –40°C to 60°C [71,75]. Clearly, the inflection point at which the isotherm begins to turn markedly upward shifts to a higher moisture content as the temperature is reduced. To illustrate this, note that the moisture content (0.674 g/g) necessary to reduce T_g to –40°C has not been attained yet in Fig. 6, and the isotherm appears quite linear over the relative humidity range shown. For further clarity, the moisture contents [$W(T_g = T)$] corresponding to T_g at 60°C, 30°C, –20°C, and –40°C were shown to be about 0.205 g/g, 0.313 g/g, 0.553 g/g, and 0.674 g/g, respectively. Oksanen and Zografi [71] reported that cellulose and elastin (a protein) exhibit similar relationships, where the glass to rubber transitions correspond to the upward inflections in their respective isotherms.

Since the viscoelastic properties of the solid undergo a significant change as the solid undergoes a transition from the amorphous to rubbery states (due to elevation of temperature at constant moisture content or to an increase in moisture content at constant temperature), one also expects marked changes in the processing properties of these solids as this transition occurs. Some properties that are likely to be affected include tablet compaction [76], gelatin capsule

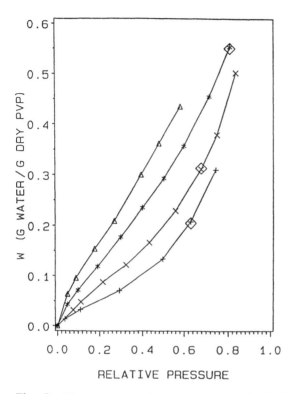

Fig. 6 Water vapor sorption isotherms for poly(vinylpyrrolidone) at 60°C(+); 30°C (×); –20°C (*); and –40°C (Δ). Data were taken from Oksanen and Zografi [71,75]. A ◊ represents the calculated water contents necessary to depress T_g to the temperature of the isotherms.

brittleness [73,77], collapse of lyophilized amorphous powders [75,78], and protein stability [79].

VII. WATER SORPTION BY PHARMACEUTICAL SOLIDS SUBJECTED TO PROCESSING

Understanding the mechanisms of moisture sorption by solids existing in either the crystalline or amorphous states allows a conceptual estimation of critical points where major changes in physical or chemical properties occur (e.g., RH_0, a crystal hydration relative humidity, glass transition temperature). Processing (i.e., milling, spray-drying, compaction, lyophilization, etc.) of pharmaceutical solids, however, often induces at least a partial conversion of most substances to a high energy form [80–85]. Such local disorder has been associated with enhanced chemical reactivity [81–85] and increased solubility [86] relative to the thermodynamically favored crystalline state. These regions have been referred to as "hot" spots of the bulk solid and, when present, leave the solid in an "activated state" [80–91].

This nonhomogeneity that exists in processed solids complicates the study of moisture sorption phenomena in these materials, as more than one mechanism of uptake must be considered. This is especially difficult, and often frustrating, for cases in which only a small amount of amorphous material is present, as experimental techniques are not readily available to measure small amounts of amorphous material in the presence of mostly crystalline substance [92]. Yet, relatively low percentages of amorphous material can absorb considerable amounts of water into their structure and act as the regions that undergo considerable change and affect the overall properties of the bulk substance [80]. This likely is especially important for low molecular weight substances that have the ability to readily recrystallize due to their overall greater mobility relative to higher molecular weight polymeric materials. This has been demonstrated for sodium chloride and sodium salicylate ground for 15 minutes in a mortar and pestle [37]. Whereas recrystallized materials exhibited no changes in specific surface areas with increasing relative humidities, the ground samples exhibited significant reductions in specific surface areas as relative humidities were increased. Figure 2 illustrates the differing moisture uptake profiles for the recrystallized and ground sodium chloride samples, normalized for specific surface area [37]. Whereas the ground material sorbed significantly more water at lower relative humidities than the recrystallized sample, the recrystallized material sorbed greater amounts at higher relative humidities. This relative reduction in sorption capacity of the ground sample is attributed to a reduction in surface area as relative humidity increased, due to the consequent recrystalli- zation of the disordered surface material [37]. Fukuoka et al. [93] have demonstrated that a variety of pharmaceutical substances indeed can be made amorphous and, furthermore, exhibit glass transition temperatures over a range from 243°K to 354°K. For example, aspirin, progesterone, phenobarbital, and sulfadimethoxine exhibit T_g values of 243°K, 279°K, 321°K, and 339°K, respectively. Although the effect of moisture content (and relative humidity) on T_g was not evaluated [93], this clearly shows that this potential mechanism of recrystallization may be important for low molecular weight substances in the temperature range to which pharmaceuticals are typically exposed.

To illustrate this more quantitatively, consider the hypothetical sucrose example discussed by Ahlneck and Zografi [80]. Assuming that all the sorbed water is taken up by the amorphous portion of material, 0.1% total moisture would correspond to approximately 20%, 10%, 4%, and 2% moisture content in the amorphous material, respectively, for 0.5%, 1%, 2.5%, and 5% of amorphous solid. The glass transition temperatures for the amorphous portions of these systems range from 9°C to 49°C, respectively [80,94]. Hence, significant changes in the solid state properties are expected at room temperature if relatively small amounts of amorphous material (i.e., <1%) are initially present. This example illustrates that even for low moisture content materials, significant changes can occur in certain regions of a solid, which may affect properties of the material influenced by molecular mobility [80].

VIII. TRANSFER OF WATER BETWEEN SOLID COMPONENTS VIA THE HEADSPACE

Combining solids that have previously been equilibrated at different relative humidities results in a system that is thermodynamically unstable, since there will be a tendency for moisture to distribute in the system so that a single relative humidity is attained in the headspace. As shown in Fig. 7, moisture will desorb into the headspace from the component initially equilibrated at a higher relative humidity and sorb to the component initially equilibrated at a lower relative humidity. This process will continue until both solids have equilibrated at the final relative humidity. The final relative humidity can be predicted a priori by the sorption–desorption moisture transfer (SDMT) model [95] if one has moisture uptake isotherms for each of the solid components, their initial moisture contents and dry weights, headspace volume, and temperature. Final moisture contents for each solid can then easily be estimated from the isotherms for the respective solids.

The SDMT model has practical utility in aiding the rational optimization of the initial moisture contents of individual components in a system to attain the final desired relative humidity. Practical applications to date have included adjustment of the initial formulation LODs prior to capsule filling to avoid gelatin

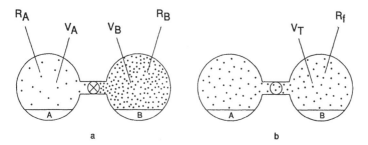

Fig. 7 Schematic representation of moisture transfer between solid components A and B with (a) headspaces isolated from one another; and (b) headspaces allowed to equilibrate. R_A and R_B = initial relative humidities above A and B; V_A and V_B = headspace volumes above A and B; R_f and V_T = final relative humidity and headspace volume above A and B. (From Ref. 95.)

capsule brittleness [77,96] and selecting the appropriate formulation moisture content and amount of desiccant to maintain the relative humidity inside a container below a defined value [97].

IX. SUMMARY

Moisture is present in all solid materials and in most processing techniques. Understanding where the water resides, its state, and the manner in which it affects the properties of individual materials, their mixtures, and ultimately, final product performance and integrity are essential for the developmental scientist to better understand the role of water in a particular system. Especially important are the kinetics of moisture uptake or loss, "equilibrium" uptake values as a function of relative humidity, whether the water resides externally or is absorbed into the material, its degree of binding with the solid, and the tendency for water to redistribute in a system consisting of more than one solid. Although water–solid interaction(s) can be extremely complex in pharmaceutical systems, application of these fundamental concepts to product development can greatly aid in understanding the role of moisture in affecting the physical-chemical properties of solid materials.

ACKNOWLEDGMENTS

The authors would like to thank Dr. Cindy Oksanen and Mr. James Conners for their technical contributions to this chapter, Mr. Conners and Ms. Linda Schweikardt for their graphical support, and Ms. Carol Della Cava for typing and editing this work.

REFERENCES

1. J. T. Carstensen, in *Pharmaceutics of Solids and Solid Dosage Forms*, J. Wiley & Sons, New York, 1977, 11–15.
2. G. N. Lewis, M. Randall, K. S. Pitzer, and L. Brewer, in *Thermodynamics*, 2nd rev. ed., McGraw-Hill, New York, 1961.
3. S. Gal, in *Vapor Sorption Equilibria and Other Water-Starch Interactions: A Physico-Chemical Approach* (C. Van den Berg, Ph.D. Thesis), Agricultural University, Wageningen, The Netherlands, 1981, p. 10.
4. S. Brunauer, P. H. Emmett, and E. Teller, *J. Amer. Chem. Soc.*, *60*, 309 (1938).
5. I. Langmuir, *J. Amer. Chem. Soc.*, *40*, 1361 (1918).
6. R. Fowler and E. A. Guggenheim, in *Statistical Thermodynamics*, Cambridge University Press, Cambridge, U.K. 1939.
7. T. L. Hill, *J. Chem. Phys.*, *14*, 263 (1946); *J. Amer. Chem. Soc.*, *68*, 535 (1946).
8. A. B. D. Cassie, *Trans. Faraday Soc.*, *41*, 450 (1945); *41*, 458 (1945).

9. S. Lowell, in *Introduction to Powder Surface Area*, J. Wiley & Sons, New York, 1979.
10. R. G. Hollenbeck, G. E. Peck, and D. O. Kildsig, *J. Pharm. Sci.*, *67*, 1599 (1978).
11. E. A. Guggenheim, in *Applications of Statistical Mechanics*, Clarendon Press, Oxford, 1966.
12. J. H. deBoer, in *The Dynamical Character of Adsorption*, 2nd ed., Clarendon Press, Oxford, 1968.
13. R. B. Anderson, *J. Amer. Chem. Soc.*, *68*, 686 (1946).
14. C. Van den Berg, in *Vapor Sorption Equilibria and Other Water-Starch Interactions: A Physico-Chemical Approach*, Ph.D. Thesis, Agricultural University, Wageningen, The Netherlands, 1981.
15. P. J. Flory, in *Principles of Polymer Chemistry*, Cornell University Press, Ithaca, N.Y., 1953.
16. A. J. Hailwood and S. Horrobin, *Trans. Faraday Soc.*, *42B*, 84 (1946).
17. C. Van den Berg and S. Bruin, in *Water Activity: Influences on Food Quality* (L. B. Rockland and G. F. Stewart, eds.), Academic Press, New York, 1981.
18. A. Venkateswaren, *Chem. Rev.*, *70*, 619 (1970).
19. A. Martin, J. Swarbrick, and A. Cammarata, in *Physical Pharmacy*, 3rd ed., Lea & Febiger, Philadelphia, 1983, 510–512.
20. L. R. Fisher and J. N. Israelachvili, *Chem. Phys. Letters*, *76*, 325 (1980).
21. G. Zografi, *Drug Dev. Ind. Pharm.*, *14*, 1905 (1988).
22. L. Greenspan, *J. Res. N.B.S.*, *81A*, 89 (1977).
23. L. B. Rockland and S. K. Nishi, *J. Food Technol.*, *34*, 43 (1980).
24. P. W. Winston and D. H. Bates, *Ecology*, *41*, 232 (1960).
25. R. H. Stokes and R. A. Robinson, *Ind. Eng. Chem.*, *41*, 2013 (1949).
26. *Handbook of Chemistry and Physics*, 67th ed. (R. C. Weast, ed.), Chemical Rubber Co., Cleveland, 1986–87, E-42.
27. M. Bergren, "Water–Solid Interactions," AAPS Short Course, Orlando, Fla. (1993).
28. C. W. Spancake, J. E. Hastedt, and A. F. Venero, *Pharm. Res.*, *10*, S-280 (1993).
29. M. J. Kontny, in *Water Vapor Sorption Studies on Solid Surfaces*, Ph.D. Thesis, University of Wisconsin—Madison, 1985.
30. L. Van Campen, G. Zografi, and J. T. Carstensen, *Int. J. Pharm.*, *5*, 1 (1980).
31. L. Van Campen, G. L. Amidon, and G. Zografi, *J. Pharm. Sci.*, *72*, 1381, 1388, 1394 (1983).
32. *Handbook of Chemistry and Physics*, 67th ed. (R. C. Weast, ed.), Chemical Rubber Co., Cleveland, 1986–87, D-189–190.
33. M. J. Kontny and G. Zografi, *J. Pharm. Sci.*, *74*, 124 (1985).
34. P. B. Barraclough and P. G. Hall, *Surface Sci.*, *46*, 393 (1974).
35. H. U. Walter, *Zeitschrift fur Physikalische Chemie Neue Folge Bd.*, *75*, S.287 (1971).
36. R. A. Ladd, *Surface Sci.*, *12*, 37 (1968).
37. M. J. Kontny, G. P. Grandolfi, and G. Zografi, *Pharm. Res.*, *4*, 104 (1987).
38. S. R. Byrn, in *Solid State Chemistry of Drugs*, Academic Press, New York, 1982, 149.
39. J. Conners, Boehringer Ingelheim Pharmaceuticals, Inc., personal communication, 1991.

40. J. S. G. Cox, G. D. Woodard, and W. C. McCrone, *J. Pharm. Sci.*, *60*, 1458 (1971).
41. M.J. Kontny, Unpublished data.
42. K. D. Ross, *Food Technol.*, *29*, 26 (1975).
43. G. Edgar and W. O. Swan, *J. Amer. Chem. Soc.*, *44*: 570 (1922).
44. J. R. Adams and A. R. Merz, *Ind. Eng. Chem.*, *21*: 305 (1929).
45. E. B. R. Prideaux, *J. Soc. Chem. Ind.*, *39*, 182 (1920).
46. M. M. Markowitz and D. A. Boryta, *J. Chem. Eng. Data*, *6*, 16 (1961).
47. C. Van den Berg, F. S. Kaper, J. A. G. Weldring, and I. Wolters, *J. Food Technol.*, *10*, 589–602 (1975).
48. C. Van den Berg, in *Water Activity: Influences on Food Quality* (L. B. Rockland and G. F. Stewart, eds.), Academic Press, New York 1981, 1–61.
49. G. Zografi and M. J. Kontny, *Pharm. Res.*, *3*, 187 (1986).
50. D. E. Wurster, G. E. Peck, and D. O. Kildsig, *Starch*, *36*, 294 (1984).
51. G. Zografi, M. J. Kontny, A. Y. S. Yang, and G. S. Brenner, *Int. J. Pharm.*, *18*, 99 (1984).
52. A. J. Stamm, in *Wood and Cellulose Science*, The Ronald Press Co., New York, 1964.
53. Y. Nakai, E. Fukuoka, S. Nakajima, and J. Hasegawa, *J. Chem. Pharm. Bull.*, *25*, 96 (1977).
54. A. Mitrevej and R. G. Hollenbeck, *Pharm. Tech.*, *6*, 48 (1982).
55. R. E. Gordon, G. E. Peck, and D. O. Kildsig, *Drug Dev. Ind. Pharm.*, *10*, 833 (1984).
56. J. C. Callahan, G. W. Cleary, M. Elefant, I. Kaplan, T. Kensler, and R. A. Nash, *Drug Dev. Ind. Pharm.*, *8*, 355 (1982).
57. J. A. Rupley, P.-H. Yank, and G. Tollin, in *Water in Polymers* (S. P. Rowland, ed.), ACS Symposium 127, American Chemical Society, Washington, D.C., 1980, 111–132.
58. N. Nagashima and E.-I. Suzuki, *Appl. Spectroscopy Rev.*, *20*, 1–53 (1984).
59. R. B. Duckworth, *J. Food Technol.*, *6*, 317–327 (1971).
60. H. J. Hennig and H. Lechert, *J. Colloid Interf. Sci.*, *62*, 199–204 (1977).
61. J. Mousseri, M. P. Steinberg, A. I. Nelson, and L. S. Wei, *J. Food Sci.*, *39*, 114–116 (1974).
62. M. J. Tait, S. Ablett, and F. W. Wood, *J. Colloid Interf. Sci.*, *41*, 594–603 (1972).
63. M. J. Tait, S. Ablett, and F. Ranks, in *Water Structure at the Water-Polymer Interface* (H. H. G. Jellinek, ed.), Plenum Press, New York, 1972, 29–38.
64. J. E. Carles and A. M. Scallan, *J. Appl. Polym. Sci.*, *17*, 1855–1865 (1973).
65. E. Hsi, G. J. Voigt, and R. G. Bryant, *J. Colloid Interf. Sci.*, *70*, 338–345 (1979).
66. M. F. Froix and R. Nelson, *Macromolecules*, *8*, 726–730 (1975).
67. B. P. Fish, in *Fundamental Aspects of the Dehydration of Foodstuffs*, Soc. Chem. Ind. (S.C.I.), London, 1958, 143–157.
68. R. B. Duckworth and G. M. Smith, in *Recent Advances in Food Science*, Vol. 3 (J. M. Leitch and D. N. Rhodes, eds.), Butterworths, London, 1962, 230–238.
69. F. M. Etzler, *J. Colloid Interf. Sci.*, *92*, 43 (1983).
70. A. R. Urquart and A. M. Williams, *J. Textile Inst.*, *15*, 550 (1924).
71. C. A. Oksanen and G. Zografi, *Pharm. Res.*, *7*, 654 (1990).

72. L. S. A. Smith and V. Schnitz, *Polymer*, *29*, 1871 (1988).
73. H. Levine and L. Slade, in *Water Science Reviews*, Vol. 3 (F. Franks, ed.), Cambridge University Press, Cambridge, U.K. 1987, 79–185.
74. L. Slade and H. Levine, *Pure Appl. Chem.*, *60*, 1841 (1988).
75. A. P. Mackenzie and D. H. Rasmussen, in *Water Structure at the Water–Polymer Interface* (H. H. G. Jellineck, ed.), Plenum, New York, 1972, 146–172.
76. A. J. Shukla and J. C. Price, *Pharm. Res.*, *8*, 336 (1991).
77. M. J. Kontny and C. A. Mulski, *Int. J. Pharm.*, *54*, 79 (1989).
78. A. P. Mackenzie, in *Freeze Drying and Advanced Food Technology* (S. A. Goldblith, L. Rey, and W. W. Rothmayr, eds.), Academic Press, New York, 1975, 277–307.
79. M. J. Hageman, *Drug Dev. Ind. Pharm.*, *14*, 2047 (1988).
80. C. Ahlneck and G. Zografi, *Int. J. Pharm.*, *62*, 87 (1990).
81. E. B. Vadas, P. Toma, and G. Zografi, *Pharm. Res.*, *8*, 148 (1991).
82. R. Huttenrauch, *Acta Pharm. Technol.*, *34*, 1 (1988).
83. R. Huttenrauch, S. Frike, and P. Zielke, *Pharm Res.*, *2*, 302 (1985).
84. M. Otsuka and N. Kaneniwa, *Int. J. Pharm.*, *62*, 65 (1990).
85. J. A. Hersey and I. Krycer, *Int. J. Pharm. Technol. & Prod. Manuf.*, *1*, 18 (1980).
86. B. Makower and W. B. Dye, *Agr. Fd. Chem.*, *4*, 72 (1956).
87. J. T. Carstensen and K. Van Scoik, *Pharm. Res.*, *7*, 1278 (1990).
88. J.-O. Waltersson and P. Lundgren, *Acta Pharm. Suec.*, *22*, 291 (1985).
89. E. G. Prout and F. C. Tompkins, *Trans. Faraday Soc.*, *40*, 489 (1944).
90. W.-L. Ng, *Aust. J. Chem.*, *28*, 1169 (1975).
91. J. Hasegawa, M. Hanano, and S. Awazu, *Chem. Pharm. Bull.*, *23*, 86 (1975).
92. A. Saleki-Gerhardt, in *Estimation of Percent Crystallinity in Milled Samples of Sucrose*, M. S. Thesis, University of Wisconsin—Madison, 1991.
93. E. Fukuoka, M. Makita, and S. Yamamura, *Chem. Pharm. Bull.*, *37*, 1047 (1989).
94. H. Levine and L. Slade, *J. Chem. Soc. Faraday Trans. I*, *84*, 2619 (1988).
95. G. Zografi, G. P. Grandolfi, M. J. Kontny, and D. W. Mendenhall, *Int. J. Pharm.*, *42*, 77 (1988).
96. M. J. Kontny, *Drug Dev. Ind. Pharm.*, *14*, 1991 (1988).
97. M. J. Kontny, S. Koppenol, and E. T. Graham, *Int. J. Pharm.*, *84*, 261 (1992).

Index